D1169116

Weird

SCIENCE

Avon Books are available at special quantity discounts for bulk purchases for sales promotions, premiums, fund raising or educational use. Special books, or book excerpts, can also be created to fit specific needs.

For details write or telephone the office of the Director of Special Markets, Avon Books, Inc., Dept. FP, 1350 Avenue of the Americas, New York, New York 10019, 1-800-238-0658.

# Acknowledgments

For one of my oldest and dearest friends, Mark.

Many people have helped me during the writing of this book, but I would like to thank especially: Jaimie Tarrell, Paul Bailey, and John Gribbin for their advice; my agents, Bill Hamilton and Sara Fisher for having the business brains I entirely lack, and John Jarrold, who is probably the most hospitable editor I've ever worked with. Finally, I would like to thank my wife, Lisa, who works harder than I do and kept me at the word processor when I would have preferred to slope off to watch TV.

Contents

# Preface

This is a book of hypotheses.

There was a time, until perhaps the beginning of this century, when the investigation of the paranormal commanded a certain respect and was conducted by professional scientists as much as by enthusiastic amateurs. Such open-mindedness ended for two reasons. First, many paranormal phenomena seemed so elusive that busy physicists, chemists, and biologists tired of trying to pin them down. Secondly, science became so overwhelmingly successful that, to many, the supernatural became almost superfluous. Nuclear physics, brain surgery, and the advancement of space travel can be as exciting as hunting ghosts or trying to prove the existence of alien life, and at the same time they are tangible with commercial and academic application.

But the paranormal remains elusive. We are no nearer to proving or disproving the existence of telepathy, psychokinesis, or clairvoyance than we were over one hundred years ago, but we do have a greater collection of scientific ideas to draw upon in an effort to reach sensible hypotheses. And that is what I've tried to do in this book.

I have always been interested in the paranormal. I went through a period of intense enthusiasm as a youth, and then into the deepest, most cynical skepticism. Now, I like to think I'm beginning to find a balanced viewpoint, the famed open-mindedness.

Like many people, I have had the good fortune to experience

the paranormal firsthand, and in keeping with almost everyone else in this position, I have no real answers that satisfy or explain completely what happened to me.

In 1974, I was a scientifically minded, nerdy fifteen-year-old at a traditional English private school. During the summer vacation, some of my friends—who shall remain anonymous to preserve their own embarrassment—began to experiment with a Ouija board. They made their own out of bits of paper and an orange juice tumbler, but where their experience really differed was not in approach but what happened next. Instead of contacting Mozart or Julius Caesar, they immediately began a conversation with something that claimed to be an alien who lived on Saturn (in a parallel universe, in which the planet was habitable), and called himself Alan Kalak 7. He told the boys that he was the leader of an alien committee who were gathering together twelve young people on Earth to form *The Group*—a collection of people who would one day alter radically the future of the planet.

Now, this was pretty heady stuff for a bunch of immature boys living through acne trauma, cricket fixation, and looming academic tests. It was even more sensational for me when the original contactees casually informed me when we all returned to school that Alan had included my name as a member of *The Group*.

Sitting at the Ouija board was a most peculiar experience, at once terrifying and exhilarating. The peer pressure not to make a complete fool of oneself was tremendous. But now, over two decades later, only one sitting sticks out in my memory, and it is the reason for this tale.

One freezing afternoon—shortly after math tests, I seem to recall—one of my friends (and the originator of the adventure), someone I'll call G., invited me to a session of the board in which just the two of us would attempt to talk to Alan.

This had never been done by any of us before. Usually there were three or four of the group, and so we were always able to convince ourselves that someone else was moving the glass and that it was all really a daft game.

G. set out the bits of paper on a glass table he had somehow commandeered for the purpose. He put out a simple Yes, a No, and a few letters and numbers. Then we sat on opposite sides of the table and placed our fingers very lightly on the glass top.

Suddenly, the room seemed very quiet, sounds from outside died, almost as though they were on a tape that had just been switched off. We could hear each other breathing heavily. Then G. said: "Alan are you there?"

For a moment nothing happened. The seconds ticked by—I could hear them clanging away at my wrist. Apart from that, silence, as we held our breath.

This was ridiculous, I began to think. Obviously, nothing was going to happen, unless of course . . . unless G. had been pushing the glass all this time. But just as I was beginning to think we were wasting our precious study time and that I ought to call a halt to the proceedings, the glass moved.

At first the movement was almost imperceptible, then it began to stagger and then to glide almost effortlessly across the smooth tabletop.

I was struck speechless. I had seen this before, but only with a group of us around the table. I looked at G.'s finger and it was almost hovering above the glass, hardly touching it. I looked at my own finger. Was I pushing it without realizing it? No, of course I wasn't. I too was hardly making contact with the glass.

G. asked Alan a few questions—something about *The Group* and the people involved—and the glass spelled out initials, answered Yes or No, as appropriate, but I was not really concentrating. I was too stunned.

Afterward, G. and I sat and talked about what had happened, and if I needed any further proof that he was as innocent of fraud as I, it came from that conversation—he was as shocked and as exhilarated as I.

I would like to report that great things came from *The Group*, that there was huge significance in the contact, that Alan Kalak 7 and his chums really were a committee of aliens from a parallel universe, but unfortunately I cannot. We all went our separate ways after we finished school, and gradually, degrees, women, and careers loomed larger than supposed alien contact. There were adventures and further mysterious happenings during what remained of our schooldays, but no cigar, and if anything, the planet is in a worse state now than it was in 1974. But, crucially, the memory of that afternoon remains and has nurtured enthusiasm even through my most empirical and skeptical moments.

I still have no idea what happened in 1974. I am not prepared to believe that we were contacted by aliens, or mischievous spirits, nor am I willing to consider that our untrained minds were capable of psychokinesis. Equally, to simply conclude that things happened via forces forever beyond our understanding strikes me as an unacceptable cop-out.

The fact is, I don't know how this and other incidents happened, but because I don't know, I continue to be curious, determined that one day I will.

Michael White
London, December 1998

 **Visitors**

> Any sufficiently advanced technology is indistinguishable from magic.
>
> ARTHUR C. CLARKE

One of the strongest ideas in occult lore is that our planet is being visited by aliens and that some of these aliens abduct humans on a regular basis. This is not simply a product of the program-makers imagination, but an idea that has become so entrenched in a range of Earth cultures as to have become almost a cliché. Whole forests have gone under the axe to help produce countless books and magazine articles on the subject, and an entire mythology has been created covering every aspect of alien visitation. The only thing missing so far is hard, irrefutable proof to support the idea.

The most active site for information concerning alien visitors is the Internet. If you want the latest on the subject, try a news group called alt.alien.visitors on the World Wide Web. Here you will encounter telepaths from the Pleiades, Reptoids from Sirius, beautiful Venusians, and a ubiquitous group commonly called the grays.

The level of material now circulating about alien visitors is staggering and has become a thriving cottage industry in itself. Magazines available at newsstands in every major city in the Western world have progressed far beyond rather tame interviews with abductees, to giving their readers detailed descriptions of alien propulsion systems and intricate anatomical studies of alien physiology.

The source of much of this material is the Roswell incident. According to enthusiasts, a UFO crashed in Roswell, New Mexico, in 1947, and believers claim the investigation into the crash has been deliberately kept from the public ever since. They also insist that U.S. government agencies have conducted experiments on both the craft and the dead aliens found in the wreckage at a top secret location called Area 51 in the Nevada desert, and that information leaked from there substantiates their claims.

There is a range of supposed origins for alien visitors. Enthusiasts who have become disillusioned with the possibilities of interstellar travel suggest that flying saucers come either from inside the Earth itself or from "other dimensions." The Hollow Earth cult has been in existence for some time, but remains little more than a fantasy. The notion of "other dimensions" is more intriguing but no less ambiguous. What are these dimensions?

It is often the case that enthusiasts of the occult use expressions like this with little or no understanding of what they mean. They then compound the problem when they try to relate such vague ideas to legitimate science, claiming that, because physicists talk about universes comprising of ten or twenty-six dimensions, this somehow accounts for their misguided theories. It doesn't—the extra dimensions currently fashionable with physicists crop up in an exotic area of physics called string theory. They are not "parallel universes," but are thought to exist only with the subatomic scales discussed in the field of quantum mechanics (sizes in the region of 10^{-33} centimeters). It is extremely unlikely that they could provide a possible location for alien intelligence arriving here in physical craft.

Consequently, in this chapter I will restrict my discussion to the notion that aliens may be coming here from other planets beyond our own solar system, and how they could possibly do this operating within the known laws of physics.

The facts of interstellar travel all revolve around distance, time, and power. Because the distances between the stars are unimaginably huge, the time needed to travel interstellar distances are correspondingly large, and any system that may have a chance of overcoming this restriction requires impractical amounts of power.

The problem begins with Einstein's special theory of relativity. First published in 1905, when Einstein was working in a Bern patent office, the special theory draws upon two firmly established scientific principles but comes up with one of the weirdest notions in the whole of science.

The first of these derives from the work of Isaac Newton, who in the 1680s, showed that the laws of physics are the same for any observers moving at a constant velocity relative to one another. The second fact, arrived at more recently, is that the speed of light in a vacuum is always constant. This velocity is represented by the symbol c and is equal to just over one billion kilometers per hour. This is true, *irrespective* of the velocity of the observer.

According to common sense, if spaceship A is moving in one direction with a velocity of 0.75c, and spaceship B approaches in the opposite direction, also traveling at 0.75c, their relative velocity would be 1.5c. But this is not actually the case. According to Einstein's equations, crews on each ship would see light from the other coming toward them not at one and a half times the speed of light, but just under 1c (0.96c to be precise).

The astonishing consequence of this is that if c is constant, space and time must be relative. In other words, if the crew aboard spaceship A or B are to see light arriving at a constant velocity irrespective of their own velocity, they must measure time differently—so, as they travel faster, time slows. Furthermore, the property of distance cannot be the same to observers traveling at different speeds. The faster one travels, the shorter any given distance becomes—a meter will be a different length depending on the velocity of the observer, and will be shorter the faster the observer moves. Finally, the faster an observer moves, the more massive he becomes. The end result of all this is that if it were possible for an observer to travel at the speed of light, he would experience three things: time would slow to nothing, he would shrink to nothing, and his mass would be infinite!

Sadly for space travel enthusiasts, this is not the delusion of a mad professor. Einstein's special theory of relativity has been proven to be true in many thousands of experiments conducted since 1905. The reason we do not notice this effect every day of our lives is that we do not travel anywhere near fast enough. A recent shuttle mission showed how minuscule the effect is at low speeds. Traveling in orbit at a sprightly five miles per second, clocks aboard the shuttle ticked less than one ten-millionth of a second slower than their counterparts on Earth. At CERN, the giant particle accelerator near Geneva in Switzerland, subatomic particles are accelerated to near-light speeds routinely and their masses seen to increase precisely as Einstein's calculations predict.

So, the law that states that no material object can travel at the speed of light is irrefutable, it is a fact of life in our universe. Consequently, the only possible ways a technologically advanced civilization could cross interstellar distances is either to travel at speeds that do not incur too many problems from Einstein's theory, but get them there eventually, or else they would have to find ways around the theory.

First, let's look at the sublight speed options.

Since Jules Verne's idea of firing a moon rocket from a nine-hundred-foot deep hole in Florida in *From the Earth to the Moon* (published in 1865), scientists and science fiction writers have come up with a range of ingenious propulsion systems to facilitate interstellar travel. These include fusion drives, antimatter engines, spaceships utilizing the properties of wormholes and space-warping devices.

All conventional space propulsion systems—by this I mean engines that do not use some exotic property of space itself, such as warping or wormholes—must work on the principle of Newton's third law of motion, which states that: "For every action there is an equal and opposite reaction." In this way a spaceship is no different from a jet aircraft—material is expelled from the back of the craft and the craft moves forward; simple. The difficulty is a question of magnitude.

The spacecraft we have developed so far all work by chemical propulsion. The greatest energy requirement has been that needed to escape the Earth's gravitational pull—to achieve an escape velocity so that the Saturn V, the shuttle, or the Ariane craft could get their payloads into orbit. All maneuvers aboard the Apollo craft traveling to the moon depended upon relatively small engines and thrusters that expelled hot gases from their exhausts and adjusted the course of the spaceship. Without these, the capsules would have been entirely at the whim of the gravitational forces at work beyond the Earth's atmosphere.

The next level of sophistication is some form of fission-powered spacecraft engine. This is the power source used in nuclear reactors and unleashed in the earliest atomic bombs. When large unstable atomic nuclei are made to decay, or undergo fission, they produce energy. The value of this energy depends upon the mass of material undergoing fission and can be calculated using perhaps the most famous equation in history: $E = mc^2$, where m equals the mass of material and c is the speed of light.

Although this is the most powerful controllable energy source we have currently, it could not provide anything like the energy needed to reach the stars, and the mass of fissionable material needed even for efficient interplanetary travel within our own tiny solar system would be so large, there would be little room left for crew or cargo.

A more powerful form of nuclear energy comes from a process called nuclear fusion. Back in 1989 there was a brief flurry of excitement when two scientists, Martin Fleischmann and Stanley Pons, claimed they had devised a technique called "cold fusion," which appeared to require nothing more than a pair of electrodes and some commonplace chemicals placed in a jar. Sadly, the excitement died when the experiments proved unrepeatable, and the hopes of scientists returned to conventional fusion. This is a mechanism by which the sun or any star is powered. In the laboratory, the process involves fusing together small nuclei such as deuterium and tritium—heavy isotopes of hydrogen—to produce large amounts of energy.*

For almost fifty years scientists have been trying to develop practical nuclear fusion—it is relatively clean because it does not use dangerously radioactive elements such as the uranium-238, which is converted into plutonium-239 in modern fast breeder reactors (isotopes that remain dangerous for hundreds of thousands of years), and it could potentially produce far more energy than fission. These are the plus points of the system; the down side has so far been the problem of containment and efficiency. In order to bring about

---

*A heavy isotope is a version of an atom that has more than its usual complement of neutrons in its nucleus. The most common form of hydrogen has just one proton in its nucleus and no neutrons. The first heavy isotope of hydrogen, deuterium, has one proton and one neutron. The heaviest, tritium, has one proton and two neutrons.

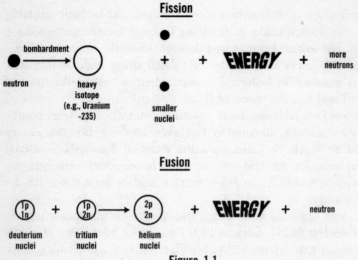

Figure 1.1

fusion, temperatures of around 10 million degrees are needed (the sort of temperatures produced at the sun's core) so that the positively charged nuclei can be forced to overcome their electrostatic repulsion. This fused material exists as a superheated plasma which cannot be kept in any form of physical container. Furthermore, the energy needed to bring about fusion has so far been much greater than the energy return, which means the system currently shows negative efficiency.

Having said that, scientists hope to crack these problems in the future, and fusion energy is seen as the most likely way in which we could save the Earth's looming resource crisis. Assuming another civilization is only a few decades ahead of us, they would almost certainly have developed fusion power, and if they are further advanced, they would have mastered the use of fusion engines aboard spacecraft. Unfortunately, this energy source could never be used for interstellar travel for the simple reason that the amount of fusible material needed to achieve even a tiny percentage of light speed would be too great.

It has been calculated that to accelerate a spaceship to just ten percent of the speed of light would require about fifteen times its mass in fuel. And this is to accelerate just once. If the craft wanted to stop at its destination, it would need to use more fuel, equivalent to fifteen times the current mass of the ship. If we assume the outward voyage has used up half the fuel—which weighed fifteen times the mass of the living quarters and cargo—a further 7.5 times the mass would be needed again. So, one start and one stop would need $15 \times 7.5 \times$ the mass of the main body of the ship (excluding fuel), or 112.5 times the mass of the living quarters and cargo.

A variation on this is the idea of the fusion ramjet. Interstellar space is not a complete vacuum, it contains hydrogen atoms, albeit distributed very finely between the stars and planets. A spacecraft could be designed with large scoops that draw in the hydrogen atoms to use as fusible material. The objection to this has always been that there is insufficient material available in space, but if the craft is moving quickly enough it would behave like a giant sea mammal drawing in plankton, or like a person running through light rain getting soaked because they are meeting the raindrops as they go.

One day, when we send people to the planets of our own solar system, we will almost certainly use fusion power in one form or another. It is a practical system, for interplanetary travel as speeds of 100,000 kph would be relatively easy to generate, allowing us to get to Mars in about three weeks. But there is no comparison between interplanetary and interstellar distances. Using fusion power to achieve speeds of 100,000 kph, we would need a thousand generations merely to reach the nearest star, and the fuel requirements to maintain even this relatively trivial velocity for so long would alone make it totally impractical.

Putting aside fusion power, there have been a number of other suggestions for ways to achieve a reasonable fraction of light speed using conventional physics. One such idea is to use the power of nuclear explosions to thrust the craft forward.

Designers of a theoretical vehicle known as *Orion* visualize using a store of thermonuclear warheads individually propelled from the back of the craft at the rate of one every three seconds. The hot plasma produced by the explosions would impact on a "pusher plate" propelling the spaceship forward. Unfortunately, to achieve a speed of just three percent of the speed of light would need almost 300,000 one-ton bombs.

A variant on this was Project Daedalus, investigated by the British Interplanetary Society during the 1970s. This theoretical system involved a craft similar to *Orion* but powered by 250 nuclear explosions per second, which could achieve some twelve percent of light speed, or 130 million kph, but again the mass of fuel required made the idea impractical.

More promising than any of these schemes could be the possibility of using exotic material known as antimatter.

All matter in our universe is made of atoms. These in turn are composed of what are called subatomic particles—neutrons and protons, which exist together in the nucleus of the atom, and electrons, which surround the nucleus. This much was understood early this century thanks to the work of such pioneers as Ernest Rutherford, James Chadwick, Max Planck, and others. Another groundbreaking physicist of this era was Paul Dirac, who in 1929 predicted that all the known subatomic particles could have counterparts with opposite properties.* These became known as antiparticles.

Protons are positively charged, and an antiproton would have the same mass and exist in the nuclei of antiatoms but would be positively charged. An antielectron, or positron as it has become known, would be positively charged and like the electron exist outside the nucleus of antiatoms. But what is most important for our purposes in designing an interstellar engine is the fact that when matter and antimatter come into contact, they annihilate each other instantly and produce energy.

In Paul Dirac's day, antimatter was merely a theoretical concept, something that had popped out of the equations when he had combined the mathematics of quantum mechanics, electromagnetism, and relativity. At that time, the existence of antimatter could not be proven, as it is not found naturally in our universe because it would disappear as soon as it came into contact with matter. Today, we can manufacture small quantities of antimatter in a particle accelerator.

---

*In those days only protons and electrons were known, the neutron was discovered three years later, in 1932.

To make an antiproton, "normal" protons are sent whirling around the accelerator ring, where they are accelerated in an intense magnetic field until they reach about half the speed of light. They are then allowed to collide with the nuclei of metal atoms. This produces pairs of particles and antiparticles along with X rays and various forms of energy. The antiprotons are then separated from the protons before they can interact and obliterate one another.

To use antimatter as a propellant, we need to allow a controlled annihilation of particles and antiparticles and to use the heat evolved to drive our spacecraft. A simple design for just such a system is already on the drawing board. The idea is to fire a tiny quantity of antimatter into a hollow tungsten block filled with hydrogen. The particles are instantly annihilated, and the energy released heats up the tungsten block. Cold hydrogen is then squirted into the center of the device, where it is rapidly heated to about 3000 Kelvin and fired out of the engine.

The great advantage of antimatter drives is that little fuel is needed to produce an effective acceleration. The great disadvantage is the difficulties of producing usable amounts of the stuff. Currently only a sixty millionth of the energy used in producing antimatter in the world's particle accelerators ends up as particles, which is one of the reasons its current market value is about $10,000,000,000,000,000 (ten thousand million million dollars) per gram.

We also have the problem of containment. Like the superhot plasma produced by nuclear fusion, special magnetic containment systems have to be used, in this case, to prevent antimatter interacting with matter before it is needed.

None of these difficulties precludes its use by advanced civilizations. Looking at our own history can provide a salutary lesson. It was only in 1919 that Ernest Rutherford discovered that the nuclei of certain atoms could be made to disintegrate by bombardment. Within just twenty-six years this discovery led to Hiroshima and Nagasaki.

Current technology may mean that antimatter is prohibitively expensive to produce, but within two or three decades this will no longer be the case. And such time spans are relatively meaningless when we look at the possibility of advanced societies developing on other worlds.

The option of antimatter propulsion systems offers hope that interstellar travel may be a possibility, but even if an advanced civilization had realized the full potential of this technology, they would not be able to circumvent the natural laws of the universe and would remain limited to sublight speeds.

What this means for interstellar voyagers is that they will either have to accommodate the consequences of traveling at close to the speed of light (which would still be very slow for the purpose of colonizing or visiting many worlds), or else travel even slower and take even longer to arrive anywhere outside their own solar systems.

Imagine for a moment a journey of fifty light-years from the aliens' home world to Earth. At 0.95c (95% the speed of light), this will take 52.5 years to complete, one way—52.5 years to the people back home, that is. Because of the consequence of special relativity—as we travel faster, relative time slows—to the crew of the spaceship this 52.5 years would only be 14.8 years.

This is still far too long to be of practical use. Even if we assume alien longevity is greater than ours, one and half decades is still a long time to be on board a spacecraft. The answer to this might be a form of suspended animation, or even cryogenics, but there are other hurdles. Crews sent out on round trips of over a century might return to their home worlds to find the political structure changed. The organization that sent them may no longer exist. Such a crew would find all their relatives either dead or ancient, and almost everything once familiar, irreversibly altered. Imagine a human able to set out on such a mission in the year 1900 returning to Earth in 2000. They may have aged less than thirty years, but the world would be almost unrecognizable to them.

If we consider, as UFO enthusiasts do, that interstellar travel is commonplace and that aliens visiting us operate as part of an organized federation or planetary authority, they could not travel at sublight speeds. First, any form of command structure would be impossible to maintain over such distances and times scales. Second, it is surely a universal rule that any endeavor must see a return for the investor within a reasonable time frame, certainly the lifetime of the investor. Who would finance any missions involving such time scales? Certainly no government we could imagine.

The only other possible option for sublight travel is the concept of the Ark. This has been a favorite of science fiction writers and space analysts for generations, and offers one system for interstellar travel that has some practical aspects.

The idea would be to create a large spacecraft on which generations of aliens would live during a mission that may last hundreds or thousands of years. Although this would require a vast spacecraft capable of sustaining a large number of crew and passengers for perhaps millennia, it would not need to travel particularly fast. If a mission was designed to take a thousand years, a distance of fifty light-years could be covered at just five percent of light speed (50 million kilometers per hour). But putting aside the technological difficulties of designing and constructing such a craft, there are other more mundane drawbacks with the scheme.

First, we return to the problem of time scales. The only possible reason a civilization would finance such a mission would be to escape catastrophe—an Ark in the biblical sense—where some or all of the population of the planet are packed into spacecraft which then head off to find a new home. We have not yet encountered a colonizing group, so it would be safe to assume that our widely reported visitors do not fall into this category. Yet, even assuming a smaller scale operation was financed and set in motion with say a few thousand passengers and crew, how would the group sustain itself psychologically?

The early generations of aliens aboard the Ark would have no hope of seeing a new world, and would be kept going only by the knowledge that their distant offspring would make it to a distant planet. It is conceivable that such a scheme would be favored by aliens who possess a very different psychological makeup than we do. They might think in a similar way to ants or bees and have an in-built instinct for the community rather than the individual. Such a scheme might work for them, but places upon it severe limitations in a general sense.

Finally, the most persuasive argument against the use of Arks is the notion of the "speed exponential curve." Using our own techno-

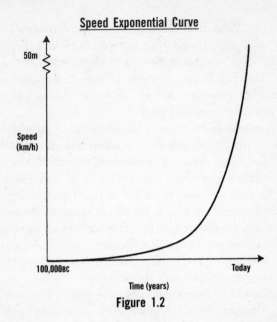

Figure 1.2

logical development as a paradigm, it can be seen that the speeds we are capable of achieving have risen exponentially with time. For the first 100,000 years of human social development, the highest speed we could reach was about twenty kph—the pace of a sprinting hunter. This was more than doubled some four thousand years ago with the mastery of horses. This same doubling occurred by the late nineteenth century with the development of trains and motor vehicles, multiplied a further three or four times within the following fifty years using aircraft and again with the advent of jets, and once more with the invention of spacecraft. At this rate, the speed exponential curve shows that it should be possible to reach one percent of the speed of light by 2070, and five percent by 2140.

The consequence of all this for our Ark voyagers is that they may arrive at their destination only to find the planet colonized long before by their own race, who traveled there in a fraction of the time.

A variation on this idea is long-term colonization. The physicist Frank Tipler has lent his support to the notion that an advanced race could "planet-hop." He bases the idea on the way the South Sea islanders

A colonizing wave by planet-hopping

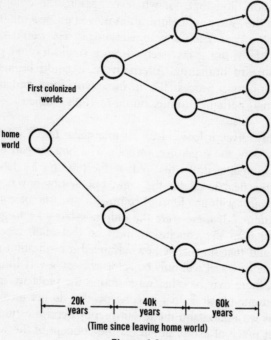

First colonized
worlds

home
world

|— 20k years —|— 40k years —|— 60k years —|
(Time since leaving home world)

Figure 1.3

spread across the Pacific Ocean by island-hopping followed by consolidation. Using this model, he believes there are two time factors to consider. First, the time needed to make an interstellar journey (t_1), the other, the time needed to establish a colony and to prepare for the next hop (t_2). Conservative estimates for the journeys would be in the 1,000- to 10,000-year range, and a reasonable period for colonizing and consolidation would be in the region of one hundred generations.

Using this system, the galaxy can be completely colonized in a surprisingly short time, because the growth would be exponential.

If we say the journey time and the colonization time combined equal an average of 20,000 years, $t_1 + t_2 = 20,000$; assuming there are one billion suitable planets in an average galaxy, these could all be reached and colonized in under one million years.

It is a sobering thought that perhaps the Earth was once the site of colonization, and that, for whatever reason, the colony did not prosper and the "wave" of colonization moved on, leaving us behind. If this was the case, such a colonization process would mean that all human life in our galaxy stems from a single mother planet, an original home of humanity. Alternatively, it could be argued that such a colonization process lies in our own future and that Earth is the original, perhaps unique, home of *Homo sapiens*.

From this survey it looks bleak for interstellar travel. Each method is either too slow, too expensive, or both. The best we can hope for is to develop antimatter drives that reduce the time for a relatively short journey to practical levels for the crew, but destroy any hope for an organized project with any form of command structure or communication with "home." If these were the only possible ways to get around the universe, then the conclusion must be that alien races are not visiting us, and that no matter how advanced a civilization might become, interstellar travel will only be achievable on a very limited scale. Fortunately, there may be other ways around the problem, using what I will call "exotic physics"; that is, concepts that do not break the laws of physics as we understand them, but merely bend the rules.

Our first piece of exotic physics is the concept of the wormhole. Like antimatter, the idea of wormholes arose as a consequence of manipulating the mathematics of physics, this time, Einstein's general theory of relativity.

Scientists have known for a long time that when a star has used almost all its available fuel, it begins to die, and the way in which it dies depends upon its mass. If it is about three times the mass of our sun or larger, it begins to shrink, setting up shock waves that result in an enormous explosion—the most violent event since the Big Bang—a supernova. But even then, because the sun was so large to begin with, some material is left at the center of the supernova which begins to collapse in upon itself again. This time, the matter becomes so dense the incredibly strong forces holding subatomic particles together, the binding forces between quarks,* are over-

---

*The most fundamental form of matter known, and constituents of protons, neutrons, and electrons.

whelmed and the star becomes a seething caldron of fundamental matter and energy. This is a black hole, so called because it is so massive and dense that even light cannot travel fast enough to escape its gravitational field.*

It often happens in science that mathematics tells us something should exist and what its properties would be before it is observed. Although the existence of black holes has not yet been confirmed, there are some promising candidates, and it is more than likely that they do exist somewhere in the universe. The chances of finding wormholes is slimmer, but there is nothing within the laws of physics that says they could not exist.

Einstein's general theory of relativity, published in 1916, is an extension of the more limited special relativity, concerned only with observers moving at a constant velocity. Einstein wondering what the situation would be for objects experiencing acceleration, imagined a lift in a state of free fall and a beam of light entering a hole in one wall. People in the lift would perceive the light traveling in a straight line. But to an observer outside the lift, the light would travel along a curved line. This bending of light, Einstein stated, was caused by the fact that the lift was experiencing acceleration, and he went on to say that because gravity is a form of acceleration, light would be bent by it.

Until Einstein, physicists saw the universe in three dimensions, with time as an extra factor. In general relativity, time is a dimension just like length, breadth, or depth; the universe actually exists in four dimensions called "space-time."

The only way we can visualize a four-dimensional universe is by representing it in three dimensions. Imagine a rubber sheet stretched flat. Now place a heavy ball in the middle—the sheet around the ball is misshapen the way space-time distorts around a massive object like a star. Roll a marble along the sheet near the heavy ball and it follows a curved path, just as light does near a star. A black hole is so massive and has such a powerful gravitational

---

*By comparison, the shuttle has to achieve a speed of 11.18 kilometers per second to escape the gravitational pull of the Earth. Light, remember, travels at 300,000 kilometers per second.

field, it curves space so much that within it lies what is called a "singularity," a point at which the curvature of space-time becomes infinitely sharp and all the laws of physics break down. Wormholes, as theorized by a number of scientists (including the physicists who first postulated them—Kip Thorne and Michael Morris at Caltech in California), are created when two singularities "find" each other and join up.

The reason wormholes are useful to interstellar travelers may be visualized from the diagram. Because of the nature of curved space-time, they offer a shortcut, bypassing the need to travel between point A and point B using the conventional route.

Now, this is obviously an attractive idea and could eliminate all the problems faced by the near-light-speed traveler at a single stroke. But, as you would expect, there are problems with the method.

First, wormholes are still pure speculation. They are not disallowed by the known laws of the universe, but neither are they certain to exist. But assuming they do, they would probably be quite rare.

Figure 1.4a

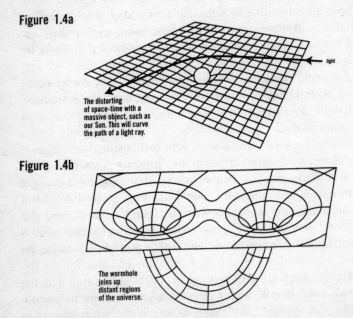

light

The distorting of space-time with a massive object, such as our Sun. This will curve the path of a light ray.

Figure 1.4b

The wormhole joins up distant regions of the universe.

The second problem is that it would be impossible to know which parts of the universe they linked until they were used. Furthermore, if they were usable, they would offer only a very limited service, linking the starting point to one fixed destination. It would be a bit like having a highway connecting London with some other mystery location with no junctions or turnoffs en route.

Ignoring this drawback, we have to consider the nature of the link, and from what we know of black holes, a "natural" wormhole would offer a very bumpy ride indeed. The inside of a black hole is probably the most inhospitable place in the universe; the gravitational forces at work there would instantly break any material object into a soup of fundamental particles and energy, and even if these forces could be resisted, once within the grip of the black hole there is no escape. So, the idea of using a wormhole created by joining two black holes at two different points in the universe does not seem very practical. The only way they could be used would be if there are certain types of black holes somewhere in the universe that do allow passage, but these might be very difficult to find.

A possible way around this difficulty is the idea of white holes. These would be the very opposite of black holes; rather than absorbing matter and energy, they might act as perfect emitters or "cosmic gushers." If a black hole and a white hole were joined, they could act as a one-way wormhole and circumvent the problem of escaping a black hole once it has been entered. Unfortunately, detailed mathematical analysis of this scenario has shown that such a system would be unstable and the white hole would rapidly decay, making the passage of a spaceship impossible.

There remains one alternative—man-made wormholes.

Since wormholes are first postulated by Thorne and Morris in a paper published in the *American Journal of Physics* in 1987, hundreds of theoretical physicists around the world have studied the concept. They have come to the conclusion that in order to construct a workable wormhole, a set of strict conditions have to be met. These include the obvious fact that the construction of the wormhole must be consistent with general relativity, and that the gravitational "tidal forces" within the wormhole be kept to a minimum. They also stipulate the shape to which the wormhole must conform, and the mass of material needed to create it. Unfortunately, the mathematics shows that in order to construct a wormhole, material known as exotic matter, which has the bizarre property of negative mass, is needed.

Although wormhole enthusiasts insist that such a nonsensical idea can be realized within the laws of physics, most scientists dismiss the notion. If they are right, it would appear that wormholes could never be manufactured, no matter how advanced a civilization might become. If they are wrong and the wormhole supporters are correct, then exotic matter has to be found and manipulated by the civilization before the wormhole could be built and used.

If wormholes look implausible, we are left with one other alternative method of interstellar travel that could facilitate a way around the light speed restriction, a concept made famous by the TV series *Star Trek*—the warp drive.

Science fiction writers since the 1940s have batted around expressions such as "space warp" and "hyperspace," but though many have been scientists themselves, they have rarely attempted to explain the concept in any detail. It has been visualized as the only possible way to circumvent the impracticalities of sublight speed travel and the nuisance of having to work within the laws of physics, but what is warping and how could it be accomplished?

Another name for warping could be surfing. This is because it is based upon the principle of manipulating space-time itself so that the space vehicle moves on a "wave." The spacecraft

pushes "home" further away

warped space-time

space-time of destination brought closer

DIRECTION OF FLYING SAUCER
Distorting space-time to create warp-drive

Figure 1.5

would have the ability to alter space-time, so that it expanded behind the craft and contracted in front of it. This means that even though the craft is itself moving relatively slowly, the departure point would be "pushed" back a vast distance and the destination "drawn" nearer.

This sounds like a cheat, but again, it is a possibility within the rules of general relativity. The difficult aspect is once again the energy requirements. For the system to work, space-time would have to be distorted significantly—or else the effect would be so small that sublight travel would probably be quicker.

Observation of our sun shows that its mass curves space-time so that it bends light by just one thousandth of a degree. For a spacecraft to utilize the expansion and contraction of space-time itself, it would have to distort the space-time continuum far more than this. In some respects the vehicle would have to behave a little like a tiny black hole. Using this as a basis for calculating the energy requirements, the result sounds depressingly familiar. To make a black hole the size of a typical spaceship, say a disk fifty meters in diameter, we would need a mass of about fifty thousand Earths compacted into the space. Expressed in terms of energy, this would be about equal to the entire output of the sun during its lifetime.

What then is to be concluded from these arguments? All forms of sublight travel restrict meaningful interstellar travel, and the options for circumventing the light-speed barrier present huge technical difficulties.

It may well be that alien intelligences have developed ways of producing enough energy to distort space-time or to create usable wormholes. To do this, their technology would need to be thousands of years in advance of our own, but this is feasible given the age of stars and the relative rates of evolution upon different planets.

Skeptics use the difficulties associated with interstellar travel as an argument against the possibility of alien visitors, but this is a narrow-minded approach and implies that other civilizations could not have developed faster or earlier than us. Far more damning are the pseudoscientific explanations for UFOs and alien visitations currently filling the Internet and the newsstands, coming from enthusiasts themselves.

It is an interesting fact that descriptions of UFOs conform neatly to the historical period in which the observation is made. At the beginning of this century, witness descriptions of alien craft often bore a remarkable resemblance to airships and elaborately decorated flying machines not unlike something H. G. Wells would have described. Today, UFOs are supposed to use warp drives not unlike those that power the fictitious *Starship Enterprise*. Yet, hundreds of thousands of sightings are reported each year. Even if just one of these is genuine, there is a case for extraterrestrial travelers paying us a visit.

Clearly, if a culture can survive long enough, it will eventually develop the technology to do almost anything within the laws of physics. But this isn't to say that we are being visited by hoards of aliens, some of whom are in league with organizations on this planet with intentions of subverting our society. Equally, the case for alien abduction appears badly flawed in spite of the fact that thousands of apparently normal people report incidents each year.

As we shall see in Chapter 13, the most common explanation for abduction is that alien beings are (a) interested in studying humans and (b) involved in genetic experiments. The problem with these arguments is that any race so advanced as to manipulate space-time to travel across the galaxy would not need to conduct physical examinations or to physically extract genetic material. Putting aside the argument that any race so advanced would probably consider such behavior immoral, both processes as described by abductees are ridiculously crude and sound suspiciously like the product of overactive but underdeveloped human imaginations.

I hope that one day our civilization will design and build interstellar space drives of one form or another, and I am sure others already have. Alien races may have once visited a rather insignificant little planet called Earth, and perhaps even pass this way from time to time, but notions of invasion forces or subversive three-foot-tall gray beings with infeasibly large eyes is suspect, to say the least. It implies that the human race is in some way special, and anyone who believes in life on other planets cannot also think in that way. In this sense, the beliefs of some UFO enthusiasts differ little from the egocentric fallacies of any other organized religion.

2 Is There Anybody Out There?

Alone, alone, all, all alone,
Alone on a wide wide sea!

SAMUEL TAYLOR COLERIDGE

Even though we cannot yet build interstellar spaceships, we may look at the stars and wonder: are we alone in an infinite universe or is the cosmos as full of life as the Earth is? This has been perhaps the biggest question facing our species since we developed the ability to think beyond our material requirements, and even now, we are only edging slowly toward an answer.

Scientists know there is certainly life on one planet—the Earth. But because we have only this one example upon which to build hypotheses, knowing for sure whether the series of events leading to life here is unique or extremely common is impossible—until we have more evidence. And, because of the almost unimaginable distances between stars, only now that we can travel outside our own planetary atmosphere and have developed machines that can see into the deepest recesses of space may we begin to hope to reach conclusions.

It has been known for most of this century that life on other planets without our neighborhood, our solar system, is extremely unlikely. Of the two planets nearest to the sun—Mercury and Venus—Mercury has extremes of temperature, with part of its surface an inferno and the other a frozen wasteland, while Venus has surface temperatures of some 800° K. Beyond the third planet, Earth, lies Mars, once thought to be the most likely contender for extraterrestrial life within our solar system. In fact as recently as 1877, the astronomer Giovanni Schiaparelli created a flurry of excitement by announcing that he had observed a network of what he called *canali* on the surface of the planet. *Canali* was wrongly translated into English as "canal" instead of its true meaning, "channel," and astronomers all over the world began to see increasingly complex canal systems as the rumors spread. Sadly, although the news inspired H. G. Wells to write *The War of The Worlds*, there are no canals on Mars—the effect was produced by a natural coloration of the surface.

The Viking probes of the 1970s found no trace of life on Mars, not a single microbe, and although some enthusiasts point to the fact that surface conditions may once have been more conducive to Earthlike animal and plant life, there is still no evidence to support the idea that life has ever appeared on the red planet.

Beyond Mars lay the gas giants, Jupiter and Saturn, with atmospheres containing toxic gases constantly churned up by powerful magnetic fields. The two largest satellites of the solar system—Titan, orbiting Saturn; and Ganymede, Jupiter's largest moon—might be more promising, and the Voyager probes that have passed close by have found what is believed to be organic molecules on the surface of Titan. But with surface temperatures around −150 degrees centigrade, and toxic atmospheres with little trace of oxygen, the chances of these molecules developing into living matter is very small. The Jovian moon, Europa, is now thought to hold the greatest promise because it is believed to possess subterranean water, which greatly enhances the chance of finding life there; but we will need to investigate this moon further before we know for certain whether it is a site for life beyond Earth.

At the outer edge of the solar system, Uranus, Neptune, and Pluto offer little comfort for those hoping to find what scientists refer to as carbon-based life-forms, because again the temperatures are either too low, their atmospheres noxious, or, in the case of Uranus, the entire planet is a single ocean of superheated water, warmed by volcanic action and covered in poisonous gases.

To find life, particularly life we can readily recognize, we must turn our thoughts and our telescopes and probes beyond the tiny confines of our solar system to the distant stars. But the problem we face then is distance. Our solar system is vast by everyday scales, some twelve billion kilometers across, but it becomes a meaningless speck and such numbers trivial when we begin to imagine contacting beings living on planets orbiting other stars.

The nearest star other than our own sun is Proxima Centauri, which lies 4.3 light-years from Earth. As we saw in Chapter 1, this is a staggering distance. What this means is that light, which travels at just over 300,000 kilometers per second, would take 4.2 years to get here. This is equivalent to a distance of 300,000 times the number of seconds in one hour (3,600) times the number of hours in one day (24) times the number of days in one year (365) times 4.2, which comes to a little under 4×10^{13} kms (4 with 13 noughts after it, or 40 million million kilometers). This is roughly equal to 100 million trips on an Apollo spacecraft to the moon. At the speed Apollo capsules traveled (about 40,000 kph) it would take about 100,000 years to cover the distance to even this, our nearest neighbor.

So, until we develop the technology to improve our speed, we can only hope to (a) contact aliens using light-speed signals such as radio waves, (b) wait for them to contact us, or (c) use telescopes of various types to try to discover as much as possible about other planets we may find orbiting nearby stars.

But just what are the chances of there being intelligent life beyond our own world?

Scientific opinion is split. There are those, like the astronomer Frank Drake, creator of the first SETI (Search for Extraterrestrial Intelligence) project, or scientist and author the late Carl Sagan who have claimed the universe is teeming with life. At the other end of the spectrum, writers and pundits, such as Marshall Savage, author of *Millennial Project*, and the physicist Frank Tipler, think we are totally alone.

The problem with trying to come up with any form of definitive answer or even an approximation is that we have no clear idea of all the variables to be considered or how these interrelate. For example, how likely is it that molecules of DNA can form given a long enough time period? How frequently do planets form around stars? How likely is it that even complex molecules can evolve into living material? We know all these things have happened at least once, but has it been only once, or billions of times?

To try to quantify the argument, in 1961 the pioneer in the search for extraterrestrial intelligence, Frank Drake, produced a now famous formula that has since become known as the Drake equation. It is very straightforward and a surprisingly powerful tool for the astronomer, except that almost all the variables can show a range of values and no one is yet sure what numbers to put in. It is the work of astronomers, biologists, and geologists to gradually narrow down each of those numbers to something more workable and to then come up with some form of answer to the Drake equation.

The equation is:

$$N = R \times f_p \times n_e \times f_l \times f_i \times f_c \times L$$

Although this might look daunting, it is actually as easy to use as working out your expenses. The letter N signifies the number of civilizations in our galaxy trying to make contact. Each of the symbols on the right-hand side of the equation represent separate factors that have to be considered in addressing the question: Is there life beyond Earth? (Each term is considered in isolation, in other words, the number assigned to say f_p is independent of that given to L, f_i or any of the others.) When numbers for all of these factors are plugged in, we end up with a figure for N. So what are these factors?

First, R stands for the average rate of star formation. A common misconception is that the universe was made at the time of the Big Bang and that was it, no change ever since. Of course this is not the case. The prevailing theory is that the universe is expanding and stars and planets are being created and dying constantly. Scientists are beginning to actually see this birth process using instruments such as the Hubble space telescope. It seems that some parts of the galaxy are more fertile than others, and the process of star birth is far slower than it was at distant points in our galaxy's past, but at a conservative estimate, astronomers think that about ten new stars are formed in our galaxy every year. So R is one of the variables that is pretty much agreed upon: 10.

f_p is the fraction of stars that are "good" and could contain planetary systems. By "good," astronomers mean suitable for forming and keeping stable, Earthlike planets in orbit around them. This is a rather complex matter. The age of the star must fall into a certain range. If it is too old, its fuel will be running down and it will emit radiation that would be unhelpful for the formation and sustaining of carbon-based life. Also, as a star gets older, the rotation of planets in orbit around it begins to slow. If the star is more than around six billion years old (our sun is about five billion years old), this will have a dramatic effect. Planets orbiting very old stars will have stopped rotating altogether and will have one face permanently turned toward the sun and the other existing in permanent night. If the star is too young, it may not have had time to allow planet formation and the mechanism that creates and evolves life-forms to run its course.

More important, planets that can sustain life-forms capable of developing civilizations could not be found orbiting pulsars or quasars—exotic stellar objects that would emit forms of damaging radiation—nor could the home star be unstable over long time periods.

Finally, many stars are binary—that is, they are made up of two stars orbiting one another. Although this system by no means rules out the formation of planets, binary stars are generally considered less likely to possess Sol-like systems than single stars.*

When Drake first suggested his equation, the value for f_p could only be guessed at, but recently astronomical findings have begun to narrow the range of numbers this could be. Back in the early 1960s, Drake placed f_p at about 0.5; in other words, half the number of stars in the galaxy were potentially able to form planets, but then, when observational techniques improved and new data was gathered, the initial results showed this figure to be wildly optimistic.

Using present-day technology, planets cannot be seen in the way astronomers observe the planets in our own solar system—the distances, as we have seen, are simply too great. It has been estimated that a telescope the size of the moon would be needed to observe clouds and coastlines on a planet within fifty light-years of Earth. And that is not the only problem. Imagine trying to detect the presence of a firefly perched on the edge of a spotlight from a few hundred miles away. The light from the spotlight would completely

---

*Sol is the name given to our sun.

swamp the effect of the firefly. In the same way, the light from an orbiting planet, which is merely reflecting light from its sun, would be totally overwhelmed by the far greater magnitude of the star.

These two problems would, you might think, stop us ever knowing if any star other than our own possesses planets, but there are other ways of knowing if a planet orbits a distant star.

The best technique we have today is observing "wobble." If you can picture a hammer thrower at the Olympic games spinning on the block and just set to let go of the hammer, the athlete, who is substantially heavier than the hammer, has a greater pull on the chain and the hammer, but the hammer (which weighs about seven kilograms) also has a pull on the thrower, who might weigh around twenty times as much. With suitable instruments this pull, or "wobble," could be measured. In the same way, a planet in orbit around a star will exert a pull on the star in an identical but much smaller way than the star pulls the planet toward it. Obviously, the bigger the planet, the greater the effect.

Even so, this is a very sensitive technique, and the difficulty of measuring the wobble of a star has been compared to using a telescope on Earth to see a man waving on the moon. Yet, within the past two years a refinement of this procedure has been used successfully by astronomers based in Geneva in Switzerland—making them the first to confirm the existence of an alien planet.

The discovery of a planet orbiting the star 51 Peg in the constellation of Pegasus, announced in October 1995 at a conference in Florence, startled the astronomy community and made headlines around the world. The planet (which is still unnamed) is half the mass of Jupiter (the largest planet in our solar system, about three hundred times larger than Earth) and orbits the star at a distance of about eight million kilometers.

Although most astronomers believe it unlikely that gas giants like Jupiter could sustain carbon-based life, it is suspected that solar systems that have any chance of containing an Earthlike planet need at least one Jupiter-type planet, which would probably be found farther away from the star than the region in which solid, cooler planets would be located. The reason for this is that gas giants act

Figure 2.1

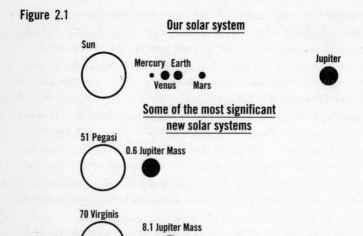

Our solar system

Sun
Mercury Earth   Jupiter
Venus   Mars

Some of the most significant new solar systems

51 Pegasi
0.6 Jupiter Mass

70 Virginis
8.1 Jupiter Mass

47 Ursa Major
3.5 Jupiter Mass

Comparison of Planets					
	Earth	Jupiter	51 Pegasi	70 Virginis B	Ursa Majoris B
Mass	0.003	1.00	0.6	8.1	3.5
Diameter	0.09	1.00	0.3–1.30	0.3–1.00	0.3–1.1
Distance from star	1.00	5.20	0.05	0.43	2.1
Day temp (°C)	15	-150	1000	85	-80

like vacuum cleaners, soaking up stray asteroids, comets, and meteors that enter the system. In this way they protect the inner, Earthlike planets, facilitating the chance of a stable environment within which life could form and a civilization develop. However, the planet found around 51 Peg is in the wrong place. It is too close to the star and it circles the star too rapidly, taking only four days to complete an orbit (Mercury takes eighty-eight days). But it is a planet, and the discovery has created a revolution in our way of thinking about the cosmos. We now know our solar system is definitely not the only one.

Within months of the discovery of the planet orbiting 51 Peg, more solar systems were discovered. In January 1996 two new planets were found around different stars, one orbiting 70 Virginis in the constellation of Virgo and the other in the constellation of Ursa Major, a star named 47 UMa. Both of these are around thirty-five light-years from Earth. Both are Jupiter-type planets, but these are in the "conventional" position around their stars. Since these discoveries, around a dozen planets have been located orbiting different stars, adding further substance to the argument that our solar system is not unique.

So what do these findings mean for the number we assign to f_p? The search for stars containing planets using the technique of "wobble" detection has so far spanned over a decade and involved choosing hundreds of Sol-type stars. Until these recent discoveries, none of the stars gave positive results, and astronomers were beginning to despair of ever finding a candidate. So, although these recent finds have greatly encouraged astronomers, they have also shown how the initial guess was too high, and at the same time illustrated how difficult it is to assign numbers to the Drake equation. Instead of 0.5, a more conservative estimate might now be 0.1, or that one in ten stars are capable of developing and sustaining a planetary system.*

Next we come to n_e, which is the number of planets per star that are Earthlike. Once again we are bound by very limited experience. In our solar system, there is really only one Earthlike planet. Mars may once have possessed a more conducive atmosphere than it does today, but now the surface temperature ranges from –50°c (223K) to 0°c (273K), and the atmosphere is so thin (with an atmospheric pressure about $\frac{1}{100}$ of Earth's), that humans would need to carry their own oxygen supply to work or move around on the sur-

---

*A caveat to this is of course that the technique can only detect large planets and the theory that planetary systems invariably contain gas giants may be wrong, so the figure may be higher.

face. At the other extreme, Venus has an atmosphere made up almost entirely of carbon dioxide (CO_2) which produces such a severe greenhouse effect that the Venusian surface is hotter than that of Mercury.

So, again being conservative, let us put n_e for our solar system at one.

f_l in the Drake formula stands for the fraction of Earthlike planets upon which life could develop, and in trying to assign a number for this, we really are in an almost totally unpredictable domain.

First, what is life? It might seem an obvious question, but biologists are quick to point out that the standard parameters are open to debate. Living beings grow and move, but so do crystals, producing regular patterns made of repeated simple units much like cells, and inanimate water or any other liquid can flow, or move. Lifeforms use energy, but so do computers, trains, rockets. Perhaps a better definition would be that a living being can *control* energy.

An alternative could be to argue that only living things process and store information, but this is the sole purpose of a computer. The debate about the possible future development of intelligence by complex computers still rages, yet the desktop computer I'm using to write this could not be described as living. So, what other criteria could we use? Could the ability to reproduce constitute life? But flames reproduce. Probably the best definition would be that lifeforms reproduce and pass on genetic material, inherited characteristics, to their offspring.

For the purposes of this book, I'm really interested in reaching a conclusion about intelligent life-forms with which we can communicate readily. It may be that any number of exotic creatures live in this almost infinite universe, but the chances of contacting them or communicating with them is even less likely than the probability of encountering a life-form with which we can communicate. There is even the possibility that we have encountered these beings and have, for one reason or another, been totally unaware of them or they of us. So, to allow for "life as we know it," we need to think in terms of carbon-based life able to communicate with us. But why carbon-based?

According to the laws of physics, which in turn give us the rules of chemistry and subsequently biology, carbon is the only element that can form complex molecules, known as organic molecules. For "life as we know it," the alien beings must operate naturally within the same narrow limitations of temperature, pressure, and radiation as we do, not least because that would be the only way we may be able to communicate with them. Within those parameters the only element to form organic molecules that constitute cells, tissue, and flesh is carbon. Carbon has the unique ability to form very strong interatomic bonds with a large number of elements such as nitrogen, oxygen, and hydrogen, as well as multiple bonds with other atoms of carbon. This allows it to form a vast range of molecules, some of which can contain tens of thousands of atoms. No other element can approach this level of versatility.

For a planet to be the home of carbon-based life, it had to possess a certain set of environmental conditions and materials in its primeval history, and a subsequent set of finely tuned conditions and materials for that life to have evolved and flourished.

Skeptics argue that these conditions are unlikely to be duplicated in the universe and that the probability of life evolving is therefore slight, but there is a growing body of evidence to oppose this view.

In 1953, just as the structure of DNA was being elucidated by Crick and Watson in Cambridge, two scientists at the University of Chicago, Stanley Miller and Harold Urey were investigating the initial conditions on Earth that produced the biochemical environment in which life began. It was known that life had originated on Earth a little under four billion years ago and that the predominant chemicals in the atmosphere at that time were ammonia, water, and methane. Miller and Urey placed these chemicals in a jar and allowed an electrical discharge to pass through the mixture. After sustaining this process for several days they found a red-brown deposit had formed at the bottom of the jar. When this "primeval soup" was analyzed, they discovered that it contained amino acids— organic molecules that act as the building blocks for all life on Earth.

Further experiments showed that a wide range of molecules essential for life could be formed in this way. The pair added another simple molecule, hydrogen cyanide (HCN), found in volcanic gases, and discovered that a number of complex molecules that are key to the formation of proteins and DNA were formed. Although these molecules are a long way from the giant structures of DNA (deoxyribonucleic acid) and RNA (ribonucleic acid), the elaborate molecules that encode the production of proteins and the day-to-day workings of cells, Miller and Urey postulated that a "soup" of life-forming molecules could have been brewed in the Earth's atmosphere during the space of just a few years. Recently, Stanley Miller has declared that these molecules could have developed in complexity and produced living cells within perhaps as little as ten thousand years. Flying in the face of critics who claim life on Earth is unique, based on his own experiments, Miller is sure that given the correct environmental conditions and the proper blend of chemicals, life could form on any planet.

Finally, the fossil record shows us that life began on Earth at the earliest opportunity. In 1980 fossils of creatures called stromatolites, or "living rocks," were found in the Australian desert, which were the simplest and probably the most ancient form of life on Earth, living over 3.5 billion years ago. We know the environmental conditions for life only became suitable around four billion years ago, and so it would seem that a period perhaps of only a few hundred million years passed before the very simplest life-forms began to appear. This does not provide evidence for the evolution of life on planets other than Earth, but it illustrates the notion that given the correct conditions, life will appear readily. The great proponent of extraterrestrial life, Carl Sagan, wrote: "The available evidence strongly suggests that the origin of life should occur given the initial conditions and a billion years of evolutionary time. The origin of life on suitable planets seems built into the chemistry of the universe."[1] What he meant by this is the principle of self-organization.

In recent years it has been suggested that certain physical and chemical systems can leap spontaneously from relatively simple states to ones of greater complexity or organization. This organizing principle, some argue, is a form of anti-entropy effect which may be, in some mysterious way, linked to life itself. Entropy is the "level of disorder" in a system, and in nature entropy always increases— an apple left to stand will gradually decompose, its cells breaking down, and the "neat, organized" form of the fresh apple will decay into a disorganized mush. The self-organization principle could, it is believed, help to reverse the natural tendency for entropy to increase in the universe. As a consequence, the chances of life deriving from a collection of complex organic molecules are greatly increased.

According to Frank Drake, "Where life could appear, it would appear."[2] He assigns a value of 1 to the parameter f_l. In other words, there is a hundred percent chance that a suitable planet will form life if it has the correct conditions. Others, such as the Nobel prize–winning chemist Melvin Calvin, and Carl Sagan, have concurred, believing life is more likely than not to form on a suitable planet.[3] For others, those who do not believe in extraterrestrial life, the value for f_l is the most crucial of all the terms in the Drake equation. They place it at zero, which would consequently make N equal to zero—no life anywhere else in the universe. The value for f_l probably cannot be anything other than 1 or zero, so, for the purposes of our discussion, I will give it a value of 1.

Next we turn our attention to f_i. This is the term that refers to the fraction of Earthlike planets where life has become intelligent. Again, when we first contemplate this expression, we are struck by the need to define. This time the question is: What constitutes intelligent life?

Many would argue that dolphins and whales are highly intelligent animals that could have formed a civilization if they had evolved on land. They can communicate with members of their own species and have been known to interact in a highly intelligent fashion with humans. Attempts have even been made to decipher the complex sequence of clicks and squeaks they use to communicate with one another.*

In a different sense, ants and bees act in an intelligent fashion when considered as a collection of individuals, each acting as a unit in a larger society, a gestalt. So, if we were to apply Drake's equation

---

*Interestingly, these efforts have so far met with only limited success, which may act as a salutary lesson for those attempting to contact alien civilizations on distant planets. Supposing we eventually do make contact, how likely is it that the two races will ever understand one another?

to our planet, we could arrive at a value for f_i between 1 and at least 4, but again being conservative, let us take the value of 1, representing the human race.

The penultimate term, f_c, represents the number of intelligent species who would want to communicate with us, and again we are faced with a highly subjective parameter for our equation. In order to use this term we have to place some limitations upon how we arrive at a value. We must first assume that an intelligent race uses some form of electromagnetic radiation with which to communicate and to interact with their universe. Most scientists would agree that it would be unlikely for an intelligent species to have developed without using any form of electromagnetic radiation. An alien intelligence may utilize extreme regions of the spectrum, they may see in the infrared or the ultraviolet because of the nature of the light emitted by their sun. Alternatively, they may live in extreme conditions such that vision is as unimportant to them as it is to some deep sea creatures, but whatever extreme situation may be imagined, they must use some form of electromagnetic radiation. If this is not the case, then such an alien race would fall out of the category of "life as we know it."

As a civilization, we utilize a range of radiation, from radio and television signals to X rays, from ultrasound to microwaves, so it is likely that any civilization at least as advanced as us would also employ similar electromagnetic waves within their technology; they may even have developed something similar to television or radio. Even if they had not created an entertainment system that leaked signals into space as our televisions have in recent decades, but were actively interested in communicating, they should be able to build equipment that would receive and decipher signals from space.

This, then, leads to a question concerning the sociological and psychological makeup of an alien intelligence: Would they necessarily want to communicate with us? It is a serious point that the signals we have been sending inadvertently into space may have presented our race in a very poor light. For some fifty years our calling card has been television signals conveying images of everything from the most violent Hollywood films to news coverage of war, famine, and torture, leaking into space in far greater quantities than any form of contrived, politically correct message we may wish to send to our celestial neighbors. Many of these signals would be too weak to reach distant stars, but this is perhaps underestimating the sensitivity of alien detection systems. Television signals are no different than any other forms of electromagnetic radiation, in that they travel at the speed of light. It is therefore conceivable that alien civilizations living on planets up to fifty light-years away could be chuckling at our antics, or perhaps battening down the hatches for fear we'll ruin the neighborhood.

So, what value do we give f_c? On the one hand, it would seem likely that any civilization would eventually develop a form of long distance receiving and transmitting system using electromagnetic radiation enabling communication, but how many races would want to make contact? There could be an abundance of races busily communicating with one another but excluding us; equally, alien civilizations could prefer to keep themselves to themselves whether or not they have been warned off. Weighing up these factors, a conservative estimate would be that ten to twenty percent of intelligent aliens would want to communicate, so f_c would be, say, 0.1.

Finally, we come to L, which represents the lifetime of a civilization (in years). And again, in attempting to assign a value, we face another complex series of permutations. The value of L puts us out on a limb, for we have to consider hypothetical sociological factors for a hypothetical race, but we have again one example to draw upon—our own experience.

It suggests an interesting synchronicity that our race developed weapons of mass destruction at almost the exact point we revealed ourselves to the universe with our electromagnetic signals, and it could be that many races are destroyed at the very point they could make contact with their neighbors.

Since Frank Drake first devised his formula in 1961, there has been much debate amongst scientists about the value of L, and during the past thirty-five years the political and social zeitgeist has altered radically. The cold war has ended, but the threat of nuclear destruction is still very much with us, and the killer instinct of human kind has not changed in the slightest. Perhaps, the ease with which human beings make war is irreducibly linked with our drive to progress and advance. It is possible the instinct that drives us to communicate derives from the same source as our aggression. If this is the case, it may be no different with other species, it may even be a natural law. In such an event, it would be likely that a large proportion of civilizations destroy themselves at around the time they develop the technology to communicate beyond their own world.

There are also a number of other ways in which a civilization based on a single planet can be destroyed. As we shall see in Chapter 19, scientists are only now beginning to realize the very real danger of planetary collisions with comets or asteroids. It is believed that a devastating asteroid collision caused a sudden traumatic alteration in the ecosystem of the Earth some 65 million years ago, resulting in the extinction of the dinosaurs, and there have even been a number of documented near-Earth collisions this century. The massive explosion reported in Siberia in 1908 which devastated hundreds of square kilometers of forest in the region of Tunguska is thought to have been caused by a meteorite exploding several kilometers above the ground.* If this had occurred over a major city, millions would have died. An object only a few times larger than the Tunguska fireball impacting with the Earth would not only devastate a wider area, but the dust thrown up by the collision could produce a blanket around the entire planet capable of destroying all life on the surface. If it landed in the sea, the tidal effects would be almost as damaging.

There is also the question of planetary resources. We as a race are perilously close to overexploiting the resources of our world, and we are already capable of severely damaging planetary mechanisms that are there to maintain an ecological balance. It is conceivable that other civilizations have followed the same path and gone further, completely destroying their own environments. Such threats as reduced fertility, AIDS, superbugs, and nuclear terrorism are all further potential civilization-killers.

One conclusion to be drawn from these considerations is that civilizations either survive little more than one to two thousand

---

*Although some UFO enthusiasts offer the alternative theory that it resulted from the explosion of a spacecraft in the Earth's atmosphere.

years, or else they continue for perhaps hundreds of millennia. It is possible that many races pass through a "danger zone" during which they have a high chance of destroying themselves, but if they come through it, they develop into highly advanced cultures capable of interstellar travel and colonization.

In an extreme case, L also depends upon astronomical factors. If we assume that life may form on a large number of planets, and that those life-forms could evolve into intelligent civilized beings, the time at which life began on their world would be a crucial consideration.

The universe is believed to be approximately twelve billion years old, and our sun is a very typical star located some two-thirds of the way along one of the spiral arms of the Milky Way galaxy—itself an "ordinary" galaxy amongst an estimated 100 billion others. In astronomical and geological terms, the Earth is quite average and life began to appear here four billion years ago, or around eight billion years after the Big Bang. But it is quite conceivable that a great many planets around other, older stars cooled long before our own planet. Astronomers have observed the death of stars far more ancient than our own. If any planet around these stars had brought forth life, any civilization that formed there would either be ancient, interstellar voyagers, or long dead.

To find a sensible value for L, we must assume a normal distribution of ages for successful civilizations. If L for a particular planet is 2,000, the race may have destroyed itself and is of no further interest. But L could be much larger. It is possible there have been and still are civilizations hundreds of millions of years old. Equally, there would be a large number of very young civilizations, perhaps no more than two or three thousand years old. Most civilizations that have survived and are able to communicate would be somewhere between the two extremes.

Drake and his colleagues have placed a value of 100,000 on L. This seems rather arbitrary, but nevertheless, if we put any figure over the two-thousand-year watershed, the equation gives us a correspondingly large number for N, the number of advanced civilizations wanting to make contact.

We assigned R as 10, $f_p = 0.1$, $n_e = 1$, $f_l = 1$, $f_i = 1$, $f_c = 0.1$,

L = a large number. Now if we put these figures into the Drake equation, we reach a very simple conclusion.

$$N = 10 \times 0.1 \times 1 \times 1 \times 1 \times 0.1 \times \text{a large number}$$

The 10 and the 0.1 cancel out, giving us $1 \times 1 \times 1 \times 1 \times 0.1 \times$ a large number, which equals $0.1 \times$ a large number.

If we call L (the average age of a civilization), say, 100,000, it means there are ten thousand civilizations sharing just this single galaxy (one of 100 billion, remember). Frank Drake believes the value of L to be much greater, which would mean that N would be correspondingly larger, and, according to some enthusiasts, N could be in the tens or even hundreds of millions. This may seem excessive, but when we consider that our galaxy contains upward of 100 billion stars, an estimate of 100 million civilizations means that there is only one such race for every one thousand stars.

So, how likely is it that we will communicate with one of these civilizations? Thus far every attempt to make contact has failed utterly; but that has not been for want of trying. The first serious scientific effort to listen in on a possible alien dialogue (or monologue) came from a paper published in the science journal *Nature* in 1959. It was written by an Italian astronomer, Giuseppe Cocconi, and American physicist Philip Morrison, who reasoned that if aliens wanted other species to contact them, they would make it as easy as possible for them to do so. What they meant by this was that an alien intelligence would broadcast a signal that would have universal meaning and be within the range that would be transmittable and detectable via radio telescopes. Cocconi and Morrison chose as their standard the frequency 1.420 GHz (gigahertz). The reason for this choice is that 1.420 GHz is the frequency at which the element hydrogen is known to resonate. Because hydrogen is by far the most common element in the universe, it was assumed an alien intelligence that had developed radio technology would know this fact and expect a recipient to be equally knowledgeable.

Enthusiasts immediately began scanning the sky, calibrating their radio telescopes to this frequency. The first SETI project was led in the early days by Frank Drake, who in 1960 established a team based at Green Bank, West Virginia, in the United States. This was followed soon after by another SETI project created at a giant radio telescope center called Big Ear, in Ohio, directed by astronomer Bob Dixon, who has been searching the skies ever since.

Today there are a number of SETI projects running concurrently around the world. NASA created a $10 million a year program in 1992, approaching the task from a different angle to the early experimenters. Whereas Drake and others had focused on the frequency of 1.420 GHz and a select group of prime candidate stars within fifty light-years of Earth, NASA decided to blitz the heavens with a broad sweep of all likely stars and a wide range of frequencies. Sadly, we will never know if the project would have garnered conclusive results because, little more than a year after it was established, funding was terminated by the interference of congressmen who thought they were paying for the indulgence of UFO cranks. One prominent member of Congress was reported to have said:

> Of course there are flying saucers and advanced civilizations in outer space. But we don't need to spend millions to find these rascally creatures. We need only seventy-five cents to buy a tabloid at the local supermarket. Conclusive evidence of these crafty critters can be found at checkout counters from coast to coast.[4]

Such ignorance does nothing to help the serious search for life beyond our planet, but luckily there are others who have a more open-minded approach. Rather aptly, Steven Spielberg, producer of E.T. and Close Encounters of the Third Kind, is funding a project based on the East Coast of the United States, and other wealthy enthusiasts are putting money into the search at a variety of sites around the world. The original NASA SETI project was scrapped but picked up by others with private funding, and in 1995 Project Phoenix was initiated. This is a pan-global operation using the world's largest radio telescopes to sweep the universe between 1 and 10 GHz, which happens to be the region where there are many natural resonances and emissions (including that for hydrogen). NASA has also developed what they call spectrum analyzers to search across wide ranges of frequencies within the chosen limits and are currently devising software to filter out noise and other interference in the signal.

Yet, despite the money and advanced technology, after forty years of searching, no conclusive evidence for what have been dubbed LGMs (Little Green Men) has been found. One famous false alarm in 1967 illustrates the difficulties facing astronomers in their search.

A Ph.D. student at Cambridge University named Jocelyn Bell detected a strong, regular signal coming from deep space in the region of the spectrum then thought most likely to contain a contact frequency. After reporting the finding to her supervisor, Anthony Hewish, they agreed they would not go public until they had investigated the signal fully. Gradually, they eliminated all possible sources both terrestrial and celestial, until they realized that the signal was actually an emission from a strange object in deep space which was sending out an almost perfectly regular pulse. The object was then discovered to be a neutron star, or pulsar, the remains of a dead star that had collapsed under its own gravitational field so much that the electrons orbiting the nucleus of the atoms making up the star had been jammed into the nuclei and fused with protons to form neutrons. This superdense matter emits pulses with such regularity that pulsars are thought to be the most accurate clocks in the universe.

Since this discovery by Bell and Hewish, other regular signals have been detected that have not originated from pulsars or any terrestrial source, but have appeared only once. A team lead by Professor Michael Horowitz at Harvard University has reported thirty-seven such signals during the past ten years, all within twenty-five light-years of Earth, but because they have not been repeated, they do not qualify as genuine signals from a race trying to contact us. They could of course be one-off leakages from specific events, but we might never know, and for scientists to analyze a signal properly, they need a repeated, strong, regular signal.

The search goes on, and with the recent confirmation that other stars possess planets, some of the thunder has been taken from cynical doubters, such as the congressmen who saw no value in the funding of SETI. If there is life on other planets, which would seem more likely than not, we do have a chance of contacting them one day. If one in every thousand stars have planets that are homes to advanced civilizations, we should have at least one such star within fifty light-years of Earth. Perhaps we are sending signals at the wrong frequency and tuning our instruments to the wrong region for that particular civilization. Maybe we are unlucky enough to be neighbors to a race who either do not want to be contacted or have developed a technology that is so different from ours that we cannot yet reach one another. It could also be that signals sent from more distant planets have not reached here yet.

Skeptics quote what has become known as Fermi's Paradox, after the famous Italian physicist Enrico Fermi who declared in 1943 that if aliens existed, they would be here. But this is not only a tautological argument, it demonstrates staggering arrogance, assuming that it is a simple matter to detect alien life, or that we are so important that aliens could want nothing other than to contact us.

Whatever the future holds for the efforts of scientists to contact alien beings, the effect of such a discovery upon our society and the mind-set of each of us would be enormous. As the physicist Paul Davies has said:

> There is little doubt that even the discovery of a single extraterrestrial microbe, if it should be shown to have evolved independently of life on Earth, would drastically alter our world view and change our society as profoundly as the Copernican and Darwinian revolution.[5]

Although this is undoubtedly true, proof that intelligent beings have visited or are still visiting us here would be exponentially more significant. In fact, it would be the biggest news story in history, and might, just might, be waiting for us just around the next corner.

3 The Mind's Eye

One touch of nature makes the whole world kin.

WILLIAM SHAKESPEARE, Trolius and Cressida

The idea that humans can communicate directly from mind to mind is probably as old as civilization itself, and a limited form of this skill may derive from a time before that. During the past hundred years countless experiments have been conducted in an effort to pin down the phenomenon and to attempt to explain how it could work. These have become gradually more refined and researchers have managed to eliminate almost every possible way in which the phenomenon could be faked, but the scientific community still sees such tests as little more than elaborate tricks.

The reason for this skepticism is that telepathy, if it does exist, is an elusive phenomenon which has been dubbed *jealous* because it often does not work properly in the presence of a skeptic, nor does it usually comply with the wishes of the experimenter. To the nonbelievers this is a symptom of what is called a nonfalsifiable hypothesis, or a hypothesis against which there can be no evidence. An example would be for me to say my pet cat, Sophie, is in fact the creator of the universe and that the universe did not exist before she created herself five years ago. You may argue that you remember the world six years ago, but I could then counterclaim that when Sophie made the universe, she put the idea into your head that you lived before that date and implanted memories and images to accompany it. There is no way such a hypothesis can be disputed using pure logic or reasoning; it is nonfalsifiable.

But with many paranormal phenomena there are other reasons

for skepticism. One of the central tenets of science is repeatability. If a scientist claims to have observed a physical phenomenon and conducted experiments to measure the effect, it is only taken seriously by the scientific community if the effect can be repeated under identical conditions by other scientists. If the experiment is unrepeatable, serious doubt is cast upon the original evidence. An example of this is the cold fusion experiments mentioned in Chapter 1. When scientists around the globe followed precise instructions provided by the originators of the theory and could not obtain the same results, they gradually realized that a mistake must have been made in the seminal experiment. The same attitude has rightly governed the approach of the scientific establishment toward telepathy and many other paranormal phenomena. But for the purpose of seeing how telepathy could work, we have to suspend disbelief and look closely at possible mechanisms.

What do we really mean by telepathy? The image from science fiction is of a being with the power to look into another's mind and to pluck out thoughts as they wish, or to manipulate the thoughts of their subject, to make them do things against their will. This, though, is an extreme form of telepathy; it may well be that all of us are capable of a type of telepathic experience based not upon supernatural abilities, but enhanced perception.

The psychologist James Alcock illustrates this idea when he says: "I was standing in a cinema waiting to buy some popcorn, and was idly recalling a conversation I had once had with the brother of a colleague . . . A few moments later I turned around, and there about thirty feet away was the man himself."[1]

The initial response to such an event would be surprise and a feeling that perhaps you had just had a paranormal experience, but that would involve mixing cause and effect. What really happened in this case was that Alcock had noticed the colleague's brother subliminally *before* he started thinking about him. This is a well-known phenomenon in psychology, although for a long time this too was scoffed at by psychologists. The effect is called "backward masking" and has been validated by laboratory experiments.

If a subject is shown an image for about a tenth of a second, he will be able to recall some of the features of the image, but if he's shown a tenth of a second flash followed by another image lasting longer, the first is forgotten, although it can influence the reported description of the second. For example, if a picture of a man is shown for a tenth of a second followed by a different man holding a knife, the subject's description of the knife-wielding figure is influenced by the characteristics of the first subliminal image.

This is a rare experience for the majority of people—hence the feeling that something supernatural is going on—but other mental skills have been studied that could help to explain a natural process, something that might be called *ultrasensory perception* or USP.

We are all now familiar with body language or nonverbal communication, but few people use it consciously. Facial expressions, head movements, body positions, tones of voice, and even odor can send us subliminal signals, the interpretation of which is usually subconscious. Head and facial movements give the most information about the type of emotion being expressed, and we are all instinctively able to interpret these, but politicians and actors are trained to pick up more subtle signals, to utilize an enhanced, but quite natural, subconscious skill that we all possess.

These skills involve just our five senses, but instead of the information being processed by our conscious minds on full-alert for signals such as those we receive when we are concentrating on something, the images are filtered into other regions of the brain and siphoned off without our knowing it. Often they are only interpreted later. And our natural senses can sometimes surprise us with their sensitivity.

Recently, the phenomenon of the cocktail party syndrome was recounted in magazines and newspapers and caused a flurry of excitement. Cocktail party syndrome is the buzz phrase used to describe everyone's enhanced receptivity to their own name. Above the general hubbub of a cocktail party or some other noisy environment, we can pick out our name, even if it is whispered on the other side of the room.

The cocktail party syndrome is nothing more than a survival mechanism left over from early human development. If our name is mentioned, it means we might be called upon in some way. It may signal the approach of an aggressor or a rival attempting to identify us, or it could be for a reason beneficial to us, something we do not want to miss out on. Our ability to subliminally notice things of which we are not consciously aware is the result of a filter system. If we were to give equal importance to everything we picked up with our senses, we could not focus on what is important and get on with our lives. Our brains are programmed to know what is important and what is not, and we can grade these sensations.

It is quite possible that modern humans are not as adept at interpreting subliminal messages as our ancestors once were. The reason for this is that we have developed certain skills to a very high degree because of the culture we have established, and neglected others, causing them to gradually fall into redundancy.

Our ancestors, some suggest, used hand signals and other forms of body language before verbal communication was developed 50,000 to 75,000 years ago. One obvious reason verbal communication was developed and adopted over sign language is that it freed the hands and does not need a visual element. If a hunter was trapped by a wild animal as he returned to the tribe, he could call for help even while clinging to the branches of a tree as the animal circled below; sign language did not offer this versatility. Sadly, although we have retained a little of our ability to read body language and to sense a range of signals borne by smell, most human cultures have not been able to evolve a twin path of development, keeping in conscious touch with our primitive instincts while adopting the sophistication of language.

Although we do not notice them, these skills remain, and utilizing them could produce humans with highly developed talents most of us would consider paranormal. Imagine the skills specially trained individuals could display if their natural talents and their ability to communicate better with their unconscious facilities were developed to the equivalent of Olympic standard. Such people would be seen as genuine telepaths.

This is what Isaac Asimov had in mind when he created the Second Foundationers in his epic books comprising the Foundation series. Asimov did not believe in thought-transference by any mystical means and had no need for it in his fiction. He instead imagined a community that had been completely isolated from the rest of humanity and selected for their innate skills. They were then trained by adepts to utilize the full range of their natural human senses, and taught to read the slightest movement in others, including involuntary muscle movements. They could analyze the meaning of every voice inflexion, could sense odors at long distances and understand their meaning. Not surprisingly, such individuals appeared to the uninitiated as superhuman and their powers viewed as paranormal.

This scenario is perfectly within the bounds of accepted science and involves no use of supernatural powers but simply the development of inherent abilities stretched to the limit. In this way, such a program would be no different to the process by which Lyndford Christie or Pele became great sportsmen.

In fact, we do not even need to draw upon such exceptions. Consider the extraordinary sensitivity of the wine-taster's pallet or the link between the fingertips and the brain of a blind person reading Braille, or of the musician with perfect pitch.

There is also the possibility that if the human body is placed in an unusual environment where the range of signals is extended, we can utilize aspects of our senses we did not realize we possessed. When astronauts were first sent into space during the early 1960s, they reported seeing strange flashing lights. One explanation for this was that they were seeing images that would be beyond those normally experienced on Earth. Their eyes detected the images, but their brains were untrained in translating the signals.

Some relatively simple animals also display skills that appear on the surface to be paranormal. The dogfish catch flatfish by picking up tiny muscle movements of their prey hiding invisible beneath the sand of the sea bed, and some species of eel possess a "net" that surrounds them, a form of electromagnetic field that can detect the presence of other creatures within range, a little like radar. Again, there is nothing supernatural about these abilities, they are the result of particular evolutionary paths along which these animals have traveled.

Within human communities, some of these skills are still to be found in a limited sense. The Cuma Indians in the San Blas islands off the coast of Panama are said to use odor as a way of helping to judge one another's mood, and clasp each other under the armpits and then sniff their palms when they meet. Our sanitized version is to shake hands, but people from developed nations are also subliminally sensitive to smells, and these can influence our feelings toward others without us realizing it.

Everyone has sensors in their joints and muscles that tell us where we are in three dimensions, and others in our inner ear to pass on information about gravity and movement. We also have elaborate systems within our bodies that regulate temperature, monitor the level of chemicals, and control our highly sophisticated metabolism. Perhaps there is no need to look beyond these to interpret most telepathic experiences.

The difference between all of these sensory effects and telepathy is a question of scale. All forms of ultrasensory perception, from differentiating closely associated smells and flavors to the ability to register an image lasting only a tenth of a second, are measurable responses. If telepathy does exist and is an alternative phenomenon—as opposed to a collection of natural abilities—it must operate in one of two ways. Either it uses an extreme form of information transmission system with which we are already familiar—most probably a region of the electromagnetic spectrum—or else it utilizes a completely unknown form of information transfer. If the former is the case, the reason we have not been able to detect it is because our instruments do not operate within the necessary range or are too insensitive. If the second possibility is true, we may never develop machines with which we could detect or measure the effect, at least not until this alternative means of conveying information is understood. Marconi, the inventor of the radio, would not have thought of constructing his prototype without being aware that radio waves existed. If he had built a radio in ignorance, finding what to him would have been hypothetical radio waves would have been a hit and miss business.

The human brain contains about ten billion neurones or nerve cells, any one of which may have many thousands of connections to other cells, making it the most complex machine known to humanity. Each neurone acts like a binary gate in a computer, switching on or off, and in this way thoughts, emotions, decisions, and inspirations are formed and transmitted through a vast network. The

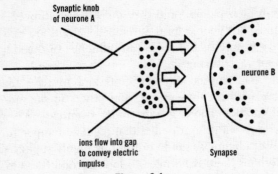

Figure 3.1

neurones are linked by axons, the tips of which do not actually touch. Instead, a signal is passed along the axon and crosses what is called a synaptic gap to another neurone in the space of about a millisecond (one thousandth of a second). This impulse has an electrical potential of about 120 millivolts and is produced by chemical means, using charged atoms called ions, which are triggered to fill the synapse and make the connection to a neighboring neurone.

It has been suggested that if telepathy is possible, the mechanism by which thoughts are transferred has to be explained at least to this level of the process. This may mean that some form of leakage occurs during the countless individual steps that constitute a thought and that a telepathic individual can somehow pick up this leakage and translate it into meaningful images.

This would be a little like using a phone-tapping device to listen in on someone under surveillance, except for one very important difference—the level of complexity involved when dealing with the human brain is several orders of magnitude greater. Phone taps work on the elementary principle of siphoning off a signal traveling along a single relatively large wire or using a remote receiver to access a signal sent between two individuals. By this analogy, the telepath (the mental phone-tapper) would access a single impulse between two neurones, but this would of course be quite useless because the simplest thought or instruction requires many thousands of neurones working in unison and a single "neurone tap" would gather almost nothing of use.

Perhaps a telepath can tap into a multitude of neurones simultaneously, but the difficulty with this idea is the decoding process. How would the telepath's receiving equipment manage to decipher all the trillions of impulses racing through the brain at any given moment? They may want to learn what their subject is thinking about a particular subject, and all they receive is the interference of signals passing on instructions to release enzymes, to scratch a leg, or to control the bladder.

Continuing with our phone-tapping analogy, a telepath would be tapping into the most sophisticated telephone exchange imaginable, trying to pluck out a few tens of millions of related conversations simultaneously, then piercing them together to make a coherent message. Of course, different regions of the human brain are responsible for different functions, so if the telepath could tune into certain regions, then the task might be a little easier.

An alternative suggestion, postulated by some parapsychologists, is that during the process of impulse transmission, the brain releases particles called *psitrons*. Although supporters speculate that these particles would be released in large numbers, they have not yet been detected. According to enthusiasts of the theory, this is because psitrons possess no mass or energy.

Although this sounds ridiculous, the notion of similarly ethereal particles is not without precedent. In the early days of quantum mechanics, physicist and Nobel laureate Wolfgang Pauli predicted the existence of chargeless, almost massless particles called neutrinos, which were eventually observed in 1956.

The psychologist Carl Jung collaborated with Pauli on a book exploring the paranormal entitled *Interpretation of Nature and Psyche*,[2] and maintained an open-minded approach to the possibility that telepathy could be explained by resorting to the esoteric fringes of known physics. He went as far as to suggest that "the microphysical world of the atom exhibits certain features whose affinities with the psychic have impressed themselves on physicists. Here, it would seem, is at least a suggestion of how the psychic process could be 'reconstructed' in another medium, in that, namely of the microphysics of matter."

But this is a terribly vague statement. It is easy to draw a hypothetical link between two disparate subjects like psychic phenomena and physics in this way, but it does not address the key facts. The crucial difference between the neutrino and the hypothetical psitron is that the former fits perfectly into the family of known particles, it plays a recognizable role, and was predicted by the strict mathematics of quantum mechanics before it was detected. Anyone can think up an imaginary particle to explain a phenomenon, give it an appropriate name, and suggest that it lies at the root of an unprovable process. Believers have even gone so far as to suggest that psychic powers do not work in the presence of skeptics because the wills of the doubters suppress the action of these particles. This is a perfect example of a nonfalsifiable hypothesis at work.

Psitrons have never been detected but this does not mean they do not exist. It is possible that some form of field or resonance or even an array of particles are produced as a by-product of brain activity; but until these are found and their properties understood, they should be considered as pure conjecture.

The brain does of course produce measurable potentials that are associated with different brain states, and their discovery by Richard Caton in 1874 raised hopes that science had stumbled upon the method by which telepathy worked. The reality is sadly much more mundane.

Four distinct types of rhythm have been identified in the human brain, and these correspond to different brain states. The waves are due to electrical activity and manifest as oscillating electrical currents.* They are detected by a machine called an electroencephalograph (an EEG) which picks up the tiny electrical impulses by attaching electrodes to the scalp and amplifying them. The signals can then be read out on a graph.

The four distinct brain waves are placed in frequency bands and measured in cycles per second, or hertz (Hz). When the brain is resting and relaxed, it produces alpha rhythms, which are detected between 8 and 14 Hz. Beta rhythms correspond to activity and predominate when the brain is working, solving problems or facilitating

---

*These brain rhythms should not be confused with the idea of biorhythms, which are thought by enthusiasts to be quite different in origin (see Chapter 7).

Figure 3.2

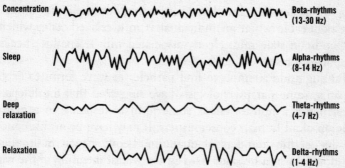

Concentration — Beta-rhythms (13-30 Hz)

Sleep — Alpha-rhythms (8-14 Hz)

Deep relaxation — Theta-rhythms (4-7 Hz)

Relaxation — Delta-rhythms (1-4 Hz)

movement such as walking or running. These rhythms occur between 13 and 30 Hz. At the other end of the brain wave spectrum are delta waves, which are produced during sleep. These have widely spaced peaks and oscillate at between 1 and 4 Hz. Finally, theta rhythms, produced when the brain is in a deep sleep or a trance state, resonate between 4 and 7 Hz.

Brain waves captured the public imagination during the 1970s, and the market was flooded with devices that were claimed to be capable of inducing alpha rhythms instantly. This was actually utilizing what yogis and Zen adepts had known for a long time: that individuals are able to control their own brain waves. It led to an awareness of what is now thought to be a fourth state of consciousness, a deep relaxation state or meditative condition corresponding to theta rhythm production.

Although research into brain waves has generated benefits for medicine, it has not led to the source of telepathic ability. EEGs are used extensively in psychiatric treatment and are particularly useful in the treatment of epileptics who exhibit disrupted brain wave patterns. The electrical impulses detected in the cerebral cortex—the outermost layer of the brain, a few millimeters thick—reflect the overall brain state; they cannot be deconstructed in order to draw off particular thoughts or even emotions.

Yet some researchers claim that certain rhythms are more pronounced when subjects are believed to be acting telepathically. These results are based upon the use of an EEG machine during telepathy tests and show that alpha rhythms accompany supposed thought transference. But this is misleading because alpha waves are most noticeable when an individual is in a relaxed state, which is also the brain state most clearly associated with telepathic successes in laboratory tests.

Putting aside attempts to find particles or wave forms to explain telepathy, some parapsychologists have suggested that the telepathic experience is a holistic effect, some form of response to a network made up of all human consciousness. It may have been this concept that one of the founders of quantum theory, Erwin Schrodinger, had in mind when he said: "I—I in the widest meaning of the word, that is to say, every conscious mind that has ever said or felt 'I'— am the person, if any, who controls the 'motion of atoms' according to the Laws of Nature."

The way in which a "human network"—or a network that includes all living beings—could operate, is little understood. Some researchers have made an attempt to clarify the concept, or to link it with aspects of biology and psychology, but the results have drawn only further controversy, and in some cases confusion.

Jung postulated the idea that there were two forms of unconscious awareness: personal unconsciousness, and what he dubbed the "collective unconscious."[3] This he saw as an inherited set of images common to all human beings. He called these images archetypes and believed each to be symbolic of a deeper aspect of human imagination, including ideas such as having parents, having children, and death.

In the hands of master manipulators, archetypes are powerful tools. The Nazi rallies at Nuremberg during the 1930s played on deep-rooted human fears and hopes by manipulating archetypes; and novelists, musicians, and painters often imbue their work with them, usually without conscious effort. It has even been suggested that the success of any form of art depends upon the artist linking with an audience via the use of archetypes. An example is the huge success of Tolkien's *Lord of the Rings* or the *Star Wars* films, which employ archetypes throughout, symbolic images such as the battle between "good" and "evil" and the "wise old man" (Gandalf and Obi-wan Kenobi). In the same way, the *X-Files* often uses the image of "the stranger" and plays on universal human fears and anxieties to great effect.

A related phenomenon is a process called "formative causation," which was first postulated and popularized by the British biologist Rupert Sheldrake in his book, A New Science of Life, published in 1981.[4] In essence, Sheldrake suggests that systems "learn" or that it is easier to repeat something if it has already been done. The mechanism for this is called "morphic resonance."

Initially, this sounds like a rather vague notion, and Sheldrake has indeed been savaged by orthodox scientists around the world from a variety of disciplines, but he has spent the past fifteen years conducting experiments which he claims verify the concept repeatedly.

One of Sheldrake's demonstrations of the principle is based upon linguistic patterns. He asked a Japanese poet to send him three similar verses. One was a meaningless string of words, the second a freshly composed verse, and the third a well-known poem, familiar to Japanese school children. He then showed the three pieces of writing to a group of Westerners, none of whom could speak any Japanese. What he discovered was that all of the subjects found it far easier to memorize the traditional poem than the other two.

His conclusion was that the traditional poem had somehow become ingrained into human consciousness via morphic resonance. Based upon this and numerous other tests, some involving living beings as well as experiments involving inanimate matter such as growing crystals, Sheldrake and his supporters believe that all things resonate with their own kind—"like resonates with like." In other words, there is a network of human interaction, and similar "fields" around other species, other inanimate objects, which influence their behavior. But it could be argued from the opposite perspective. Using the example of the Japanese poem: Could it not have been that it had survived and become familiar to school children because it was easy to learn? Could not certain functions become easier with repetition because those doing them have a natural affinity toward them and avoid tasks that do not come naturally?

If morphic resonance or the collective unconscious are realities, they may point the way to alternative ways in which minds may communicate. The traditional image of telepathy is that it occurs via pseudophysical means, facilitated perhaps by rays of particles, but the truth may be far more subtle. In a sense, any artist may be communicating with his public telepathically by tuning into archetypes or using morphic resonance to cross the barriers of space and time. Perhaps there are special people (and other animals) that have a greater sensitivity toward these resonances, are more adept at manipulating archetypes in far more sophisticated ways then we usually experience.

Whatever mechanism lies at the root of the telepathic experience, parapsychologists have been obliged to follow the traditions of science in attempting to demonstrate psychic phenomena. This is really the only way in which they can hope to convince a skeptical scientific community and to develop an understanding of what is happening, if anything.

Researches into paranormal phenomena began during the nineteenth century, but it really came of age with the American parapsychologist, Joseph Rhine, who summarized his findings in his 1934 book, *Extrasensory Perception*. Rhine worked at Duke University in North Carolina and pioneered the use of what became known as Zener cards, after their inventor, Karl Zener. There are five designs on the cards, a circle, square, star, plus sign, and three wavy lines.

These experiments involved the experimenter taking the top card from the pack and attempting to transmit the information on the card to the "reader" or subject, to see if they could identify the symbol on the card. According to probability, there is a twenty percent chance of simply guessing correctly, but in some trials subjects obtained remarkably high scores. On one occasion a participant got 588 "hits" from just over 1,800 trials—a success rate of 32 percent. This does not sound like a great improvement on the average, but the likelihood of achieving such a score by chance is astronomically high. In a variation of the tests, Rhine offered a subject a hundred dollars for every success. They produced a run of twenty-five successes, netting $2,500, a result calculated to have odds of just under three thousand million million to one against.[5]

After Rhine's experiments hit the headlines, other researchers followed his lead and rapidly brought the study of parapsychology

Zener cards

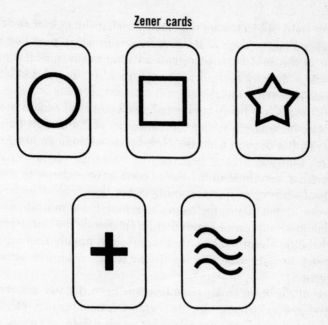

Figure 3.3

into disrepute with a series of infamous fakes. Since then parapsychologists have expended great efforts in attempting to develop fraud-proof experiments to demonstrate what they believe to be a genuine and measurable phenomenon.

Modern experiments rely upon random number generators which churn out numbers that are supposed to be completely without pattern (a little like the lottery machines wheeled out each Saturday evening). These experiments are seen as more reliable than the early Zener card tests, but the objections of the skeptics, the self-styled *psi cops*, have driven the researchers to new heights of sophistication.

The latest experiments are known as Ganzfeld or blank-field studies. These involve participants undergoing sensory deprivation in a form of isolation tank. Ping-pong balls are placed over their eyes and white noise is played through headphones. The experience has been compared to staring into a formless fog. After about fifteen minutes most subjects experience hypnagogic images, those often experienced on the edge of sleep. A sender, usually a friend or close relative of the subject, is placed in an acoustically shielded room, from where they try to send an image, usually a one-minute video sequence or a static image.

This research is being conducted at a number of centers around the world, including a site at the University of Edinburgh founded in 1985 with a bequest from the Nobel-prize-winning author Arthur Koestler, who was a great believer in paranormal phenomena. So far, there has been little in the way of conclusive evidence to support traditional ideas of telepathy coming out of these experiments. Like Rhine during the 1930s, the teams have found rare individuals who have achieved impressive scores that lie far outside the normal range of probability. Unfortunately, these results are usually unrepeatable and therefore cannot be deemed in any sense conclusive by ortho-dox science.

One of the most striking implications from the vast number of experiments conducted during this century is the idea that telepathic ability can apparently be enhanced by a wide variety of factors. The example of the man who produced twenty-five hits in a row when given the incentive of financial reward is a mundane one. Experi-menters have become interested in the idea that if other senses are suppressed, then telepathic powers can come through more readily. This is the reason for isolating the subject in the Ganzfeld experi-ments, but it has also formed the basis of experiments linking sleep with telepathy. Once again the results show that a small number of impressive but unrepeatable events suggest more pronounced tele-pathic ability if the other senses are dampened. One researcher has correctly pointed out that "if psychic powers exist, everyday experi-ence shows that they must be very weak for most people.[6]

Other anomalies could have an influence upon telepathic power. It has been found that children with mental defects score higher in telepathy tests, and in one set of experiments a child known simply as "the Cambridge boy," who had been born with physical and mental disabilities, achieved well above average scores when his mother was present.[7]

The explanation for this is that if telepathic abilities really do exist, they may be more useful to individuals who cannot communicate in the conventional manner or have their other senses suppressed in some way. There have also been a number of unconfirmed cases of telepathic powers becoming apparent during life-threatening situations. These have been dubbed by parapsychologists as "need-determined" cases or "crisis telepathy," but dismissed by orthodox science as apocryphal.

According to some researchers, this form of telepathy could be explained by its survival value and might even be a genetically favored trait. Humans, they argue, have submerged this talent with other more readily developed and utilized abilities, but some rare individuals are more in tune with this power, and it comes to the surface in an emergency.

It has also been supposed that other species display this facility. During the 1970s, Soviet parapsychologists attempted to demonstrate this effect experimentally. They took a set of newly born rabbits away from their mother and killed them at set, recorded times. The mother was wired up to an EEG and her brain patterns monitored. According to official reports, the mother rabbit displayed sharp electrical responses at the precise moment each of her offspring were killed. Unfortunately, because news of this experiment leaked out from Soviet Russia, it is difficult to verify, and nobody in the West has so far reinvestigated it.

Skeptics continue to pour cold water on the entire phenomenon of telepathy. One of the most usual arguments is to ask why telepathic individuals do not use their skill to win lotteries or to chalk up staggering success at the race track? They also wonder why in lab tests the talents of the claimants mysteriously vanish.

The problem with telepathy is that a century of investigation has turned up little evidence that complies with standard scientific practice. But during this same period, science has performed quite apparent wonders, from curing diseases to placing humans on the surface of the moon.

A study by the highly skeptical National Research Council in the United States found in 1988 that there were what they called "problematic anomalies" in some experiments that could not be explained; in other words, incidents of success that could not be accounted for merely by chance. And, despite the apparent lack of evidence, many people believe in telepathy. In one survey, 67 percent of people questioned said they had experienced ESP.

Psychologists have noticed that "a sense of deep personal conviction may be the key to achieving good results,"[8] but this is surely not the whole story. There is still no satisfactory scientific explanation for what supporters claim has occurred during a growing number of experiments, but it would be unscientific to conclude from this that telepathy is imaginary. It may just be that we don't know how it works.

4 Moving Heaven and Earth

O the mind, mind has mountains; cliffs of fall
Frightful, sheer, no-man-fathomed.
GERARD MANLEY HOPKINS

Telepathy is really only the tip of the psi powers iceberg. If we think of telepathy as being "mind communicating with mind," then what are the chances of mind interacting directly with the physical world, with matter?

Psychokinesis or PK is defined as: "The apparent ability of humans to influence other people, events, or objects by the application of will, without the involvement of any known physical forces."[1] Whereas telepathy could be described as an interaction between two psychic "fields" or "forces," PK involves an interaction between the mental and the material, so it is one stage further down the psychic route. To the skeptical, this means a marriage from hell, but from the perspective of parapsychologists it is merely a natural progression from the more prosaic telepathy.

PK has been the subject of even more experimental work than thought transference, and according to how you view this evidence, it has either been proven beyond doubt to be a genuine, natural process or else all the tests and experiments conducted during a period of almost a century have been faked or may be explained by other factors.

The earliest serious attempt at trying to quantify the concept of PK is attributed to the parapsychologist J. B. Rhine, who worked on telekinesis experiments concurrently with his attempts to pin down telepathy during the early 1930s. He was led into the subject when a young gambler told him he could influence the fall of a die by willpower alone. Rhine immediately set about conducting thousands of tests on scores of subjects, in an attempt to reach a statistically meaningful conclusion that would show if there was any such effect.

He based the test upon participants trying to will two dice to produce a score of more than seven. As with the results of his telepathy tests, Rhine and others found that most people achieved scores that deviated little from the values expected by chance, but once in a while he turned up individuals whose scores did not fit the normal pattern. On a few occasions he found someone whose scores deviated from the average to such a degree that the probability of it happening by accident was sometimes placed at millions to one against.

In one set of experiments Rhine conducted 6,744 tests. These should have produced 2,810 successes by chance alone, but one subject achieved a score of 3,110, a deviation calculated to occur by chance only once in a billion tests.

Rhine attracted criticism of his experimental methods almost immediately, but as with his telepathy experiments, he went to great lengths to debug his tests from any chance of trickery or unintentional influence. After nearly thirty years of such experiments, he reached the conclusion that "the mind does have a force that can affect physical matter directly."[2]

Since Rhine's seminal investigations, literally hundreds of groups around the world have conducted other forms of PK tests. During the 1970s researchers led by Helmut Schmidt at the Mind Science Foundation in San Antonio, Texas, replaced Rhine's die with a Geiger counter. The reason behind this was that Geiger counter readings derive solely from radioactivity, which is produced by the breakdown, or "decay," of radioactive nuclei. This decay process is completely random and as close to fraud-proof as parapsychologists could hope to get.

Schmidt had participants attempt to influence the readout from a Geiger counter—usually a display showing a series of flashing lights or an oscilloscope screen showing a wave pattern. Variants on this theme were developed in the 1980s using white noise patterns generated electronically.

One of the leading parapsychologists currently working with PK is Robert Jahn, an engineering professor based at Princeton University in New Jersey. He developed the white noise experiments and has gone on to try different versions of what he calls a random event generator—a machine that produces random displays or number sequences—the electronic equivalent of tossing a coin thousands of times. The generator is fitted with a collection of safety devices to detect any change in temperature, the influence of external magnetic fields, or any physical disturbance such as tilting or weighting the machine in any way.

Ignoring the cynicism and sometimes open hostility of his orthodox colleagues, Jahn has spent the past fifteen years conducting millions of tests, using over a hundred subjects to see if there is any deviation from chance. His findings are still inconclusive. Taking the collection of experiments and subjecting them to statistical analysis, he has found that there is some effect, which he judges to be about a 0.1 percent deviation from pure chance. In other words, on average, one thousand tests throw up one significant deviation from what was expected.

If this all sounds unconvincing, there are other more worrying aspects to Jahn's experiments. He conducted a large number of trials where the subject was placed far away from the random event generator in his laboratory. Some of the subjects were asked to attempt the test from as far away as Africa and England, but confusingly, Jahn found that their success rate did not vary at all with distance.

In another set of experiments, he asked his subjects to make their attempts at psychokinetic influence up to several days before the test, and again found no difference in the quality of the results. Others have discovered the same odd anomaly. In a different collection of tests conducted by Helmut Schmidt, he disconnected the random event generator and substituted the "live" readout with a recording of the signals from the previous day, but didn't tell his subjects what he had done. He found that, if anything, the results were better than they had been in the usual experiments.

PK enthusiasts have produced a very odd explanation for this. The subjects, they claimed, were practicing what has quickly been dubbed *retroactive psychokinesis*. This involved them sending thoughts back in time and space to the previous day and effecting the readout.

Astonishingly, many parapsychologists subscribe to this explanation. Cynics merely call it a nonsensical explanation for an obvious and telling flaw in the entire theory and practice of parapsychology. Yet, there is actually another far more mundane explanation that appears to have escaped both camps. If the results on Day 2 are affected by the subject even though the display from Day 1 was being shown to them, perhaps their minds were interacting with the playback device. If we are trying to demonstrate PK, there is no reason why the subject might not control some machine feeding the fake display, any more than they might alter the pattern from a random event generator. After all, the playback device is certain to be some form of tape recorder or digital device, and as susceptible to PK as any other material system.

So, putting to one side the views of the extremists from either camp—the ardent skeptics and the wholehearted believers—what scientific conclusions can be drawn from the vast range of PK experiments?

Enthusiasts point to the rare, outstanding anomalies and conclude that something odd is happening and that this must point to proof of PK. But they seem unperturbed by the fact that those cases that deviate greatly from chance are quite exceptional and very rare. More common is a tiny perturbation from the expected, which may perhaps point to a weak force at work or could be due to a number of other anomalies.

One such anomaly could be stray magnetic fields or electric currents. It has been found recently that some people who live near power cables sometimes experience physical illness and depression. This is thought to be due to the close proximity of the brain to powerful electrical currents. All electrical impulses have associated magnetic fields, and those around carriers such as national grid power cables interact in some way with the similar but far weaker magnetic fields created by the electrical signals produced in the brain. Via some unknown mechanism, these disturbances manifest as physical and emotional instability. In a similar way, it is possible that magnetic fields some distance from the test center could have a weak effect on the machinery used for the tests. Another source might be electrical interference by leakage from equipment elsewhere in the lab or even beyond the building. Experimenters have tried to negate this anomaly by encasing the test equipment and the subject in a special container called a Faraday cage, which shields them from electromagnetic disturbances.

Such a tiny effect as that observed by parapsychologists could be produced by other natural sources. We all live on a giant magnet. Like most planets, the Earth has its own magnetic field which fluctuates naturally due to movements hundreds of miles beneath the crust. It is also disturbed by fluctuations outside the atmosphere. Sunspots, which are cooler regions on the surface of the sun, are able to disrupt the sun's powerful magnetic field, and this in turn can alter the Earth's associated field. Such magnetic disturbances might conceivably affect electronic machinery and, indeed, the fields each of us produces by electrical activity in our brains and from each nerve impulse passing constantly through our bodies.

Other factors to consider are currents of warm or cool air and minute geological disturbances such as microearthquakes. Although these factors are almost undetectable, they could be powerful enough to disturb PK experiments.

These objections may sound pedantic, but if parapsychology is to be taken seriously by scientists, it has to play by the same rules as orthodox science. Many critics of PK and other paranormal phenomena cite undue care or unprofessional attitudes to research as the most likely source for their apparently impressive results.

The only reasonable conclusion to be drawn from the millions of PK experiments conducted since the 1930s is that if the results do demonstrate a genuine effect produced by the human brain, then the effect is very small. And because it is so small, it is incredibly difficult to measure. Parapsychologists have dubbed this effect *micro-PK*, and there is a growing body of evidence to substantiate an anomaly of this type.

In the mid-1970s a psychologist, Gene Glass, came up with a revolutionary approach to the study of experimental results from parapsychology experiments, called *meta analysis*. He realized the problem with PK was that the effect was so small, usually in the region of 0.1 percent over chance, that a large number of results would be needed to show up the anomaly created by any genuine paranormal activity. Furthermore, the smaller the effect, the more results would be needed.

An analogy would be the effort put into tuning a radio. If a signal is strong, such as say a BBC transmission picked up in southern England, it would be easy to tune into it. If, on the other hand, you were trying to pick up a weak signal from a pirate or independent broadcaster, you might have to spend some time fine-tuning the radio to detect the signal. This fine-tuning is equivalent to conducting a large number of samples or tests.

So Glass's idea was to somehow pool the data from all the tests that had been conducted over a long period. The problem was, the tests carried out since the 1930s were quite different from one another. Some experimenters had investigated the possible effects of PK on falling die or wood blocks, and others followed the altering of light displays or the random decay of unstable isotopes using a Geiger counter. But Glass found eventually that using suitable mathematics, results from disparate tests could be merged, which means that the parapsychologist has access to a far larger sample— tens of millions of results taken by scores of experimenters over a period of some sixty years.

One of the best examples of applying meta-analysis to PK experiments comes from the work of a psychologist, Dean Radin, at Princeton University's Psychology Department, and Roger Nelson, a member of the Princeton Engineering Anomalies Research program (PEAR). They did not conduct their own experiments, but instead tracked down over 150 reports summarizing almost 600 separate studies and a further 235 control studies by 68 different investigators, each of whom had been researching the influence of consciousness on microelectronic systems—experiments where the subject was asked to disturb the workings of an electronic random event generator.

To their amazement, they found the probability of the net result deviating from the normal pattern merely by chance was 1 in 10^{35} (1 with 35 zeros after it).

Again, this result does not say for certain that PK is a genuine phenomenon, but it does suggest there is some factor or combination of factors that alters the behavior of matter other than by the visible, conventional means. Whether this is the effect of human consciousness or sunspot activity, microearthquakes or thermal currents, is another matter.

Those convinced that the aberrations shown up by meta-analysis are of human origin believe the effect is generated by micro-PK and that this phenomenon is with us all the time. They suggest that many incidents we might think of as coincidence are a result of this force. When we drop a book and it opens to exactly the page we wanted; when we look through a filing cabinet and put our fingers right on the piece of paper we've been looking for; when we net the basketball without even looking. These, according to the supporters of micro-PK, are all examples of a subconscious ability to influence the way matter behaves. Furthermore, it is thought that this phenomenon works best when the subject is not deliberately trying to make it work, when the individual is concentrating on something quite different.

We have all experienced moments when things have either gone extremely well for us or extremely badly—good days and bad days. We've all experienced beginner's luck, moments when you can do no wrong. It is on these occasions, the believers say, that micro-PK is working at its best, and these are exactly the occasions when we are the least likely to be trying to make it work.

Rex Stanford, a psychologist, working at St. John's University in New York, has conducted an interesting variation on the usual PK experiment to illustrate the effects of micro-PK. He places his subjects in a locked room and gets them to perform a series of very dull tasks. In the next room is a random number generator. What he does not tell the subjects is that they can only be released from their task and allowed to leave the room when the generator produces a sequence of numbers that appear only once every two or three days under normal circumstances. Yet on several occasions some subjects have managed to get out of the room within forty-five minutes.

The problem with accepting micro-PK is that the force producing the effect is the same as that involved in macro-PK. Any system that allows us to subconsciously control the way a book lands or the movement of a ball in a game of billiards, football, or cricket is the same phenomenon that could allow us to move objects at will. They are on the same scale and would presumably operate by the same wave form, particle stream, or other inexplicable force. And it is not whether these effects occur occasionally or frequently that bothers the scientist, it is how they could occur even once. Because, at the root of the dilemma remains the fact that no form of psychic force has been detected, yet we are asked to accept that the mind can interact with matter—the marriage from hell I mentioned at the start of this chapter.

To illustrate the problem, let us consider the physics of PK. What are the energies involved, and is there any compatibility between what is needed and what could be reasonably generated by the brain?

Let us imagine an experiment in which a subject who claims the ability to perform psychokinesis is asked to move an object along a table using just the power of his mind. Suppose our object weighs a hundred grams, the mass of a spoon or a pair of glasses. Now assume the participant is to accelerate the object to the modest velocity of ten centimeters per second, and to maintain this velocity for a few seconds. If we add a small contribution from friction, the energy needed to do this comes to approximately 1×10^{-4} Joules (one ten thousandth of a Joule).

This is a relatively small amount of energy, roughly equivalent to that stored in one millionth of a gram of sugar. But equally, to produce even this much energy from a force that has so far remained undetected by any conventional means requires a conscientious suspension of disbelief. Consider the figures.

As we saw in the last chapter, the brain has associated electric and magnetic fields. Now, we could suppose that the electrical impulses from the action of neurones is responsible for creating the force with which the object is moved. But this field must interact very weakly with the material world, because we cannot pick up the force or any form of tangible interaction with any known instrument. But let us be liberal and say that one thousandth of the power of the electrical impulse penetrates the skull, reaches across space to the object, and accelerates it to a speed of ten centimeters per second for a short period of, say, three seconds. The voltage produced in the neurones is approximately 100 millivolts, so this would mean that the human brain would have to produce a current in excess of 0.25 amps to provide the necessary energy. To put this into perspective, a current of little more than half of this (0.15 amps) passed through the heart would kill a person.

A less spectacular but equally intriguing form of PK involves a mind-matter interaction of a different type. This is a process known as thoughtography—the art of generating images on photographic paper or on film without the use of chemicals. A famous example of this talent was investigated by the parapsychologist Jule Eisenbud during the 1950s. Her subject was a man named Ted Serios, who could produce spontaneous images on film merely by looking at a camera. He became a celebrity for a time and produced a book called *The World of Ted Serios*. Although he was never caught cheating, he later attracted suspicion when he was found to be holding a tiny device in his hand whenever he performed his trick. He refused to let anybody analyze the device, but could not make thoughtographs without it, claiming that it focused his energies.

This form of PK is another example of the power of the mind apparently altering the physical world. This time, some form of electromagnetism may have been able to change the chemical structure of the photographic paper or the materials coated on the film. Under normal circumstances, light of particular wavelengths would impart sufficient energy to the chemicals on the paper or the film to break interatomic bonds and stimulate the creation of new chemical arrangements which would collectively produce a photograph. If Serios, or any other thoughtographer, were genuinely creating an image, it must have been through a similar mechanism. Perhaps it is possible for the very weak fields created by the brain to be amplified and focused at a precise set of wavelengths that in some way duplicate the action of sunlight.

So far in this chapter, I have concentrated on what might be termed "peripheral PK effects"—the results of micro-PK. However, in the early 1970s the world awoke to what appeared to be an example of full-blown psychokinesis, and it was captured on film and broadcast to the world on television. Uri Geller had arrived.

Geller's contribution to the paranormal evokes strong feelings both for and against. To some he is a talented psychic, to others he is little more than a showman, a trickster masquerading as a mystic. During almost twenty-five years in the public arena, he has reworked his career many times, but has done little in recent years to top the splash of publicity and sensationalism that surrounded his first appearance in Britain in 1973.

At that time a number of scientists took Uri Geller seriously enough to expend effort and resources studying his claims and trying to reproduce his stunts under laboratory conditions. John Taylor, a professor of mathematics at King's College, University of London, overcame the ridicule of some of his colleagues and the resistance of the college authorities in trying to quantify Geller's apparent skills. Although he was initially intrigued by Geller's performance under test conditions, Taylor soon found he could not reconcile the evidence with electromagnetism and was forced to either dismiss the phenomenon or accept that the bed rock of science was wrong. He chose the former.

"When science faces up to the supernatural," he concluded, "it is a case of 'electromagnetism or bust.' Thus we have to look in detail at the various paranormal phenomena to see if electromagnetism can be used to explain them."[3]

Within a few months of starting the tests at King's College, Taylor had switched from being open-minded to denouncing Geller as a conjurer. In his 1976 book *Superminds*, he wrote:

Uri Geller appears to have posed a serious challenge for modern scientists. Either a satisfactory explanation must be given for his phenomenon within the framework of accepted scientific knowledge, or science will be found seriously wanting. Since such an explanation appears to some to be impossible, either now or in the future, they argue that the Geller phenomenon is incompatible with scientific truth, and that the value of reason and the scientific point of view is therefore an illusion. Will the gates of unreason then be allowed open and drown us in a world inhabited by aetheric bodies, extraterrestrial visitors, spirits of the dead and the like? Will reason then wholly give way to superstition?

To be fair to Uri Geller, some of his stunts have been impressive. He has been tested by a dozen different research groups around the world, beginning with a team at Stanford Research Institute in California in 1972 before he became an international celebrity. Since then he has traveled around the world and subjected himself to analysis in laboratories in Japan, France, and several in Britain and the United States. He has been videotaped materializing objects apparently from thin air, stopping machinery, and of course distorting a vast assortment of cutlery—not only on film, but before large audiences on live television shows.

Geller himself conveys the impression that he does not fully understand from what source his talents spring, and appears to be totally convinced that what he is doing is paranormal. In the final analysis, he may be the only one who will ever know for sure, because if he is a fraud, he will certainly never admit as much.

The researcher Arthur Ellison has said of him:

For some strange reason Geller has a different image of truth which doesn't quite agree about the strengths of materials taught in university. He believes that if he gently strokes a spoon between finger and thumb without putting a terrific pressure on it then it will bend. And sometimes it does . . . I've seen Geller take a Yale key . . . [he] stroked it and the end came off . . . Geller's image of truth just doesn't involve Yale keys being forever rigid.[4]

In one test, a one gram weight was placed on a sensitive scale and the apparatus covered with a glass bell jar. The weight on the scale registered as a line on a chart recorder, and Geller was asked to alter the reading paranormally. During a test lasting half an hour, he was able to deflect the chart recorder twice, each for one-fifth of a second. The first deflection was equivalent to increasing the weight to 1.5 grams, and the second to 2 grams.

On another occasion, he was able to dramatically alter the readout of a Geiger counter that had been sealed in a unit some distance away across the lab. Then, when he was allowed to pick up the device, he concentrated hard and the beep accelerated to a wail that stopped the moment he put down the unit.

In one sense, at least, Geller is an accomplished showman. This has been part of the reason for his success. He does not merely produce impressive tricks, but is happy to employ gimmicks such as spoon-bending, which caught the public's imagination in the early 1970s. When he first appeared on British television and bent an assortment of spoons before the cameras, he became an overnight sensation and people throughout the country and from all age groups were rushing for the kitchen drawer as soon as the show was over.

Even today, more than a quarter of a century after his first shows, scientists have still not been able to show how he creates the effect. The magician James Randi, a vehement critic of Geller, has declared publicly that spoon-bending is nothing more than a vaudeville trick and has himself duplicated Geller's performance quite convincingly by sleight-of-hand. Others have taken a more analytical approach, studying the objects at the heart of the controversy—the spoons themselves.

A team in France has taken electron microscope photographs of spoons before and after bending experiments. They tested the crystalline structure of the metal and measured the weight and dimensions before handing them to spoon-benders and again after the experiment. These tests revealed little except a local hardening around the area of the distortion said to be similar to that caused by a considerable pressure applied externally and concentrated on a small region. Clearly, no observable external pressure had been used, so the experimenters were forced to conclude it had been produced "internally."[5]

Other tests have shown that as the bending occurred, the metal of the spoon took on the consistency of chewing gum in the region of the distortion. Interestingly, this effect can be caused by using corrosive chemicals, but that always results in weight loss and visible corrosion of the metal surface, neither of which has been observed in these PK tests.

So what are the possible mechanics of spoon-bending? If it is not a conjurer's trick, how could it be done?

As John Taylor points out, scientific explanation is limited; it's a question of electromagnetism or bust. So if Geller and other spoon-benders are able to focus electromagnetic waves into the spoon, what sort of energy would they need to produce?

Metals are substances in which the atoms are arranged in a uniform lattice. The shape and characteristics of this lattice depend on the type of metal. If we consider, for the sake of simplicity, that a spoon is made of iron, we need to look at the way in which the lattice of iron could be distorted by PK or any other method.* To do this we will need to relate the energy requirement to those used to distort the metal more normally—by the use of heat.

Iron melts at 1808K (around 1,500°c), but of course we don't want to actually melt our spoon, just soften it enough to distort its shape. A better value to consider is something called the latent heat of fusion of the metal, which is the energy needed to convert solid iron into liquid iron—to disrupt the metal lattice. If we imagine a spoon weighing in the region of 100 grams, perhaps only twenty percent of its mass is involved in the distortion, so we only need to consider the energy needed to alter 20 grams of iron.

Obviously we do not need all the energy required to convert iron from a solid to a liquid, because all we wish to achieve is the "chewing gum" consistency observed by some experimenters. Let's be conservative and say that the energy needed to create this state is only ten percent of that required to melt the spoon. So, the energy needed is equivalent to ten percent of the latent heat of fusion of 20 grams of iron, a figure that turns out to be around 600 Joules.

Earlier we saw that the energy needed to move a 100-gram object along a smooth surface was a tiny fraction of this (one ten thousandth of a Joule). The electrical energy Geller would need to generate in order to provide 600 Joules using only the electrical potential of neurones in his brain would require a current in the region of 150 amps—the sort of current used to power heavy-duty electronic machinery.

The empirical analysis of spoon-bending, and indeed all PK ac-

---

*They are more typically made from stainless steel, which is in fact much tougher and resistant to heat and pressure than pure iron.

tivity, does not paint a rosy picture, but that should be expected—
it is a marriage from hell. Using the forces we know and understand,
matter cannot be manipulated by the energies associated with
thought, or at least the outward, measurable manifestations of
thought—electrical currents in the brain. So we are left with explanations that lie outside the usual scope of the physical and natural
sciences.

At this point parapsychologists often resort to drawing in ideas
from the fringes of physics, such as quantum mechanics, but these
do not offer a satisfactory explanation and are often used willy-nilly.
Because quantum mechanics is itself a weird area of physics, it is
also vulnerable.*

One of the most telling aspects of PK research is the fact that,
like telepathy, psychokinetic effects have no respect for distance.
The set of experiments conducted by Robert Jahn at Princeton University, in which subjects are apparently able to alter readouts up to
several thousand kilometers from the laboratory with as much ease
as when they sat a few meters from the detector, illustrates this point.

Almost all known physical forces operate according to the inverse
square law. What this means is that as distance is increased, the
effects of force diminish by the square of that distance. For example,
Isaac Newton described in his masterwork *Principia Mathematica*,
published in 1687, that the force of gravity operated via the inverse
square law.

By Newton's reasoning, if planet A circles the sun at a given
distance, it will experience a certain gravitational attraction toward
the sun. If planet B with identical mass orbited at double this distance, it would experience only a quarter of the gravitational attraction experienced by planet A (the inverse of 2 squared). An
identical planet C, orbiting three times as far away as planet A,
would feel only one-ninth the gravitational attraction experienced
by A (the inverse of 3 squared).

All known forces operate by this inverse square law, but support-

---

*I'm of the opinion that quantum mechanics has little to do with PK or telepathy but
may be an explanation for the phenomenon of prediction and synchronicity, which is
discussed in detail in Chapter 6.

ers of PK argue that there are other forms of energy transmission which do not. The most popular argument is that the intensity of radio signals does not weaken very much with distance. From this, enthusiasts draw an analogy between radio and "thought waves."

The first part of this statement is true. If radio signals are generated by a powerful enough transmitter, and fine-tuned using what is called a signal-optimizing system, their intensity does not diminish to any large extent over reasonable distances. But this is not really the point. Radio signals convey information, they do not move objects or enable the bending of spoons; some form of force is needed to do these things, and all known forms of force operate by the inverse square law. Furthermore, if PK operates by electromagnetic radiation, of which radio signals are a small part, where in the spectrum is this radiation to be found? As discussed in Chapter 3, it might operate at extreme ends of the electromagnetic spectrum, but the range has been thoroughly searched and no trace of a psi wave discovered.

If PK is produced by a genuine form of mental energy or acts via some as-yet-unknown force, then the only conclusion to be drawn is that this force has nothing in common with others we have so far experienced in the universe. Depending upon your viewpoint, this fact either strengthens or weakens the case for a paranormal explanation for PK. To the skeptical, it merely reaffirms the claim that because PK cannot be explained by recourse to electromagnetism, it must be put down to trickery or sleight-of-hand. To the believer, it confirms that the phenomenon is not governed by the normal laws of physics, cannot be measured by humans wielding electronic gadgets, and lies above and beyond us. Perhaps PK is an adjunct to some odd quantum mechanical effect, but if this is the case, an explanation for the way it works may be a very long time coming.

The Fire Within

Billy, in one of his nice new sashes,
fell in the fire and was burnt to ashes;
Now, although the room grows chilly,
I haven't the heart to poke poor Billy.

HARRY GRAHAM

The earliest reported case of what has become known as "spontaneous human combustion" (SHC) dates from early in the seventeenth century when a carpenter named John Hitchen died along with his wife and daughter during a thunderstorm. Recovering his body from the wreckage of his house, neighbors were astonished to discover that the man's body was burning from the inside but there was almost no outward sign of fire. He continued to burn for three days until his entire body had been reduced to ash.

Of course, reports like this from so long ago are, by their very nature, unreliable, and it is interesting to note that until this century, all victims of SHC were recorded as having been heavy drinkers, even if they were not. Although the official line taken on SHC by the medical profession has changed little since the first case—that all victims are in some way associated with fires or were alcoholics—there is an impressive body of evidence to suggest there is more to this rare and frightening phenomenon. Coroners do not refer to victims dying from SHC even if all the evidence points that way and correlates with many other similar cases; and perhaps because of its rarity, spontaneous combustion of human beings is still viewed as paranormal and therefore fantasy.

The most famous case of SHC was indeed fantasy—it sprang from the fertile imagination of Charles Dickens, who dispatched the character Krook in *Bleak House*, published in 1853, using the device of spontaneous combustion. But in recent times several baffling cases have been witnessed, sometimes by scores of people, and the intriguing medical details analyzed by skeptical experts as well as paranormal investigators. In 1986, Viennese pastor Franz Lueger is reported to have burst into flames and exploded before an entire congregation during a particularly impassioned sermon. In Los Angeles in 1990, Angela Hernandez blew up on an operating table at UCLA. And in one of the most thoroughly documented cases in Britain, a seventeen-year-old student at Halton College in Cheshire burst into flames in front of her friends in 1985 and died after fifteen days in the hospital.

At first glance, many cases of SHC seem to be easily explained. Victims are frequently associated in some way with naked flames or flammable liquids. Also, a disproportionately high number of victims are homeless people who, upon further investigation, are found to have been alcoholics. Yet, as with many phenomena categorized as "paranormal," there is a significant percentage of incidents that completely defy these standard explanations. Furthermore, some aspects of SHC are extremely difficult to accommodate using accepted medical knowledge, and may require a reappraisal of some treasured notions of what the human body is capable of when exposed to extreme conditions.

First, let us consider the anomalies associated with SHC.

The most striking aspect of spontaneous human combustion is that in almost all cases the body of the victim has suffered the results of intense fire, but the surroundings are relatively unaffected. In many incidents the body has burned from the trunk outward, but the fire has not reached the extremities. Some bodies are reduced to ash, but fingers, and on occasion entire limbs, show almost no trace of combustion. Frequently, clothes and shoes are undamaged and items lying close by are unmarked.

The second surprising aspect of SHC is that those parts of the body affected by the fire, including bones of the rib cage, the pelvis, and the backbone, are often completely reduced to ash. This is significant because the temperature required to powder human bone is relatively high—in excess of 1000°c. Incinerators used in crematoria operate up to 950°c, and even at these temperatures bones often have to be crushed mechanically after cremation.

So, explanations for SHC have to account for the very high temperatures required to cause the sort of damage reported, and suggest ways in which a human could burst into flames from a source that appears to be internal.

One of the earliest attempts to produce an official explanation for spontaneous human combustion relied on the questionable claim that the vast majority of cases involve an external flame or heat source. It is true that in a large percentage of reported cases there is a local heat source, but certainly not all. And in a significant number of those incidents where fires or ovens were found nearby, they were not in use at the time of the incident. The best example of this is the death of the student Jacqueline Fitzsimon at Halton College in 1985, who burst into flames walking down the stairs leading from her catering class. Her lecturer, Robert Carson, swore that all the rings on the ovens had been turned off an hour before the group left the room, and a Home Office chemist, Philip Jones, was unable to make a smoldering catering jacket burst into flames.[1]

Concerning the argument that heat sources are always present, skeptics explain that combustion is facilitated by victims either falling onto a fire or becoming unconscious and leaning against a heat source. The victim then experiences what has been called the "wick effect," with the human body acting like a candle, burning slowly from the inside outward.

Dr. Dougal Drysdale, a fire safety engineer, described how this worked on a television documentary called, A Case of Spontaneous Human Combustion, broadcast by the BBC in 1989.

"In a way, a body is like a candle—inside out," he said. "With a candle, the wick is on the inside and the fat on the outside. As the wick burns, the candle becomes molten and the liquid is drawn onto the wick and burns. With a body, which consists of a large amount of fat, the fat melts and is drawn onto the clothing, which acts as a wick, and then continues to burn." He then went on to demonstrate the effect with a lump of animal fat wrapped in cloth.

It has since been claimed that the demonstration was falsified using time-lapse photography, making it appear that the fat had combusted furiously and with little effort. But such criticisms aside, the demonstration did not begin to address the most significant facts surrounding cases of SHC. Humans may be a little like candles in one respect, but in other, important ways, they are quite different. First, viewers were not told what sort of fat was used in the experiment. Second, most obese humans are composed largely of water rather than pure fat. But, crucially, the question of temperature was conveniently sidestepped, and no serious attempt was made to explain how the wick effect could generate temperatures even close to those required to destroy human bones.

Given, then, that the wick effect offers little assistance in trying to explain the majority of SHC cases, whether initiated by a local heat source or not, are there any better explanations?

Perhaps a more productive approach would be to look at the type of fire witnessed in many cases of SHC. Although there is a spectrum of effects, one common factor is the presence of a blue flame. Another is the surprising fact that water has little effect in quelling the fire, and in some cases actually makes it worse.

The appearance of a blue flame suggests the presence of methane gas, the major component of natural gas. Cattle often experience a build-up of methane in the digestive tract, which bloats the stomach and sometimes requires a vet to make an incision to release the pressure. It has been suggested that a similar process can occur in humans, creating pockets of gas which under normal circumstances can be released eventually, but in rare cases meet with some intense heat source—with explosive results.

The fact that water facilitates rather than extinguishes the fire is more curious. Fires of this type are usually linked with the presence of the elements magnesium or titanium. The best example is aircraft fires, because of the high proportion of magnesium and titanium used in the framework of airliners and military craft. Magnesium burns with a very intense, very hot flame, and in the presence of water it releases hydrogen gas, which is itself highly flammable. Magnesium is also one of the trace metals found in the human body.

Yet, these links do not in themselves point the way to how SHC occurs. Magnesium is present in the body, but only as a trace element. It could conceivably act as a "fuse" to initiate a fire which would then be fueled by more common materials in the body, such as methane. But so far, no one has provided a demonstrable mechanism how this could happen.

The chemical reactions that occur in our bodies all the time are varied and incredibly complex. The human body is an elaborate machine that can absorb energy in the form of food, perform a variety of specialized as well as general functions—both voluntary and involuntary—and then emit the waste after all the usable energy has been extracted. Whether it is the movement of a leg, the mechanism behind the creation of thought, or the systems governing breathing, all the functions of the body eventually come down to biochemical processes. Although these processes involve a complex array of biochemical devices, each comprises a set of simple, step by step reactions.

It was discovered in the nineteenth century that chemical reactions are either exothermic—that is, they evolve heat—or else they are endothermic, drawing in heat from the environment. Biochemical processes are simply chemical reactions involving biomolecules—large organic molecules found in living things. Some of these reactions are exothermic, others are endothermic. Perhaps, in cases of SHC, these chemical systems become distorted in some way and the energy produced is channeled into precipitating an uncontrolled process.

Whether caused by the ignition of methane or by some strangely distorted exothermic processes, the common objection to any form of fire starting in the human gut has been the presence of large amounts of water, which constitutes about seventy percent of body mass. But recent studies conducted by chemical engineer Hugh Stiles has shown that the presence of water does not actually affect the chances of ignition and the sustaining of a fire.

The mechanism he describes begins with a heat source oxidizing the carbon in organic molecules found in the cells of the gut to carbon dioxide. This is an exothermic reaction and generates a large quantity of heat which then evaporates the water in neighboring cells. Calculations show that the trunk of an average human being burning in this way would produce in the region of 200 megajoules (200 million joules) of heat, which is more than twice that needed to evaporate the entire water content of the body. It then follows that the leftover energy feeds the process, resulting in an intense fire from within or else an explosion sparked perhaps by the ubiquitous flammable gases in the gut.

The problem with this theory is that such a reaction would produce large amounts of water vapor, which would then condense on every surface near the site of the incineration. According to documented reports, this is not usually the case. Many cases of SHC occur when the victim is alone, and bodies are often found long after the fire has run its course, by which time any water may have been removed by secondary fires. But in cases where the body has been found soon after the event, or when the conflagration has been witnessed, there have been no reports of unusual condensation near the corpse.

Another chemical that is a candidate for sustaining an internal fire is phosphine. This is a poisonous gas that combusts spontaneously with air at room temperature. It is sometimes seen as a glowing vapor in marshy areas and given the name "will-o'-the-wisp." It is not normally found in the human body, but could be produced by an unusual reaction involving naturally occurring phosphates.

According to retired industrial chemist, Cecil Jones, phosphine could account for some reported cases of SHC and might have been responsible for the incident at Halton College.

"It is possible that the victim passed wind containing phosphine," he has suggested. "The mixture would be confined by her clothing, but would rise up her back, partially igniting in the restricted amount of air available, giving the initial sense of heat, then escaping at her shoulders and bursting into flames with the phosphorescent glow seen by the two witnesses. This would account for the excessive burns to the back and buttocks, but nothing to the front of the body or legs."[2]

This effect has even gained an official name—the *phosphinic fart*—and it may account for some cases, but clearly not all.

For a more general source of heat, we need to return to biochemistry, but this time to look at the energy exchange processes that take place in the cell.

Humans gain energy from food by a series of biochemical processes. The first step is to convert food into glucose. This is then subjected to a process called *glycolysis*, during which it is converted to a molecule called pyruvic acid. The pyruvic acid is then changed into a variety of molecules, and along the way energy is released. Although this is a complex process overall, each step involves a simple chemical change.

An example is the famous Krebs' cycle, named after the biochemist Hans Krebs, which takes place in a part of the cell known as the mitochondria. As the name implies, Krebs' cycle is a "loop," or a cyclical system in which pyruvic acid enters a reaction cycle and is converted into a series of other molecules. At each step in the cycle, other molecules enter and leave the system, siphoning off energy that is continuously made available by more molecules of pyruvic acid entering the cycle.

The crucial point about this and other biochemical systems is that each step involves the production of a tiny amount of energy, but collectively they provide all the energy required by the body. In order for this source to be used to produce SHC, the energy generated in each step in each cell would need to be utilized in a short space of time.

According to orthodox medicine, there is no mechanism by which this has been shown to operate. Each stage of the process is physically separated from the others by tiny membranes within the mitochondria in each cell, but it is theoretically possible for these membranes to be broken down and for some form of "initiator" to instigate the process.

A possible candidate for the role of initiator is the chemical species known as free radicals—exceptionally reactive atoms in which one of the electrons in the atomic structure exists for a short time in an unnaturally high energy state. Because this state is not normal for the electron, it tries to return to its usual level by whichever means becomes available. This usually involves a reaction with another chemical.

An example of this is the breakdown of ozone in the Earth's atmosphere. Here, free radicals are involved in a series of reactions between the ozone and hydrofluorocarbons (HFCs) from aerosols and industrial refrigerants, which eventually lead to the decomposition of the planet's natural protective layer of ozone.

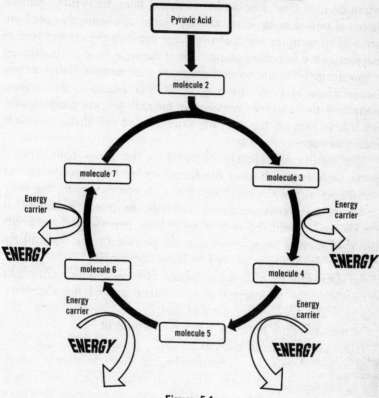

A simplified representation of the Krebs' cycle

Figure 5.1

Free radicals are also suspected of playing a part in the aging process, initializing reactions that produce the degeneration of tissue. It is not too farfetched, therefore, to suggest that free radicals can operate under special conditions within the body. Whether or not they can initiate the spontaneous release of energy from metabolic processes remains open to conjecture, but the energy produced by such a process, if it were to occur, is more than enough to account for the sort of heat and dramatic effects witnessed in cases of SHC. To put it into perspective, one gram of glucose releases 38,000 joules of energy.

But just how likely is it that something could go wrong with a system such as this? The answer is that it does, frequently. All biochemical processes are controlled by elaborate chemicals called enzymes. The complex series of steps that produce the correct type of enzymes in the necessary proportions at the right time are controlled by specific genes. Sometimes these genes malfunction. If this occurs, a wide range of problems may result. For example, if the gene controlling the enzymes responsible for cell division malfunctions and fails to turn off the enzyme, uncontrolled cell division occurs, leading to cancer.

If a similar malfunction occurred in the genes controlling a system such as glycolysis or the Krebs' cycle, it is conceivable that it could get out of control. Instead of a cancerous growth, the body would witness greatly increased metabolic processes. Taken to an extreme and conducted over a short time period—perhaps hours instead of months or years—this could provide the mechanism for generating the energies needed to bring about SHC.

An alternative to using the metabolic processes within cells might be to utilize the fat stores that accumulate around the abdomen. This may also account for the fact that in most cases of SHC, the fire is seen to come from the trunk or the midriff.

Hibernating animals, such as squirrels, possess a particular form of fat called "brown fat," which is broken down to supply energy during the hibernation phase. This is an extremely efficient energy source, providing in the region of 500 watts of power in the form of heat for each kilogram stored. This is enough to power a small electric fire.

Humans do not possess brown fat, but when the body is starved of nutrients, "normal" fat may be broken down. This is a relatively inefficient process, but an obese individual would have a large stock to draw upon, and again, a free radical initiation may possibly enable a form of chain reaction to be sparked off that would readily supply the sort of energy needed to bring about spontaneous combustion.

As I mentioned earlier, the pattern of fire in the victims suggests the combustion begins in the gut, and victims are invariably found with their extremities unscathed by the fire. This arrangement would fit with both the mechanisms described—the cellular and the body fat method. The reason depends upon which theory you adhere to.

Consider the cellular mechanism. In order for the mitochondria to continue functioning, they need oxygen. After the victim has died, the heart stops beating, but parts of the body live on until the oxygen supply is used up. By the time the fire reaches the knees, the oxygen level in the mitochondria would certainly have dropped below a critical minimum, and the cells would die and combustion would stop.

If, on the other, we say that body fat is responsible for sustaining combustion, the fire would probably burn itself out by the time it reached beyond the trunk, because the extremities of even the most obese person are usually less fatty than the abdominal region.

As in other possible mechanisms, some recent studies have proposed the phenomenon of cold fusion. This strange process was mentioned briefly in Chapter 1 as a possible energy source for interstellar spaceships, but it is seen by most of the scientific community as little more than fantasy.

In a set of experiments conducted in 1989, Professors Fleischmann and Ponns devised a system that appeared to produce energy by nuclear fusion at room temperature. The process seems to share some of the characteristics of what are called electrochemical systems. In these, reactions on metal surfaces immersed in special solutions provide electrical energy: this is the basis for the common battery. Rather than employing the more usual chemical reactions that take place in electrochemical devices, the two scientists proposed that, under the right circumstances, metal surfaces could act as facilitators for the fusion of atomic nuclei.*

Professor Fleischmann maintains to this day that his system works, and his attempts to create a workable and practical energy source from cold fusion have been funded by Toyota for the past few years. How this mechanism could create SHC is unclear, but

---

*Chemical reactions involve the electrons of an atom rather than the nuclei. Atomic processes, both fission and fusion, employ the nuclei and have nothing to do with these extranuclear electrons. Because far less energy is needed to keep electrons in atoms compared to that required to keep neutrons and protons in nuclei, chemical reactions are much easier to facilitate.

in their book, *Spontaneous Human Combustion*, authors Jenny Randles and Peter Hough quote an anonymous physicist as saying:

> It seems likely that of the elements found in the body, potassium or sodium are the most likely to be involved in a fusion reaction. The tissues with the highest potassium levels are in the brain, the spinal cord, and skeletal muscles. This would suggest that the feet and the hands would be the least affected, and the head the most.
>
> It would be very hot, maybe incandescent. This would hardly affect surrounding materials, as a slow burn seemingly would. It would all be over in seconds, or a few minutes at most. There just might be residual radioactivity, it would be surprising were there not, but not necessarily a lot.[3]

This is an interesting proposition and sounds convincing, but the problem is, if we accept this as a cause, we have actually traveled further away from an explanation. Not only is SHC unproven, cold fusion is regarded by the majority of scientists as nothing more than a hypothesis that has been unrepeatable by almost every worker in the field.

There is really no need to rely upon explanations from eccentric regions of science to account for ways in which SHC could conceivably happen—there are plenty of possible anomalies within the biochemistry of the body and the natural forces around us to generate just-plausible mechanisms.

One SHC researcher, Larry Arnold, has offered up the "strange particle" hypothesis as a source, suggesting the involvement of an imaginary particle he calls the *pyrotron*, which could in some mysterious fashion initiate combustion. Although, like the psitron, the pyrotron has a cute name, there is no experimental or theoretical support for its existence and it is, I believe, an unnecessary complication.

So far I have concentrated on ways in which the body could sustain a form of combustion once it has been initiated, but what sort of event could spark off the process?

The first place to look should be the body itself. Although this sounds like an unlikely source for the sort of energies needed, it is not beyond the bounds of reason. We saw in Chapter 3 that the human brain is a mass of electrical circuitry more complex than any computer so far built. Each neuronic impulse operates at a voltage of 100 to 120 millivolts. It is possible that a suitable current, perhaps produced by a surge of energy from some unknown freak activity somewhere in the brain, if rerouted, could initiate combustion in a minute area, which could then lead to a series of increasingly disruptive processes ending in an explosion or an internal "fire."

The biochemical processes taking place in cells also involve electrical potentials. A voltage higher than that found in the brain is produced across the inner membrane of the mitochondria—a potential difference of 225 millivolts. According to one calculation, this amounts to 45,000 volts per cubic centimeter.[4]

But we should not get carried away with such figures. Adding together voltages from millions of neighboring cells does not actually mean very much. It is a little like saying that each computer in a building handles a voltage of 240 volts, and if this malfunctioned, we could create an electrical surge of tens of thousands of volts by its spreading to hundreds of other computers nearby. It is conceivable that such a peculiar circumstance might arise, and in the same way, it is possible that a freak event could occur inside a human body, but it would be a very unusual anomaly.

Looking beyond the body itself, there are powerful natural and man-made forces at work. It has been estimated that sixteen million thunderstorms occur on the planet each year, and at any one time there are 1,900 of them raging in different parts of the world. Surprisingly, only a handful of people are killed each year by lightning in Britain (although in America, with a population only four times larger, 100 to 150 individuals are killed by lightning annually).

Lightning is produced in thunder clouds, which are churning masses of air containing large numbers of negatively charged ions. These ions repel negative charges in objects over which the cloud passes, disturbing electrical fields in all living creatures, including humans. This is why animals and some sensitive people know when a thunderstorm is approaching. Some individuals have been known to feel depressed or even physically sick as the negatively charged cloud moves overhead. The brain chemistry and the electronic balance in the cells of the body are disrupted by the repulsion of negative charges, and in some unknown way, this causes emotional and physical responses.

According to reports, there is no apparent correlation between thunderstorms and spontaneous human combustion. When people are struck by lightning, they do not burst into flame. In fact, most victims show no outward sign of having been burned. The reason for this is that the intense heat produced by a thunderbolt—something like 30,000°c—touches the body for only a few microseconds, which is not long enough to cause a burn. Instead, the victim dies from a complete destruction of the nervous system—the electronic circuitry of their bodies is literally fused.

Although there is no direct correlation between lightning storms and SHC cases, there are similar natural weather effects that could account for some reports. St. Elmo's Fire, a ghostly glow sometimes seen around the masts of ships and tall buildings, is produced by ionized air.

A similar phenomenon is ball lightning. Used frequently to account for a collection of paranormal activity, from UFOs to crop circles, it is a little understood weather effect, and so rare its very existence is still disputed by many meteorologists. If it is a genuine weather feature, the consensus from occasional sightings is that it takes the form of a ball of energy that is generally spherical, from one to more than a hundred centimeters in diameter, which lasts for a few seconds. The balls are reported to move horizontally at speeds of a few meters per second, and to decay silently or with a small explosion. It may be a form of condensed energy, perhaps a small cloud of ionized gases that sometimes breaks away from a thunder cloud, imagined by some to be a "thunder-bolt-in-a-ball."

Ball lightning does present itself as a serious candidate for initiating SHC. It is very rare, possesses large amounts of energy, appears, from some accounts, to be able to penetrate material objects, and lasts for a very short time. Furthermore, it leaves no clear trace of its presence.

Another possible initiator is the increasing number of electronic fields produced by the machines we use. It is easy to forget that every piece of electronic hardware in our homes, offices, and cars have associated magnetic fields, and each, to some degree, interfere with one another.

I remember as a child how the television picture in our living room would become disrupted when the neighbor turned on the lawn mower or the food mixer (it was always in the middle of *Star Trek*, I seem to recall). Today, electronic machinery is better insulated and this cross interference is rare, but even the best insulation cannot prevent a certain amount of leakage, and as we fill our environment with ever more sophisticated machines, the chances of stray fields and the possibility of power surges altering the electrochemical makeup of our bodies should be considered seriously. Much has been made of the eye strain resulting from the use of computers, yet little research has been conducted on the possible detrimental effects they have upon our nervous systems.

Static electricity is also with us on a daily basis, and some people are far more susceptible to it than others. There have been many cases of people who act as stores for static electricity and have even unconsciously disrupted electronic devices nearby. It is conceivable that victims of SHC are particularly vulnerable to electrical fluctuations in the environment, generated not by natural weather conditions such as thunderstorms but by the machinery around them.

Beyond these relatively mundane sources, there have been a collection of what might be called "paranormal explanations" to account for the initial "spark" that triggers SHC.

Many of these musings have no theoretical grounding and appear to be unnecessarily mysterious, yet, one interesting fact has arisen from them. It would appear that spontaneous combustion only occurs in human beings.

It may be that SHC is such a rare phenomenon that we have simply not observed it in the wild, but reports would be expected from farmers through the centuries, especially given the propensity for cows to store large volumes of combustible gas. The cellular mechanisms described above are almost identical in all mammals, so why should they only malfunction in humans?

One suggestion is that the mood or the psychic state of the victim can somehow influence the chances of SHC. There is no solid evidence to support this, only a few incidental facts that could point to some form of emotional trigger. Two reports from the 1950s tell of apparent SHC victims who were in the process of committing suicide at the time of the combustion.[5] Some researchers have also suggested that the large number of SHC cases in which the victim was found to have been drunk might be linked to mood rather than biochemistry.

From the collected reports taken over three centuries, there emerges no real pattern to pin down a type of person or any particular peripheral event that may be the root cause of the event. There is no particular time of day when these conflagrations occur more or less frequently, although more cases are reported during the winter months than the summer. Skeptics would say that this is the time of year when there are more likely to be open fires roaring in the grate or ovens left on. Believers counter with suggestions of a link between SHC and SAD—Seasonal Affective Disorder. But again, there is no hard evidence to support this connection.

So, what conclusions may be drawn from the supposition, theorizing, and meager collection of facts surrounding SHC?

There seems to be little need to search for any supernatural cause. The human body is a startlingly complex and formidably versatile machine which still turns up surprises for biologists, chemists, and physicists. It is possible that an environmental factor could trigger a completely unknown biochemical disturbance, causing the body to turn against itself or become overloaded in some odd fashion.

It is conceivable there are many types of spontaneous combustion. The incidents involving explosion may be fundamentally different to the slow-burn cases, or the cause may be the same, but different victims may respond to the initiator in different ways, perhaps dependent upon the quantity of gases in their gut or their physical characteristics. There certainly seems to be a category of SHC that may not be that at all, but due simply to the build-up and eventual release of gases such as phosphine, which then combust upon contact with air.

Because SHC is so rare, it is difficult to construct a definitive explanation. The glib responses of traditionalists are certainly unsatisfactory and account for only a small number of cases, ignoring aspects of many others. In this respect the "scientific" explanations are not scientific at all. They merely sweep under the carpet any uncomfortable facts—those that defy explanation by conventional means. Sadly, these "difficult" aspects are also the ones that offer the best opportunity to discover something strikingly new.

6 Visions From a Future Time

For we know in part, and we prophesy in part.
1 CORINTHIANS 13:9

When we consider the full range of paranormal phenomena, one of the most difficult for the scientist to accept is precognition. Ironically, it is also one of the most ancient metaphysical ideas and appears in some form in almost all cultures and eras, adopting many guises, from Shakespeare—"Beware the ides of March"—to end-of-the-pier novelty attractions.

Before we look at the ways in which precognition may operate according to some rather elaborate ideas from modern physics, let us first consider the more mundane ways in which some forms of foretelling the future can be explained.

In his book *Supernature*, Lyall Watson lists over a dozen exotic forms of prediction, ranging from aeromancy (divination via cloud shapes) to tiromancy (a form of prophesy involving cheese), but there are more traditional forms of fortune-telling and wide-scale prophesy.[1]

In ancient Egypt the priests of the sun god Ra were prophets, and the Greeks consulted oracles such as the one at Delphi, to determine the most auspicious time to wage war or to broker peace. In more recent times, the most famous western prophet was Michel Nostradamus, and methods such as the tarot, crystal ball gazing, and palmistry have become part of cultural history, embodied in the literature and folklore of many countries and eras.

Nostradamus was a mystic who lived in France between 1503 and 1566 and composed thousands of prophesies in the form of four-line verses called *quatrains*. Some enthusiasts see accurate prophesy within his writing, but Nostradamus is open to the same criticism as almost all traditional forms of divination—the problem of interpretation. What Nostradamus predicted has only been fitted with events *after the fact*, which diminishes their impact, and, like many other forms of prophesy, his statements are extremely vague. Although he has been credited with predicting many technological discoveries, wars, dictators, and catastrophes of the twentieth century, the original prophesies hardly offer clear-cut pronouncements. Take the following quatrain for example:

> *They will think they have seen the Sun at night*
> *When they will see the pig half-man:*
> *Noise, song, battle, fighting in the sky perceived,*
> *And one will hear brute beasts talking.*
>
> CENTURY I, VERSE 64

This is meant to be a prediction of atomic weapons and fighter aircraft, but it is easy to superimpose such images after the events have transpired. Commentators of the nineteenth century may have placed a totally different interpretation upon it, and if the work of Nostradamus had not been discovered until, say, the twenty-second century, what would people living then think of it? Like many occult phenomena, prophecy or precognition is judged by the standards and universal viewpoint of the time.

A quite different form of precognition involves an individual who is interacting with a system; for example, practitioners of the Chinese divination method called the I Ching, or the European version, the tarot. Nostradamus was a lone prophet, sitting in his study, supposedly gazing into the future, but tarot or the I Ching are a form of ritual which, believers claim, focuses certain mental powers and bypasses the conscious mind. Enthusiasts of these techniques suggest it is our subconscious minds that can break the shackles of time, and that if we allow it free reign, we have the power to access the future. By using any system, whether it be a set of cards or the yarrow stalks of the I Ching, the everyday clutter of the conscious mind is filtered out so that the subconscious can take direct control.

In a similar way, many individuals who claim to have the power of precognition say they receive images from the future while they sleep. If there is any truth in the phenomenon, then it would make sense that the best time to receive messages would be during sleep, as this is when the subconscious takes over from the conscious mind.

Whether prophetic images come via a ritualistic system or by drug or sleep-induced states in which the conscious mind is bypassed, a surprising number of people claim to have had precognitive experiences. Most of us forget the images long before we might try to interpret them—they are crowded out by the conscious mind and the confusion of everyday life. But according to believers, some of these images remain with us and pop up again in our conscious minds when triggered by some subliminal event—this is what is usually meant by a feeling of déjà vu.

Some people who retain images from dreams can consolidate them, and a few have even been so bold as to make public their predictions. Surprisingly, these have, on occasion, proven disturbingly accurate. Worse still for the scientist, once in a while these precognitive images are almost impossible to dismiss with mundane, mechanistic explanations.

After the Aberfan disaster in 1966, when a coal slide destroyed a school, killing over one hundred people, many claimed they had "seen" the event before it happened. According to one psychiatrist who studied the cases of precognition associated with this disaster, some sixty predictions appeared to him to be genuine. One of the children who perished in the tragedy told her mother two weeks before the disaster:

"No, Mummy, you must listen. I dreamt I went to school and there was no school there! Something black came down all over it!"[2]

One William Klein foretold the sinking of the *Titanic*, and was so impressed with his own vision of events to come that he warned a friend not to travel on the ship. The friend ignored his warning and met an icy death. A well-known researcher of the paranormal at the turn of the century, W. T. Stead, was warned by two psychics not to travel on the *Titanic*, but surprisingly, he ignored both of them and drowned.

These claims could of course be put down to coincidence. It is not unusual for children to have anxiety dreams linked to school, and this could take the form of something black engulfing the building. The *Titanic* was a ship of revolutionary design for the time, and there was a great deal of publicity surrounding the maiden voyage of what was claimed to be the first "unsinkable" ship. It is not surprising that many would have thought the worst. When the worst happened, it seems as though they had performed a miracle by foreseeing it.

However, some researchers and believers in precognition claim there is no such thing as "coincidence." Instead, they claim that human beings have an innate ability to roam the highways of time, and that the barriers of "past" and "future" may be elastic.

Carl Jung coined the term *synchronicity* and wrote a book about the idea called *Synchronicity: An Acausal Connecting Principle.*[3] In a further collaboration with the physicist Wolfgang Pauli, they created what became known as the Pauli-Jung theory, in which they described the unprovable but nevertheless attractive idea that the collective unconscious could somehow influence the world. They visualized a system that opposes the randomness of Nature by imposing a "unitary world," or an *unus mundus*, guided subconsciously by a universal intelligence (that of the human race and other sentient beings). The archetypes mentioned in Chapter 3 are a manifestation of this enveloping "force," and a way in which it leaks into the material world via our subconscious minds.

This is an appealing idea, rooted in humanism and transcending the mundane world in which we live and work, but is there any substance to it? How could such an all-pervading awareness operate within the rules of science?

If we restrict ourselves to the matter of predicting the future, there are two possible mechanisms through which this might happen. Both require the use of ideas from the very fringes of modern physics, but which are not illogical or, in theory, beyond the limits of recognizable science. The first of these depends on that useful theoretical construct—the wormhole.

We saw in Chapter 1 how wormholes might be used to travel large distances across the galaxy or beyond without having to traverse the distance between two far-flung points. There are huge practical

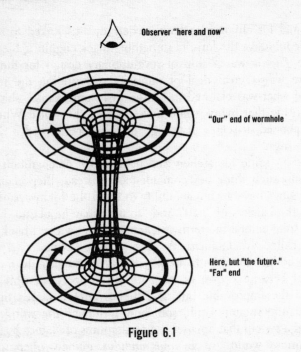

Observer "here and now"

"Our" end of wormhole

Here, but "the future."
"Far" end

Figure 6.1

problems involved with this method of transport, and I showed how it could only offer, at the very least, a limited form of communication between two fixed points. But as well as offering a passage between two physical points within the same space-time framework, wormholes may also offer an opportunity to access information from the future.

The wormhole in figure 1.4 linked two points in space, but if you now picture one end of the wormhole moving at close to the speed of light (figure 6.1), the whole system suddenly takes on the characteristics of a time machine.

One of the consequences of Einstein's special theory of relativity is that the measurement of time is dependent upon the velocity of an observer. If one end of the wormhole is moving at close to the speed of light, while the other is moving much more slowly, then observers at the two ends would be in different times.

For our precognition scenario, we need to have the far mouth of the wormhole positioned in space not far from where we are at the other opening, so as to "see" events that have meaning for us, here on Earth. But our end of the wormhole would initially have to be moving at a velocity close to the speed of light for there to be any viable difference in the measurement of time at the two ends. But once the distinction between the ends is established, the velocity difference can be zero.

If the other, far mouth of the wormhole was relatively stationary compared to ours, it might follow a circular path once in seven days, but to us it would seem to take far less time, let us say one day. So effectively, we have a time machine—the far end lies in our future. An observer at "our end" looking through the wormhole to another observer at the far end would see that their date was six days in our future and we would exist in their "past."

So how could this facilitate precognition?

Prophesy or precognition requires the passage of information from a future time to our time. To do this there is no need for the complexities of transporting a spaceship or people through treacherous wormholes with their associated gravitational fields and devastating forces. If information can pass through the wormhole and be detected by a sensitive individual, then precognition is a possibility.

In his book, *The Physics of Star Trek*, Lawrence Krauss analyzes the possibility of developing a matter transporter as used in the television series *Star Trek*.[4] He concludes that the transportation of a material object is practically impossible, but that the system might just work if we view the transporter as moving *information* rather than bodies.

Even then the mathematics is enough to induce a migraine! It appears that the information content of a human being—all the data needed to reproduce a single human—is in the region of ten million billion times the total information contained in all the books ever written. But a way around this restriction is to say that information conveyed from a future event through a wormhole to a receiver in our time would not need to be so detailed.

Documented cases of precognition demonstrate one unifying characteristic: the images from the future are almost always nebulous and ill-defined; often subjects receive a mere outline describing the event. This, then, may require a far less sophisticated information transfer system. Psychics who "see an outline of a ship sticking out of the water" and later claim to have witnessed the sinking of the *Titanic,* may have received information from the event which somehow leaked through a wormhole linking the future with the present; they then processed that information and "saw" the catastrophe before it occurred.

When outlining the use of wormholes for space travel, I highlighted the fact that they would be of limited use. This is also the case when using them to explain precognition. The far mouth of the wormhole would have to be in just the right place at the right time in order for information about an event to be passed back to the receiver. This in itself seems highly improbable.

Believers in this theory have suggested that perhaps mini-wormholes are present within our environment all the time, and that in some inexplicable way, catastrophic events trigger them to act as conduits for information. They go on to suggest that human consciousness can somehow interact with these mysterious, theoretical objects, and when the human mind faces catastrophe, or perhaps death itself, information can be passed through a wormhole linked to our present. But until we know more, this explanation is actually no explanation at all, merely an hypothesis used in an effort to add substance to another. Instead many believers in precognition are beginning to show an interest in the idea that prophesy may be linked with that much-maligned area of physics known as quantum mechanics (QM).

Quantum mechanics began life during the early part of the century as a completely revolutionary theory that overturned the prevailing (classical) ideas of Victorian physicists.

The earliest model for the atomic world held that the atom was composed of a nucleus around which electrons orbited—like a solar system in miniature. Electrons were known to have about one two-thousandth the mass of a proton (one of the constituents of the nucleus) and to possess a negative charge to counter the positive charge of the proton.

But during the first decades of the century it was realized that this model could not possibly work. For a start, mathematics demonstrated that electrons could not be sustained in their orbits like planets, and their orbits would decay so that they merged with the protons in the nucleus. As this was clearly not happening in the universe we live in, it was assumed correctly that the model must be wrong.

Through the pioneering work of physicists such as Planck, Bohr, and Schrödinger, a far more sophisticated model of the nature of the subatomic realm emerged, and with it a number of counterintuitive consequences that have caused confusion for the nonphysicist ever since. One of the pioneers of quantum mechanics, Niels Bohr, even went so far as to say that "anyone who is not shocked by quantum theory has not understood it."

The problems really began when particle physicists realized the electron was not a ball of matter with a negative charge but could only be described in terms of probability. In other words, there is a high probability that an electron will be found a certain distance from the nucleus and a low probability of it existing much farther or much nearer the nucleus than this.

Linked with this notion is the Uncertainty Principle developed by Werner Heisenberg in 1927. This shows that there are limits placed upon the accuracy to which pairs of physical quantities can be measured. For example, if we try to measure, say, the position *and* the momentum of a subatomic particle, the very act of measuring these quantities disturbs the particle so much that they cannot be said to have an exact position and an exact momentum at the same time. We can only assign *fuzzy* values for the two factors. This fuzziness is described by the *wave function*—meaning a description based entirely upon probabilities.

Now at first glance this might seem like a trivial matter—so what if subatomic particles cannot be pinpointed to an exact location? In fact this is the whole essence of QM and lies at the root of all the problems it presents for the nonphysicist. It is also the very reason why quantum mechanics could conceivably help to explain precognition.

If we cannot define the universe at the most fundamental level, then it must mean the universe is constructed upon probabilities. There can be no certainty, no clear-cut definitions, no pure yeses or no's. From this springs some very strange quantum mechanical ideas.

The first of these is that the universe can only be studied on a statistical level; if we probe too deeply, or try to single out individual particle transactions, we end up with nonsensical results. An alternative way to consider this is to say that the universe really only demonstrates an apparently logical framework if viewed holistically.

This we can just about accept intuitively, but modern quantum mechanics has gone much further. Heisenberg himself suggested that the fuzziness of the quantum world could mean the traditional notion that *effect* must always follow *cause* could break down. Worse still, the experimenter or observer may be able to interfere with the experiment, that human consciousness might control the processes of the universe in some way.

To illustrate this, consider a famous thought experiment devised by one of the founders of quantum mechanics, the German physicist Erwin Schrödinger during the 1920s.

Schrödinger imagined a box containing a cat and a radioactive source. If this material decays, it triggers a poison that kills the cat, but because the radioactive material decays randomly, there is a fifty-fifty chance it will happen and kill the cat and a fifty-fifty chance it won't. The only way the experimenter will know what has happened is to open the box to see if the cat is alive or dead. This means that until then the cat is both dead and alive. The probability will only become a certainty by the action of the experimenter opening the box, so the observer controls the outcome, or in technical terms, "collapses the wave function."

Although there is nothing illogical or mathematically false about this description, it does not feel "right." Indeed, it leads to a whole series of paradoxical arguments. For example: What if we replace the cat with a human? Presumably they would be as able to collapse the wave function as readily as the experimenter. What would they experience inside the box? Would they be able to override the effect of the experimenter?

Now imagine the experiment had become a media event. What would happen after the experimenter had opened the box? He may find either a live human or a corpse, but the cameras and journalists outside the lab, unaware of the events inside, would not know what has happened, so is the subject alive or dead? Equally baffling is the question of what would happen if the cat or the human was replaced by a computer, or indeed if the experimenter was replaced by a computer? How would these changes affect the outcome?

As mystifying as Schrödinger's cat experiment may be, it is based upon sound theory and decades of reasoning within the discipline of quantum mechanics. It does not feel comfortable because it appears to contravene the logical processes we have been educated to appreciate and some that may be instinctive to us as humans. Yet, these principles may be right and our intuition wrong.

These bizarre notions have been interpreted in a number of different ways. The most traditional approach is called the Copenhagen Interpretation, which was developed in 1927 and suggests that the indeterminacy of individual events within the subatomic world cannot be extended to the macro world, and that the large-scale world of everyday experience is only comprehensible on a broad statistical level. Critics claim this side-steps the issue and have come up with some lateral ways of looking at the quantum dilemma.

One suggestion, known as the Everett-Wheeler (after the creators of the idea, Hugh Everett and John Wheeler), postulates that all possible outcomes of a process (in the case of Schrödinger's cat — two: dead cat, alive cat) are observed somewhere. In our universe only one outcome is observed, perhaps the cat survives; but in a parallel universe the opposite is observed—the cat is dead.

This would mean that whenever anything happens, there are at least two alternatives which are observed by observers in two totally separate universes that can never meet. Furthermore, with each passing second, from the dawn of time, the number of possible "futures" has increased and is still increasing to a near-infinite variety of outcomes.

This idea is taken seriously by some members of the scientific community. It has been the subject of several papers and used as a platform for further, even more convoluted schemes. Yet it is quite untestable and will probably never be validated. Ironically, its cocreator, the eminent physicist John Wheeler, has attacked parapsychologists, declaring that the scientific community should throw out the "weirdos from the workshop of science."[5]

Enthusiasts of the paranormal see this as arch hypocrisy and claim that the ideas of Wheeler and many other physicists are far more extreme than the most imaginative notions of the parapsychologists and researchers of the occult. The quantum physicists counter this with the argument that their theories are founded on mathematics and a self-consistent body of knowledge stretching back the best part of a century. They also point to the many aspects of quantum mechanics that have been shown to work and the technological developments this discipline has offered the world.

And they have a point. Quantum mechanics is the foundation of the science of lasers, advanced electronics, and telecommunications. Without an understanding of this exotic area of physics, there could be no television, advanced computing, space travel, CD players, no global telecommunications, no Internet, no laser surgery. By comparison, parapsychology has continued to be one of the most evasive aspects of human exploration and remains impossible to pin down.

A third interpretation, which seems less controversial than the Everett-Wheeler model but is nonetheless still beyond scientific verification, is the idea that human consciousness can interact directly with the wave function. This suggestion is credited to the American physicist Eugene Wigner and offers the idea that the human mind can subconsciously manipulate the universe at a fundamental level.

Here we have echoes of several ideas discussed in earlier chapters. The parallels with Carl Jung's unitary world or *unus mundus* and Rupert Sheldrake's Morphic Resonance are obvious. Such an explanation may account for PK and types of telepathy, but above all, the Wigner interpretation offers a solution to how precognition might occur.

Central to this explanation is the idea that the interaction between consciousness and fundamental processes is not restricted to the here and now. In other words, human consciousness has the ability to transcend distance and time in the way it manipulates the universe. In technical jargon, it is said to be *spatially* and *temporally invariant*. But what is surprising is that this is not merely the eccentric imaginings of overenthusiastic believers. It may be supported by

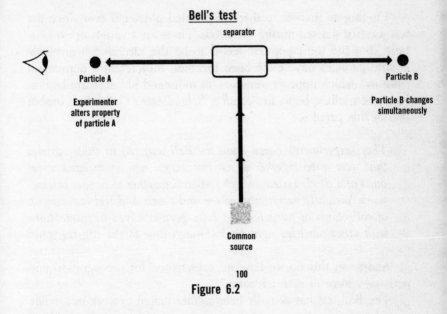

Bell's test

separator

Particle A

Particle B

Experimenter
alters property
of particle A

Particle B changes
simultaneously

Common
source

100

Figure 6.2

a convincing set of tests based upon a thought experiment devised by the American physicist John Bell.

In the experiment, two particles from a common source are fired at a device, which then sends them in opposite directions at the speed of light.

One of these particles is then "altered." What this really means is that the particle can be changed in a very limited number of ways. For example, its *spin* can be altered, or if it is a high-energy electron, it can be allowed to return to its normal "ground state" energy level. But the truly remarkable consequence of this is that the second particle which, remember, has been traveling in the opposite direction at the speed of light, is also altered *simultaneously* by the change made to the first particle.

For the quantum physicist this presents an exciting but worrying paradox. The speed of light is a finite value, an irrefutable universal constant that cannot be negated; so how could one particle alter the state of the other when any communication between them is seemingly impossible?

Finding an answer to this has puzzled physicists ever since the test was first devised during the 1960s. There are a variety of explanations, but the most popular seems to be the idea that if any subatomic particles have once been together, they retain a permanent "affinity" which appears somehow to transcend physical limitations. In his best-selling book, *In Search of Schrödinger's Cat*, John Gribbin says of this paradox:

> They [experiments based upon the Bell test] tell us that particles that were once together in an interaction remain in some sense parts of a single system, which responds together to further interactions. Virtually everything we see and touch and feel is made up of collections of particles that have been involved in interactions with other particles right back through time to the Big Bang.[6]

And so in this do we have an explanation for precognition and perhaps a form of race telepathy?

The Bell test has actually been demonstrated to work in a range of real experiments conducted since it was first contrived over thirty years ago. But the problem with linking it to the paranormal is that no information is transferred in the connection between the two particles, A and B. If particle A has a certain spin and is "flipped" into a different state by the experimenter, then particle B is altered in the same way, but crucially, no information need be transferred in order to do this. This change is merely a shift from one spin to another, and each spin state is dependent upon a set of what are called *random quantum fluctuations*. Changing from one random pattern to another requires no information. Furthermore, it is found that if an experimenter tries to repeat the test by changing a parameter that does require information to be transferred, there is no simultaneous effect upon particle B.

Theorists of the paranormal say this is actually no barrier because psychic effects do not involve information as we know it, but they cannot support this claim with hard facts, and such comments are just the sort that justifiably irritate physicists like John Wheeler.

So, what may be judged from this collection of ideas and theories? Several eminent scientists have been convinced that, within certain limits, quantum mechanics can explain the esoteric world of the paranormal. Men such as Jung and Pauli were fascinated with a possible link between the two and constructed elaborate explanations for a marriage of QM and psi. Today there are many scientists around the world who are continuing the tradition. Henry Stapp at the University of California, Berkeley, is utilizing the ideas of Eugene Wigner to reach a mathematical interpretation of how QM could lie behind psychic phenomena, and Nobel laureate Brian Josephson has said that if psi phenomena had never been noticed, quantum mechanics would have predicted them.

Unimpressed by apparent restrictions such as the limit to information transfer, parapsychologists suggest a mechanism for how QM and precognition could relate. They propose that human consciousness is able to influence the past, present, and future because, at its most fundamental level, the law of causation can be manipulated. They claim the universe is a single unit and that each human mind is part of this vast network. Human consciousness, they believe, can "see" events that have not yet happened because we are all one, always have been and always will be. This may seem supportable using experiments like the Bell test, say their opponents, but it flies in the face of the Copenhagen Interpretation and ignores the math, while at the same time conveniently extracting the kernel of verifiable science. Furthermore, no one has even begun to offer a way this vague but intriguing concept could actually work.

In some respects it is easy to sympathize with the skeptics who say that such interpretations come from ignorant nonscientists who are trying to manipulate something they don't understand to their own ends. To many physicists, the ideas of the parapsychologists are insulting and an affront to the decades of study and research they have committed to their subject. But on the other hand, quantum mechanics does offer itself as a sacrificial lamb.

To be accepted, even the most unorthodox scientific hypotheses have to be supported by rigorous mathematics and must be consistent with a body of scientific knowledge stretching back to the seventeenth century. Yet, many aspects of modern physics remain unprovable.* For instance, there has never been absolute proof of

---

*Although there is very, very good circumstantial evidence.

the existence of a black hole, and many advanced theories from particle physics remain untestable because the equipment needed to demonstrate a range of effects is beyond our technological capacity.

To the enthusiast of the paranormal, science at the limit can sometimes seem as esoteric and otherworldly as their own ideas. The difference lies in approach: Science is based upon mathematical integrity and, where possible, experimental verification—no theory is accepted until proven by experiment. Parapsychologists, the physicists claim, are too easily led into assumption and neat ideas that appear to fit the facts without proof, mathematical rigor, or a sufficient degree of self-criticism.

One of the greatest scientists who ever lived abhorred quantum mechanics despite the fact that he helped to create it. Albert Einstein saw the avenue along which QM progressed as a voyage into the absurd, and commented famously that "God does not play dice," meaning that the universe was not merely a collection of probabilities. On another occasion he commented: "Quantum theory reminds me a little of the system of delusions of an exceedingly intelligent paranoiac concocted of incoherent elements of thought."

Erwin Schrödinger was as dismissive of modern QM as Einstein, once claiming: "I don't like it, and I'm sorry I ever had anything to do with it." By a delicious irony, he created his famous thought experiment as a method of demonstrating the apparent absurdity of many quantum mechanical implications.

Yet, many elements of QM have been tested and shown to work. Many of the more advanced ideas spring from the same well as those that have given us much of our modern technology—so how far can we go in accepting or denying the more imaginative theoretical offshoots? If we dismiss aspects of QM, we are risking throwing out the baby with the bathwater.

It may be that modern physicists have gone along a path that has led to ridiculous and quite wrong theoretical explanations for practical and workable phenomena. If this is ever shown to be the case, the parapsychologists will have scored a victory, but I suspect the answer to this lies some way in our own future, and to acquire it, we will probably have to wait.

The Agony and the Ecstasy

When you buy shoes measure your feet.
CHINESE PROVERB[1]

Pain is a perfectly natural, even necessary physical process, but there is growing evidence that the human mind can train the body to negate pain and control a range of other bodily functions, bending the physical to the power of the will.

The incredible abilities of certain individuals, known in India as fakirs or yogis, have been documented since ancient times. Descriptions of fire-walkers, self-mutilators, and tales of people capable of the most amazing feats of endurance are mentioned in the Bible, in the writings of the ancient Greeks, and in Oriental texts. Today, displays of these talents may be witnessed by tourists around the world.

I have experienced such skills at very close quarters and in the most unlikely place — the Boardwalk of Venice Beach. Amongst the robotic dancers, limbo artists, buskers, and chain-saw jugglers, I spotted a small crowd gathered around a terrifyingly thin white man who was busy preparing a bed of glass made from a collection of bottles. He was joking with the crowd as he smashed the bottles and arranged the pieces of a sheet laid upon the ground, and when he spotted me taking pictures, I was called over and asked to help in his act.

Five minutes later, the glass heaped high, three-inch jagged shards protruding from the pile, the performer, who told me his name was Larry, had me steady a chair as he climbed onto the seat. A moment later he jumped feet first onto the pile of glass. The crowd fell silent, but Larry was totally unharmed. He then proceeded to leap up and down on the glass, rubbing his feet among the slithers as though he was paddling in a stream. When he had finished, he showed me his soles—there was no blood, not a single scratch.

In many cultures such acts of courage and control are treated as a religious ritual, a cleansing process, and in ancient times facing such terrors was used as a form of judgment—the traditional trial by fire. According to one story, in 1062, the Bishop of Florence was accused of corruption by the saintly Peter Aldobrandini. Peter declared that he and the bishop should both walk on fire if the clergyman wished to prove his innocence. A corridor of hot coals was prepared and a bonfire placed at each end. Peter Aldobrandini entered the flames, walked the path of flaming coals and through the fire at the far end totally untouched. When it came to the bishop's turn, he wisely declined, opting instead to resign his position.

Such stories led people to believe that these seemingly miraculous feats could only be achieved by the pure of heart, or through deep meditation and rigorous fasting, and today such beliefs are still common among those who perform the acts or watch agape. Modern research has shown that religious observance is only helpful because it imbues confidence. The ability to walk on fire, pierce the body with hooks, or lay on a bed of nails unharmed is a question of applied science, training, and belief in oneself.

One of the most dramatic examples of human endurance is the art of fire-walking. The Old Testament contains a vivid description of such an event in the tale of Nebuchadnezzar's attempt to execute Shadrach, Mesach, and Abednego. According to the biblical account, the fire was so hot it killed the executioners taking the accused to the flames, but left the victims miraculously untouched. "The princes, governors and captains, and the king's counselors, being gathered together, saw these men, upon whose bodies the fire had no power, nor was a hair of their heads singed, neither were their coats changed, nor the smell of fire had passed on them."[2]

In isolation this could be treated as just another farfetched Old Testament drama, but the same, seemingly miraculous exhibitions are witnessed every day, and within recent years some of these have been subjected to serious scientific study.

In 1980 a team from the University of Tubingen in Germany traveled to the annual festival of St. Constantine at Langadhas in northern Greece to investigate the phenomenon. As the festivities reached a climax and the fire-walkers prepared themselves by chanting and performing ritualistic dances, the scientists set up their apparatus. They attached thermocouples to the walkers' feet, and electrodes linked to an EEG, which measures brain wave patterns, were placed on the scalps of the yogis. The team noted the length of the fire pit at four meters and that it was filled with coals and embers to a depth of five centimeters. The surface temperature was measured at 495°c (932°F).

After the display, the investigators measured the temperature of the soles of the walkers' feet to be 180°c (356°F), but could detect no sign of blistering or scorching. At the same time, the EEG recorded that the subjects had shown significantly increased theta rhythms during the walk.

At another ceremony in Sri Lanka, the psychic investigator, Carlo Fonseka, visited several hundred fire-walking demonstrations where he took measurements of the fire pit and monitored the walkers. He found that the fire paths were usually between three and six meters in length and eight to fifteen centimeters deep, with fire temperatures lying between 300°c and 450°c (580°F and 850°F). But not contented with watching the performers in action, Fonseka decided to set up a laboratory test to quantify the ability of subjects to resist high temperatures.

The apparatus was simple, a forty-watt lightbulb inside a metal cylinder. Volunteers were then asked to place the soles of their feet on the top of the cylinder for as long as they could. The control subjects, people untrained in fire-walking, felt some heat after between six and ten seconds, and after thirty to forty seconds they experienced extreme pain. The fakirs from the fire-walking ceremony felt no heat for an average of 29 seconds, and could keep their feet in position for up to 75 seconds.

Fonseka's conclusion was that the fire-walkers all had soles with far thicker epidermal layers. They acquired this protective covering as a result of wearing shoes or sandals only rarely, and it acted as a very efficient thermal insulation.

But this was not the only reason for their success as fire-walkers. The other crucial element, Fonseka claimed, was speed. He conducted a study of over a hundred walks, and found that the fakirs spent an average of only three seconds in contact with the coals during the entire walk, with each step requiring 0.3 seconds.

Even so, three seconds is plenty of time to receive severe burns, and the secret, according to this set of results, seems to be a combination of conditioned soles and speed. This is borne out by frequent cases where ill-prepared walkers, who insist upon "having a go," end up in the hospital with third-degree burns.

But this still cannot be the entire picture. There are several other factors to consider in accounting for success with the art of fire-walking, because it has been demonstrated safely by westerners with soft feet and with very little preparation. Hardened soles and a lightness of touch are obviously helpful, but some physiochemical aspects must be of equal importance.

Dr. Jearl Walker, professor of physics at Cleveland State University, has proposed something he calls the *Leidenfrost effect.* Johan Leidenfrost was the first person to notice that liquid exposed to sudden intense heat produces an insulating cushion of steam. Walker concluded that perspiration produced on the soles of the feet because of the excitement of the ceremony evaporates as soon as it comes close to the hot coals. This creates a layer of protective steam on the soles of the runner, which lasts long enough to cushion the impact of the heat.

Another, purely empirical theory comes from physicist Bernard J. Leikind at U.C.L.A., who suggests that part of the explanation derives from the difference between temperature, heat, and internal energy.

The words "temperature" and "heat" are often misused; it is easy to forget that temperature is merely a measure of the heat content, but heat is the result of molecular vibration within an object. Take for example the hot coal in the fire path. This is made largely of carbon, which has a very regular chemical structure. But unless a substance exists at absolute zero (−273°c) its molecules will be in a state of constant vibration, and the more internal energy they possess, the faster they vibrate. When one object warms up another, it happens because some of the energy from the hot object has been transferred to the cooler one, causing the molecules of the cooler object to vibrate faster.

Complications arise because different materials at the same temperature possess different amounts of energy. This seemingly odd result comes from the fact that different materials have different *specific heat capacities*, which signifies their relative ability to store heat. Perhaps more important still is the ability a substance has to convey or pass on its thermal energy. This is called the *conductivity* of the substance, and the carbon in the coals of a fire path has a surprisingly low conductivity. So, although the embers may measure a temperature of some 500°c, this energy is not efficiently transferred to the walker.

To draw an analogy, imagine yourself walking barefoot across a collection of different surfaces on a hot summer day. Sand and metal would feel very much hotter than, say, rush matting or a wooden surface, because the sand and the metal are better conductors and transfer their thermal energy to your feet with greater efficiency. This, then, is one very good reason why fakirs have not updated their display and replaced the hot coals with a length of heated metal!

Collectively, these effects go some way to explaining how firewalking can be achieved without burning, and how those who have prepared succeed over those who do not. But there remains one other requirement that in many cases can make a significant difference.

The power of positive thinking or self-confidence is not a supernatural quality, but merely a matter of applying instinctive abilities. The fact that the team of scientists from Tubingen found greatly increased theta rhythms in their subjects confirms the idea that firewalkers are able to overcome pain by adopting a rarefied mental state. Theta rhythms are associated with meditation and deep relaxation, so it would appear that to prepare for a fire-walk or any other extreme feat like this, the yogi must channel his brain patterns and calm his body and mind.

Professor Leikind from U.C.L.A. and his colleague, William J. McCarthy, both ventured into the fire during their experiments into the physics of fire-walking and emerged unscathed. They do not have hardened soles and had no special training before the event, but are convinced that part of the reason they could do this was because they had prepared themselves mentally. They did their fire-walk at the home of Californian self-help prophet Tony Robbins, who led fee-paying courses instructing participants how to allow their theta rhythms to dominate their brain patterns.

Several techniques were employed to do this and were noted by the investigating scientists. The first trick was to ensure that the walk was scheduled for the early hours of the morning when people would normally be asleep. Staying up like this had the effect of depressing sensitivity. It also allowed them to be more suggestible and to channel their theta rhythms with greater ease. Other pain-blocking techniques included special breathing patterns and chanting. Repeating such expressions as "cool moss, cool moss" seemed to help the participants to focus their thoughts, to concentrate on achieving their goal and blocking pain.

What surprised researchers initially was that these techniques did not just inhibit pain, but the majority of walkers showed no physical signs of heat. There was rarely a blister to be found among the devotees who enrolled in the course.

It may be that some participants have a greater talent for controlling their bodies than others. It is certainly possible that fakirs who have spent their lives training and are convinced of the religious aspect of the technique are able to manipulate some of the biochemical processes within their own bodies. This is known as biofeedback, and a number of laboratory tests have shown that gifted or well-trained individuals can control such functions as heartbeat, skin sensitivity, and muscle tension. However, for the majority of "amateurs" who have managed to walk on fire after only a few hours preparation, self-confidence may allow a relatively painless journey, but the Leidenfrost effect, conductivity, and speed have probably played the most significant role.

The ability to adopt a trancelike state induced by chanting and special breathing is actually a civilized version of a natural survival technique. Often prey fall into a cataleptic state when they are caught but not killed by a predator. Wildebeest sometimes stand motionless with glazed eyes as a pride of lions bites at its legs, and some tarantulas readily submit to the attacks of tarantula wasps.

The reason for this is twofold. Paradoxically, it offers the possibility of escape, because by not struggling, the predator is less likely to deliver a coup de grace. The other, more relevant reason is that by adopting a trancelike state in which theta rhythms dominate, the animal suffers less pain.

The stories of individuals slipping into an involuntary relaxed state and performing tasks they would otherwise consider unimaginable are surprisingly common. A friend of mine drove her car over a cliff and managed to crawl from the wreckage that had squashed her into a space a meter square. Despite sustaining a broken arm, a severe concussion, and facial injuries, she managed to walk half a mile to the nearest telephone, where she calmly called her father and waited for an ambulance to arrive. Today she has almost no memory of the interval between escaping the wreckage and reaching the hospital.

Part of my friend's ability to do this came from a flood of chemicals in her bloodstream. The hormone adrenaline constricts the blood vessels, stemming blood flow and providing emergency energy supplies, and a mechanism known as the *extrinsic mechanism* comes into play under extreme conditions to help clot blood. But of key importance may be a natural ability to regulate the brain wave patterns and, in the case of extreme emergency, to create a balance of patterns that allow us to do what is necessary in order to reach safety.

In the late 1970s, Dr. Wolfgang Larbig from the University of Tubingen carried out a series of experiments on an Indian yogi who was wired up to a collection of monitors, and his brain responses compared to a group of control subjects.

The experimenters applied painful electric shocks to the fakir and the volunteers and monitored their heart rate, skin conductivity, and brain wave patterns. They found that after a short time, the volunteers could take no more and looked exhausted, whereas the fakir had remained impassive throughout. They also discovered that the conductivity of his skin at the end of the tests was very different from all the other subjects, that his heart rate was much slower, and that he had produced strong theta rhythms throughout the experiment.

In order to do this, the yogi was thought to be practicing a form of self-hypnosis, and some psychologists are beginning to use this technique to help people overcome physical problems such as chronic pain or psychosomatic disorders. According to one enthusiast, the hypnotist Leslie LeCron: "Essentially all hypnosis is self-hypnosis. The operator is merely a guide and the subject produces a result."[3]

LeCron believes that the basics of self-hypnosis can be taught very easily, but the ease with which an individual can allow themselves to slip into a hypnotic state varies from person to person. The technique requires concentration on a simple fixed image—a candle is a favorite—which then allows the subject to focus their minds and channel their brain wave patterns into a slower rhythm. These are usually either alpha waves (normally associated with sleep and found at 8 to 14 Hz) or the ubiquitous theta rhythm, at 4 to 7 Hz.

A similar practice is Transcendental Meditation. This became popular in the West during the late 1960s thanks to the Beatles' well-publicized flirtation with the art. At its simplest level TM is used by many people as a relaxation technique, and by others as a cure for insomnia, but for the real enthusiasts it is an avenue into some of the more advanced practices displayed by yogi and holy men. Central to the concept is the use of a simple word chanted silently over and over again. This is really a mechanism to send the brain into a relaxed state and to reinforce a single simple rhythm, a substitute for the candle or the simple visual image employed by hypnotists such as Dr. LeCron.

TM was actually known about in the West long before the Fab Four popularized it. The nineteenth century English poet Lord Tennyson was a keen user, once describing it as:

A kind of waking trance—this for lack of a better word—I have frequently had, quite up from boyhood, when I have been all alone. This has come upon me through repeating my name to myself silently, till all at once . . . individuality itself seemed to dissolve and fade away into boundless being, and this not a con-

fused state but the clearest, the surest of the surest, utterly beyond words.[4]

Fire-walking is not the only form of demonstrating resistance to physical pain. A speciality of Sri Lankan yogis is hook-hanging. This involves hanging by half a dozen wires tied to hooks passed through the flesh of the back.

For many practitioners it is a form of religious observance, and they see it as a way of atoning for their sins in much the same way that Catholics count rosary beads or pay penance. For the uninitiated, it is extremely dangerous, and even trained fakirs sometimes suffer infections as a result of their devotions.

Like fire-walking, the ability to hang by hooks is a combination of mental control and physics. The suppression of pain is again a result of controlling brain wave patterns using breathing techniques and accentuated by the participants by a self-belief in what they are doing. Physics plays a role in that the hooks have to be arranged in such a way that the fakir's body weight is distributed evenly.

When hook-hanging was first witnessed by western scientists, they were surprised to note that the yogis rarely bleed from their wounds. To explain this we return to the extrinsic mechanism and the adrenaline rush experienced when an ordinary person is placed in an extraordinary situation. Like my friend's experience after her car crash, the body of the fakir responds to the self-inflicted crisis. The difference is that in the case of an accident, the response is automatic and involuntary, but in the case of the hook-hanger, the same physical responses can be turned on and off and controlled at will.

A famous image of pain endurance is the yogi lying on a bed of nails. Once more this is explained by a combination of mental preparation, physical training, and science. Like the fire-walker and the hook-hanger, the yogi performs this act for religious reasons and so he is motivated to train, to be influenced by ritual, and to allow positive thinking to dominate. This assists the production of enhanced theta rhythm activity which can then be channeled to control a range of physical characteristics. Also, the fakir has been trained how to lay on the nails and how to lower himself and raise himself from the spikes to avoid penetration of the skin. Finally, the distribution of the nails is such that none of them has to bear too great a weight, which lowers the risk of piercing.

Other yogis show a talent for piercing the body with spikes, wires, and even rapiers, yet they rarely bleed despite rupturing blood vessels and tearing tissue. Training has enabled them to pierce themselves in specific places, carefully avoiding internal organs, major arteries, and muscle. Those who perform this trick frequently have also produced areas consisting of unusual amounts of scar tissue, which allows the wound to be reopened more readily, but with the blood vessels sealed. Aside from these precautions, the fact that the fakir can control bleeding is possibly due to a highly developed ability to monitor and manipulate many of the body's natural functions.

Although the vast majority of fakirs and yogis live and perform in India and the Far East, there are examples of westerners who have managed to acquire many of the skills of the holy man of ancient tradition. Probably the most famous of these was Harry Houdini, who enthralled audiences with his sensational act from the turn of the century until his death in 1926.

Houdini regularly escaped from straitjackets in seconds, but he also survived being immersed in freezing rivers with his hands and feet manacled, claiming that he had defied death this way over two thousand times during his career. On one occasion he survived in a sealed coffin submerged in a swimming pool for an hour and a half, beating the record set by an Egyptian fakir. He claimed he could take any blow anywhere on his body as long as he had time to prepare. Ironically, he died from a ruptured appendix after a fan punched him in the stomach before he had time to tighten his abdominal muscles.

So what conclusions can the open-minded scientist draw from these acts of self-mutilation and pain resistance? Clearly, the mind can, to a degree, control the natural processes of the body, but just how far this ability could extend is difficult to judge. Some religious sects take the development of these powers to an extreme, and followers devote their lives to honing their natural skills. There are apocryphal tales of holy men fasting for twenty years while sitting impassive in a full lotus. Such stories have never been substantiated and defy all the laws of science. But lesser demonstrations are quite common and could be attributable to a form of "mind over matter" as well as sensible application of technique and sound science. What research and everyday experience highlights is that we all have the ability to come through the most extreme conditions imaginable involuntarily, which shows just how adaptable and responsive our bodies are. It is these natural talents the yogi exploits with dramatic and often startling results.

8 A Chance of a Ghost

From ghoulies and ghosties and long-leggety beasties
And things that go bump in the night
Good Lord deliver us!
ANONYMOUS

According to one recent report, one in six people believe in ghosts, and one in fourteen claim to have seen one.[1] But, the problem is: What do we mean by the term? More than any other paranormal phenomenon, ghosts engender a range of emotions from terror to cold ridicule, and to many they lie at the extreme edge of acceptable supernatural oddities. The primary reason for skepticism is that in all but a few rare cases, ghosts are thought by believers to be spirits of the dead. To accept this we must assume humans have souls and that the soul survives death. Beyond that we have to concede that the ghost has returned to the world of the living for a purpose that implies a strong link between this world and the next; for why else would a dead person care what happens here and now?

All of these steps in accepting the notion that ghosts are anything other than hallucinations or tricks of the light, self-delusion or hoaxes, means we have to throw away every science book written during the past three hundred years. It also implies that we as individuals are very much more important than everyday experience tells us we are, and that we have personalities that survive physical death and then interact with the mundane dramas of earthly existence.

In spite of these criticisms, ghosts are perhaps the oldest and strongest held occult belief and have appeared in almost all cultures and all historical periods. According to their legends, the shamen of tribal cultures can act as conduits through which the living and the dead may communicate. The ancient Egyptians charted the progress of the individual after they had passed beyond this world in their *Book of the Dead,* and in our own culture, the Bible is full of apparitions, ghosts, and visitations. However, no one has so far produced irrefutable proof that ghosts are supernatural entities, nor can anybody provide a satisfactory theory linking apparitions with the spirits of those departed.

In view of this, we must turn to science to give a reasoned and logical analysis of what ghosts are and how they have become so ingrained in our race consciousness.

There are two distinct phenomena that come under the umbrella term of "ghost": apparitions and poltergeists. Although they may both flow from the same psychological spring, for our purposes they can be distinguished: apparitions are largely *passive,* whereas poltergeists are *active.* Apparitions are usually witnessed as images, which impart information or respond to living beings only rarely, whereas poltergeists are frequently reported to interact with the living and are often said to be malevolent, even murderous.

The first analytical study of ghosts was produced toward the end of the nineteenth century by a dedicated group of researchers of the paranormal calling themselves the Society for Psychical Research (SPR). The group formed around three scholars from Trinity College, Cambridge, and endeavored to investigate the current vogue for spiritualism using purely scientific means. Most active were Frederic Myers and Eleanor Sidgwick, who wrote several academic books about paranormal phenomena and helped to conduct thousands of tests and investigations exposing fraud and hoaxes. By the end of the century the society had amassed eleven thousand pages of reports later distilled into two important books, *Phantasms of the Living* and *Human Personality and the Survival of Bodily Death.*[2, 3]

Early on it was realized that apparitions could be classified into three distinct categories. The first type are ghosts that appear to a single person and are thought by the "sighter" to be an image of a dead person, perhaps a relative or a close friend. The second type is an image that appears to several people simultaneously. The third type (which may also occupy the second group) are visions of people who are still alive. This last variety are called "crisis apparitions."

The majority of ghosts appear to individuals, invariably at night or in darkness, and have little or no interaction with the environment. There have been countless cases of these "simple apparitions," but absolutely no proof that such images are produced by a dead person deliberately trying to communicate with the living. Furthermore, photographs taken of ghosts are invariably found to be fakes or to show quite natural aberrations of light or photographic anomalies.

Frederick Myers defined ghosts as "a manifestation of persistent personal energy, or as an indication that some kind of force is being exercised after death which is in some way connected with a person previously known on earth."[4]

This is a purely empirical definition and has done little to satisfy those who believe ghosts are the spirits of the dead. What it suggests is that there may be some natural mechanism by which the energy of an event or a person could be recorded in the environment, and it is interesting to note that some cases of apparition show common characteristics. These might add weight to this idea without the need for supernatural causes.

The first thing one notices about ghosts is that they seem to be remarkably simple-minded. As the writer Colin Wilson has pointed out: ". . . a tendency to hang around places they know in life would appear to be the spirit-world's equivalent to feeble-mindedness; . . . one feels they ought to have something better to do."[5]

Simple apparitions usually perform a basic set of movements (and occasionally sounds) within a limited frame of reference. They appear to be tied in some way to a particular building or even a specific room, and to follow through repeated, identical movements—for example, a walk along a corridor or across a room—and many appear only at certain times or under special conditions.

One example out of many thousands of such cases is that of the "New Year's Eve Nun." This ghost was first reported during the 1930s at a girls' school in Cheltenham, where on New Year's Eve a nurse and the headmaster both saw a nun in a sitting position at the edge of the school playground. There was no chair in the vicinity. The following New Year's Eve the pair saw the same woman in the same position.

Cynics have suggested there is no coincidence in the date except that perhaps the school staff had enjoyed an end of year tipple, but there are several interesting aspects to this and the majority of such cases that also match Frederick Myers's definition. This image did not interact with the modern environment at all and appeared to be quite unaware of the presence of the "sighters." The nun had been sitting on a chair in her own time, probably in a room that no longer existed, and appears to have been some form of played-back image, a tiny, isolated segment of some previous event.

In one respect we see ghosts every day. If we watch a film such as *The Thirty-Nine Steps*, made over sixty years ago in 1935, it is almost certain all the adult actors will now be dead (actually Robert Donat died in 1958 and Dame Peggy Ashcroft in 1991), yet we see them moving and talking, acting out their roles. It can be even more disturbing to see interviews of deceased celebrities, watching them talk about their plans and aspirations, laughing and joking with the interviewer.

Imagine how someone who had never heard of film or video would feel if they were shown a recording of someone they knew who had recently died. They could be forgiven for thinking they were seeing a ghost, and might only accept it as a mechanically reproduced image if the process was explained to them.

Unfortunately we do not have a mechanism to explain how recordings of images could be made and played back without the use of a machine such as a television camera or a video system, but the idea has been around in one form or another since the early days of the SPR.

At the end of the last century it was postulated that some form of photograph of a scene or an individual could be taken and somehow projected at a future date. When cinema became popular, it was a logical step to upgrade the technological comparison and suggest that a "film" record was somehow made and played back. Today we think in terms of video, but how could such a process work?

All forms of recording rely upon an imprint being made in a specific medium. A photograph is produced when light activates chemicals to produce photochemical reactions that create a black-and-white or colored product, depending on the type of film used. In this way, the pattern of the original scene is transferred to the negative and reproduced on specially prepared paper. A musical recording is made by creating a series of very specific patterns in a plastic disk. When the stylus travels along these patterns or grooves, the original sound is reproduced. As we improve our technology, the means by which this information transfer and retrieval is achieved changes, but the basic principle is the same. A CD reads a pattern on a disk using a sensitive laser, a video player reads a pattern on a tape produced by fluctuations in the material of the tape sensitive to variations in an electromagnetic field. Parapsychologists suggest this principle of transfer and retrieval can also be applied to explain the appearance of ghosts.

However, there are two problems with this comparison. First, there appears to be no device by which this could happen. Second, in many cases the preserved energy seems to be linked with human emotions.

Enthusiasts of this theory suggest a number of ways in which a recording device could be produced by the surroundings themselves. One idea is that buildings—or rather, the material used to build walls—contain chemicals that can act as receivers and holders of the information needed to produce a ghost image. Research is currently underway into the abilities of different materials to record images, and it has been claimed that buildings containing unusual quantities of quartz are more common sites for the generation of apparitions.[6]

In addition to considering the nature of the material acting as a storage and playback medium, we need to look at the energy needs of the system. A photograph is different than a film in that it is merely a static image. Ghosts invariably move (even if within a very narrow frame of reference), so there must be something in the environment to reproduce a mechanical effect. Using a video player, or even a hand-cranked cine projector, energy is expended in order to create a moving image. Walls do not usually move, so we would require some alternative method of producing an active image.

One possibility is that the sighter provides the energy. Alternatively, unusual atmospheric conditions could unlock a stored image. One common feature of ghost sightings is a sudden drop in temperature immediately before an apparition. It has been speculated that energy is taken from the environment in order to process and project the stored information. A further alternative is that several walls are involved in the projection in the same way a holographic image is constructed from a collection of superimposed two-dimensional recordings.

The second intriguing aspect of this system is the link with human emotion. One argument for the case that ghosts are action replays suggests that the initial image is produced within a specific region of space-time during a moment of intense emotional activity.

Support for this comes from the fact that apparitions often signify some intensely dramatic scene or traumatic personal event. One such case tells of a woman named Elizabeth Dempster who moved into a flat in London and immediately began to feel a brooding, unhappy presence. In an effort to dispel the morbid atmosphere, she decorated the flat in bright colors. It was then that she began to see in the bedroom the image of a mournful woman dressed in Victorian clothes. After researching the history of the house, she discovered that soon after it was built it had been occupied by an Italian woman and her husband. When the woman learned that her husband had been killed suddenly, she locked herself in the bedroom and stayed there until she died of starvation and self-neglect.

Another example of an apparition thought by supporters to be linked with emotional impact are ghost battle scenes. These sometimes appear on specific days, perhaps when the environmental conditions are appropriate. Such spectral battles are often witnessed by several people simultaneously and can come complete with sound effects and even smells.

A famous report from the seventeenth century tells of a group of shepherds who claimed they witnessed a reenactment of the Battle of Edgehill. This was one of the decisive conflicts of the English Civil War in 1643, when five hundred men were brutally killed within a small area during a few hours of one another.

But how could emotions be recorded in a medium such as a building, or, in the case of a battle scene, by nothing but thin air?

Supporters resort to the concept that energy produced by the brain could be transferred, and suggest that during particularly traumatic incidents, the human brain produces unusual brain wave patterns and is capable of projecting this energy.

Humans certainly do produce unusual brain wave patterns during moments of trauma or emotional anguish, but even during these times, the energy associated with such disturbance is many orders of magnitude too weak to have any impact upon the material world. However, it is perhaps possible that the combined energy from hundreds or thousands of simultaneous deaths or shared anguish could create a gestalt that imprints itself upon the environment.

In Hong Kong the greatest number of ghosts are reported in buildings that were occupied during World War II, and in particular those in which Chinese prisoners were tortured to death. The second highest population of ghosts in the city is to be found in the hospitals. But if humans were able to imprint emotion onto the environment, why is Hiroshima or Nagasaki not haunted by powerful afterimages of the thousands who died there simultaneously? Why is Belsen or Dachau not a region of space in which intense visible projections are seen on a daily basis? Could it be that the environmental conditions in those places were not right at the time to imprint images, or that the conditions to play them back are not suitable today? One answer might be that in the case of the atomic bombing of Japan, the energy released during the explosions not only killed thousands instantly but stopped the mechanical process by which an image could be stored.

A variation upon this theme of recording and replaying images has been postulated by several modern-day researchers of the paranormal. "Ghost-hunter" Andrew Green, who made headlines in 1996 after he was called in to find a ghost spotted several times in the Royal Albert Hall, claims ghosts are projections created by the sighter themselves.

Ghosts, he believes, are "forms of electromagnetic energy between 380 and 440 millimicrons of the infrared portion of the light spectrum . . . If I was told my wife had been killed, this kind of information shocks me to the core. I picture her in my mind's eye. And that image is transferred to where I last saw her. It may be fifty miles away or upstairs in the bedroom, it doesn't make any difference, that image is there suspended till someone else picks it up."[7]

So, by this reasoning ghosts are telepathic projections, and we again come up against the problem of the energy needed to create such an image or to transfer the necessary information to another individual who sees it. There is also the fact that ghosts would only be seen if someone who had known the person were still alive themselves and able to create the projection.

If we reject the idea that ghosts are either recordings or projections, we have to consider a limited range of other options, some of which might at first appear obvious. One of the original investigators of the paranormal, Eleanor Sidgwick of the SPR, produced a list of mundane explanations for ghosts that should always be considered before entertaining inexplicable sources. She listed these as: (1) hoaxing, (2) exaggeration or inadequate description, (3) illusion, (4) mistaken identity, and (5) hallucination.

Hoaxing was rife at the end of the last century, when the SPR was setting up in business. The craze for spiritualism had started earlier in the century and had become a booming cottage industry with an increasing number of mediums and spiritualists ready to fleece the gullible. It was an interest that crossed all social divides and enticed the superficial interest of artists, writers, and curious scientists. Charles Darwin once attended a séance with other scientific colleagues, academics from Cambridge, and the writer George Eliot, but came away from it even more skeptical than before.

In recent times there have been some famous hoaxes. "The most haunted house in England," Borley Rectory in Essex, which burned to the ground in 1939, was supposed to be the site of some five thousand paranormal incidents during a period of a few years. It was made famous when the researcher into the paranormal, Harry Price, investigated the building and wrote a best-selling book about it called *The Most Haunted House in England*, published in 1940. During the 1950s the SPR investigated Price's claims and revealed that they were entirely faked.

Another example is the "Amityville Horror" house in Amityville, New York, which became the subject of a famous book and a Hollywood horror movie. The entire story was fabricated for purely commercial reasons by the onetime owners, George and Kathy Lutz, and a lawyer, William Weber.

During the nineteenth century there was money in spiritualism and mediumship, just as there is today in successfully faking sensational paranormal experiences. People want to believe, they want to talk to their deceased loved ones, and, in an age before the advent of television and virtual reality, people were more impressionable, they would pay for the privilege. One of the functions of the academically minded SPR was to expose cheats and frauds, and during the latter half of the last century there were even prosecutions and imprisonments as a result of exposés helped by the investigations of groups like the SPR.

Today, hoaxes account for only a small percentage of ghost sightings, and improved technology can usually spot obvious fakes, but exaggeration and inadequate description account for a significant percentage of apparitions. Often the witness is sincere but has misconceived an experience unintentionally, or else they have been subject to an optical illusion. Fear can distort, and illusions under stress are surprisingly common. If a lone witness sees what he thinks is a ghost in bad light or under unusual environmental conditions, he can often be forgiven for misinterpreting what his eyes tell his brain.

However, the most interesting of all so-called *trivial* explanations for ghosts is hallucination, and many researchers—both enthusiasts of the occult and empirically minded skeptics—accept that the vast majority of apparitions can be explained in this way.

Hallucination is an intensely researched psychological state that is surprisingly widespread. Back at the turn of the nineteenth century one of the most useful pieces of data gathered by the SPR was a survey conducted on 17,000 people to determine the incidence of hallucination. They found that 2,300 of those asked had experienced a hallucination sometime during their lives, and according to a modern-day survey of American college students, seventy percent claimed they had experienced the auditory hallucination of hearing voices while awake.[8, 9]

In his book, *Fire in the Mind*, the psychologist, Ronald K. Siegel has said of hallucinations:

> In the past, hallucinations were often regarded as the exclusive domain of the insane. Through the research and cases in this book, we begin to understand that anyone can have them. They arise from common structures in the brain and nervous system,

common biological experiences, and common reactions of the
brain to stimulation or deprivation. The resultant images may be
bizarre, but they are not necessarily crazy. They are simply based
on stored images in our brains. Like a mirage that shows a mag-
nificent city on a desolate expanse of ocean or desert, the images
of hallucinations are actually reflected images of real objects lo-
cated elsewhere.[10]

The most common time to see a ghost is late at night and usually
at the point of going to sleep. The "ghost at the end of the bed" is
the stuff of legend and the mainstay of horror films, yet there is a
well-understood reason for this. As the body switches from the voluntary nervous system—the system that allows us to function in our
everyday lives—to the involuntary nervous system, we commonly
experience what are called *hypnagogic hallucinations*. One interpretation of these is that "our wires get crossed," the brain is momen-
tarily confused by the switch from one nervous system to the other,
and images are dredged up from either deep in the conscious memory or from the subconscious. This, it is believed, accounts for the
vast majority of apparitions.

These visions or hallucinations can seem very real and may be
accompanied by auditory sensations or even smells. A similar experience is sleepwalking. Many people have at one time or another had
the odd experience of suddenly coming to in the bathroom or sitting
in front of a blank television screen in the den. Often these experiences seem very real at the time but are almost totally forgotten by
the following morning.

Another common form of sleep-related hallucination is what has
been dubbed *hypnopompic hallucination* and occurs when we
awaken. Again, this is due to the body switching nervous systems,
this time from the involuntary to the voluntary, and may account
for a large number of cases of apparition.

Both hypnagogic and hypnopompic hallucination may also be
used to explain what have become known as "hitchhiker apparitions." Since the beginning of the car age, an increasingly common
phenomenon within ghost mythology describes accounts of drivers
seeing ghostly figures in or at the side of the road. These sightings

often occur along stretches of road famous for particularly grizzly accidents or known in the region for apparent spectral activity. Sometimes drivers have even reported knocking down people, feeling the bump of the body under the car, and when they have gone to see what had happened, they're left staring at empty pavement.

Even more dramatic incidents tell of drivers tending someone they believed they had hit, covering them with a blanket, only to find the body had vanished by the time the police arrived. In 1979 a driver named Roy Fulton claimed he picked up a male hitchhiker along a stretch of road in Stanbridge, Bedfordshire, late at night. The young man opened the door and sat silently on the backseat, ignoring Fulton's attempts to begin a conversation. Only when Fulton turned to offer the hitchhiker a cigarette did he realize the boy had vanished.

Such incidents are open to ridicule and to claims of hoaxing, and perhaps a large proportion of them are deliberate frauds, but in some cases they could be put down to either hypnagogic or hypnopompic hallucination. Drivers sometimes fall asleep at the wheel, and it is quite possible hallucinations could occur as they lose consciousness or wake suddenly. These brain-generated visions are then amplified by environmental effects and individual circumstances. Driving alone along narrow country roads in the dark can induce suggestive images in the mind, and speeding along a seemingly endless stretch of featureless highway is often almost mesmerizing.

A related phenomenon is crisis apparition, where sighters sense the presence of someone they know either close to or at the point of death. There are many documented cases in which people have apparently seen projections of close relatives or friends who were in a crisis situation at the time, often immediately before their moment of death. In their book *Phantasms of the Living*, Edmund Gurney, Frederick Myers, and Frank Podmore documented 701 cases of apparent crisis projection and admitted they could not explain many of these incidents.

One of the most famous stories of crisis apparition comes from the 1930s. One stormy, freezing night in the mid-Atlantic, a one-eyed English pilot named Hinchliffe with his female copilot was attempting the first east–west crossing of the ocean when suddenly their biplane hit bad weather. The high winds tossed it around, the compass was disturbed by magnetic interference, and without a reference point for hundreds of miles in any direction they were soon hopelessly lost. The plane began to nose-dive toward the waves, its engine screaming in protest, and a moment later it hit the water, killing pilot and copilot instantly.

The same night, two friends of Hinchliffe's, Squadron Leader Rivers Oldmeadow and Colonel Henderson, were steaming toward New York aboard an ocean liner several hundred miles away from the scene of the crash. Neither of them had seen Hinchliffe for some time and they were totally unaware he and his female colleague had attempted the flight. It was in the middle of the night, just when, according to later corroboration, Hinchliffe's plane hit the storm, that Colonel Henderson, dressed in his pajamas, burst into his friend's room shouting:

"God, Rivers, something ghastly has happened. Hinch has just been in my cabin. Eyepatch and all. It was ghastly. He kept repeating over and over again, 'Hendy, what am I going to do? What am I going to do? I've got the woman with me and I'm lost. I'm lost.' Then he disappeared in front of my eyes. Just disappeared."

Supporters of paranormal explanations for ghosts and apparitions have proposed that this incident and others like it are due to a form of emergency or crisis telepathy, that at the moment of death the human brain is capable of transmitting an image or news of a person's situation, perhaps as a final survival attempt. But it is also possible to see such events in a far more prosaic light.

First, there is the strong possibility of hallucination. The sighter in this story, Colonel Henderson, may well have enjoyed a pleasant evening at the captain's table before turning in for the night, and could have experienced an alcohol-induced hypnagogic hallucination.

But, say the enthusiasts, how does this account for the fact that Henderson had no idea his friend was in the middle of a risky flight?

The answer is, he almost certainly did know about it subconsciously. It is possible he had been reading a newspaper and noted subliminally an article about his friend attempting an Atlantic crossing without reading the piece or even realizing consciously that he had spotted it. This could then have enhanced his hallucination, providing the subconscious image around which he produced the vision.

An alternative suggestion for this and many other cases of crisis apparition is that people have a subconscious knowledge of an event but need to create an hallucination in order to process the information through their conscious mind. The usual reasons for this inhibition are fear and guilt. These emotions could force the conscious mind away from analyzing or thinking about a situation, and the brain then has to resort to filtering the information through an alternative system where it does not meet the same resistance.

A final explanation for hallucinations appearing as crisis apparitions is wishful thinking or comfort thinking. When this is the source, the visions are called *need-based hallucinations.*

Most people hope there is an afterlife, and many, especially those who feel insecure or fearful, can experience this desire so strongly they produce "evidence" to support their wishes. To them, a ghost is proof of an afterlife, and so their subconscious mind is empowered to conjure up an appropriate image. In other situations people can imagine they are being visited by a comforting, supportive figure who either warns them of imminent danger or gives them extra impetus to fulfill a difficult task. The record-breaking driver, Donald Campbell, claimed he'd been visited by his father Sir Malcolm Campbell on many occasions, and believed his father had been sent to warn him of impending danger.

Of a quite different order to apparitions is the phenomenon of haunting, especially poltergeist activity. Hauntings are usually witnessed by several people and encompass a wide range of apparently supernatural activities—materializing and dematerializing of objects, noises, smells, and, on rare occasions, violent, even life-threatening incidents.

When a group of people all witness the same set of experiences, it is very difficult to explain them as hallucination, hoax, or any other natural process. But many do yield to rather mundane causes if the investigators probe deeply enough.

The first stage in any investigation of a haunting is to eliminate natural sounds and smells. These may take some searching, and enthusiasts of occult explanations are fond of listing cases in which months of investigation into particular hauntings have been wasted and no link to natural causes detected. These cases are rare and do not offer proof of supernatural activity, they simply show the researchers did not investigate thoroughly enough.

After hoaxes, natural causes, and illusions have been ruled out, we can again turn to hallucination. Surprising as it may seem, it is possible for a group of individuals to experience the same induced images. This phenomenon is called *mass hallucination*, and is brought about when one of the group is a stronger personality than the others and creates a convincing suggestion which is then adopted by the others. This explains many cases of hauntings involving parents and children where one of the adults—or in rare cases, one of the children—unwittingly implants the idea, after which fear and anxiety take over.

Enthusiasts of the occult take this idea and add an element of the supernatural to create what they call the *infectious hallucination theory*. This proposes that one of the group experiences an hallucination which is then transmitted telepathically to the others. But psychologists have shown through experiment that this mechanism is actually quite unnecessary. If the correct blend of personalities are put together in an atmosphere of perceived danger and fear, an hallucination created by one of the stronger personalities is infectious enough to spread to the others without the need for telepathy.

The final explanation for poltergeists, and one supported by some enthusiasts of the occult, is the idea that emotional disturbance in human beings can be projected into dramatic physical events. The theory suggests that people with the ability to project telepathic images may, if they are placed in an emotionally challenging situation, produce enough psychic energy to move furniture or throw objects across a room.

As we saw in Chapter 4, the energy needed to do this is of a quite different order to that generated by the human brain, and, according to the known laws of physics, is completely impossible. A less problematic explanation may be that a human source of hallucination is planting images into the minds of the witnesses. Again there is no need for telepathy in this situation. If the creator of the image is a strong enough character, they may be able to induce hallucinations in the other witnesses. They may even be able to convince those on the receiving end of any violent activity that they actually feel pain or have cuts, burns, and bruises.

Poltergeist activity has been shown to center around children and teenagers more frequently than adults; in particular, pubescent girls appear to be a common source. This has led occultists to the idea that hormonal and emotion imbalance enhances PK abilities in these young women. A more logical explanation might be that during a time in which body and brain chemistry is in a disturbed state, the subjects could be capable of encouraging hallucinations in those around them by suggestion and emotional manipulation. Mothers placed in highly stressful situations, thanks to the growing pains of their teenage girls, might be particularly susceptible, especially if the notion their home is haunted has already been "seeded" in their minds.

For at least a century, believers in the paranormal source of ghosts and apparitions have tried to provide clear evidence for their claims. During the nineteenth century the fashion for mediums and spiritualism succeeded in convincing large numbers of people that ghosts are supernatural, and it was not until skeptics exposed many practitioners as frauds that the popularity of séances diminished.

Photographic evidence is also very weak. Most of the examples that have entered popular culture show what look like pantomime ghosts with sheets draped over their spectral heads. Those that do not show cavaliers in full regalia or bug-eyed demonic creatures, look like faults on the negatives—bursts of light or optical aberrations. As far as I am aware, there is no convincing film or video footage of ghosts or apparitions. If ghosts are visitors from an afterlife and are as common as people believe, then this is rather surprising.

One striking development that has been claimed as evidence for the supernatural origin of ghosts is *electronic voice phenomenon*, EVP. This is the name for the process where background recordings apparently throw up the sound of a dead person speaking.

This phenomenon first came to public attention in 1920, when the October issue of *Scientific American* carried a feature about the famous inventor, Thomas Edison. In the piece, Edison claimed he was working on a device that could be used to communicate with the dead. Most of his contemporaries believed he had finally gone senile, and his ideas were ignored by the scientific establishment. Not surprisingly, by the time of his death in 1931, he had still failed to deliver the promised machine, but a few months earlier, an American psychic named Attila von Szalay claimed he had heard the voice of his deceased son and later "trapped" these voices using a 78 rpm record cutter. Three decades later, in 1959, a Swedish opera singer named Fredrich Jurgenson was listening to tape recordings of bird song when he thought he heard human voices speaking in Norwegian far back in the sound mix.

Soon, others were following the lead of Edison, von Szalay, and Jurgenson, and books about the phenomenon began to appear. The most famous was *Breakthrough*, published in 1971, which documented over seventy thousand recordings, many of which were made by tuning a radio between stations and recording the resultant white noise.

It is difficult for the scientist to accept that these recordings carry messages from the spirit world. The first and most obvious criticism is that the interpretation of the recordings is totally subjective. Different listeners hear different things on the tapes. Only when a stronger personality insists a certain voice is talking in a specific language and delivering what they conceive to be a sensible message do others begin to hear the same sentences.

The second criticism is based on the fact that human beings seem to have an in-built inclination—in many cases a need—to find order within chaos. This is thought to be a deep-rooted survival tool, an ability to analyze and find patterns in seemingly random events. Such a skill enables an individual to be better prepared for unpredictable outcomes—a form of psychological preparation. It also demonstrates an instinctive need for security: order equates to stability, which means safety. Those who hear distinct voices amidst the random signals and recordings are deluding themselves through a deep psychological need.

A third explanation for this effect has nothing to do with the human subjects but is based upon a fundamental universal law—chaos theory.

A chaotic system is defined as one that shows "sensitivity to initial conditions." What this means is that a slight variation in the initial state of identical processes can lead to very different end results. For example, a speck of dust floating on the surface of a pair of oscillating whirlpools displays chaotic behavior. The particle appears to move randomly and its course becomes increasingly unpredictable, but the way it moves during one experiment will also be very different in another, depending on the initial conditions of the test.

Chaos is observable in everyday situations. The world's weather systems are chaotic, which makes forecasting notoriously difficult. A dripping tap behaves chaotically, and it is even suspected that global financial systems show chaotic behavior, which means long-term speculation is riskier than many prospectors realize.

The concept of chaos can be visualized by a very neat example known as the *butterfly effect*. This says that the flutter of a butterfly's wings in England could eventually create a thunderstorm in Australia. Although this would appear to be fantastic, it is possible because the tiny disturbance created by the butterfly is an example of altering the initial conditions. Because weather systems are sensitive to initial conditions, this minuscule effect can become amplified so that through a long and complex chain of events it can be seen to produce a thunderstorm on the other side of the world.

So what has this to do with "ghost voices" captured on tape?

One of the striking characteristics of chaotic systems is that within chaos there is also order. This, scientists believe, is how an apparently chaotic universe throws up "islands" of order. Our earth, our civilization, we as individuals, could all be short-lived, local anomalies of relative order in an ocean of randomly shifting particles we know as the observable universe. As James Gleick says in his international best-seller, *Chaos*: "The simplest systems are now seen to create extraordinarily difficult problems of predictability. Yet order arises spontaneously in those systems."[11]

White noise from a tuned-out radio, or background sounds from a recording of bird song, can readily create a chaotic system. But within that chaos there may be elements of order; shifting, fleeting fragments of organization. When an enthusiast, a believer in the supernatural source of ghosts, listens to these patterns, they may pick out momentary parcels of order and amplify them into coherent but nevertheless imaginary sentences.

Ghosts certainly exist, but there is no hard evidence to suggest any form of personality or soul survives the death of the brain. Therefore ghosts appear to derive from other sources. These sources are plentiful and varied, ranging from deliberate fraud to mass hallucination. It may also be possible that a form of energy representing the physical characteristics of a person could be trapped by an unusual confluence of environmental conditions. It seems to me unnecessary to add to this the idea that some form of psychic energy created by extreme emotion is also trapped. Many ghosts proven not to be fakes, illusions, hallucinations, or freak environmental effects appear to be limited in their movements and merely reenact a specific scene endlessly. Occasionally, the sighter feels depressed by the apparition or is sensitive to an oppressive mood, but this is almost certainly an effect created by the mind of the individual, not the ghost. If such a record and playback system exists, it is a purely physical phenomenon, perhaps generated initially by extreme emotion, but one that science has so far failed to quantify.

Until physicists and biologists find a way to study the possibility that apparitions are replayed images from the past, we can only assume ghosts derive from the minds of those who see them, projections of our own desires and fears, and that they have no material form in the physical world.

The Lost and the Lonely

> But first on earth, as vampire sent,
> Thy corse shall from its tomb be rent,
> Then ghastly haunt thy native place
> And suck the blood of all thy race.
> LORD BYRON

Creatures of the night, bloodsucking beasts and sea dragons, haunt all of us subconsciously. Such monsters stalk our nightmares and play on our imaginations: They are universals, race images, Jungian archetypes. Some may also have their roots in scientific reality.

The umbrella term *mythical beasts* has been used to describe a vast collection of creatures from all parts of the world, creatures that have appeared in the history books of almost all cultures and wander the pages of literature and art. But recent investigations have begun to show that the term "mythical" may not be so appropriate as some skeptics believe. These animals are certainly shy in the extreme, yet collectively, the witnessed cases, photographs, and filmed archives show that our world is almost certainly home to a host of creatures about which we have only begun to speculate. These could range from isolated remnants of prehistoric creatures to rare genetic aberrations.

There seem to be two distinct groups of creature that have not yet been fitted into any genus accepted by science. I'll call these *human variants* and *evolutionary cul-de-sacs*. The first of these are all creatures that approximate to human beings, but possess extraordinary characteristics and include vampires, mermaids, and werewolves. The second category consist of beasts that would appear to have evolved outside the avenues observed by modern biology, creatures such as sea monsters, the Loch Ness monster, the yeti, and the bigfoot.

Of human variants, the most intriguing and the most frightening is the vampire. According to the theologian Heinrich Zopfius, writing during the early eighteenth century:

Vampires issue forth from their graves in the night, attack people sleeping quietly in their beds, suck out all their blood from their bodies, and destroy them. They beset men, women, and children alike, sparing neither age nor sex. Those who are under the malignity of their influence complain of suffocation and a total deficiency of spirits after which they soon expire.

In Europe, people have known about vampires since at least the early Middle Ages, and they have persisted in the modern imagination partly because of the abiding fascination they have held for writers. The most famous vampire in literature is the eponymous antihero of Bram Stoker's *Dracula*, written in 1897. Although the public image of a vampire derives largely from this creation, by the time the book was published, vampires had already become a mainstay in Gothic horror literature. *The Monk* by Matthew Lewis, written in 1796, was one of the original vampire novels, and a massive work called *Varney the Vampire* was published in the 1840s and ran to over a thousand pages.

All of these books convey the image of vampires as bloodsucking humanoids who cannot endure sunlight and prey on their victims in order to survive. They all have a vulnerable aspect to their characters and convey a heavy sense of eroticism. Since the eighteenth century, when these stories first became popular, the image of the vampire has remained remarkably consistent and appears to be an almost timeless image of corruption, power, and sadness.

Amazingly, there are today a growing number of people who claim to be vampires. The films *Interview with a Vampire*, the 1990s remake of *Dracula*, and Tarantino's *Dusk Till Dawn* have all helped to encourage a neo-Gothic subculture that appears to have inspired some fragile mentalities to buy themselves coffins and shun the sunshine (most especially the lemon glow of Los Angeles).

The Internet increasingly plays host to a surprisingly healthy community of vampires, and the confessions of blood-obsessives have appeared in several recent books, including *Chaos International* in 1992 and Rosemary Ellen Guily's *Vampires Among Us*. True-life stories of modern-day vampires are becoming almost commonplace within certain communities. Jack Dean, a self-confessed human blood drinker, developed his taste for the stuff after he was injured in a car crash, and another, Philip Hine, cultivated his fetish when he was sprayed with blood after a friend cut himself badly.

According to one report, there are currently thirty-six registered human blood drinkers in Los Angeles, and at the last count seven hundred Americans claimed to be vampires. The fact that none of these are likely to be the genuine article but are victims of an obsession or induced fixation makes the craze no less alarming. The *X-Files* episode "3" delved into this chilling underworld and came closer to actuality than almost any other episode so far broadcast. But at a time when AIDS is rife, the fringe club scene in which "love nips" are exchanged like kisses may be seen as a telling sign of the level of psychological damage some people live with, especially at the frenzied heart of the end-of-millennium metropolis.

Yet, in spite of all the hysteria and sensationalism surrounding the concept of vampires, it is now believed that some vampiric characteristics have sound medical foundations and that the legend derives from a series of genuine genetic anomalies.

The mystery of the vampire probably began in Dark Age Bohemia. In those times, small communities often lived in complete isolation, and entire populations could lead their lives without ever traveling beyond the valley in which their village or town was located. Quite ignorant of a larger world beyond the encircling mountains, such communities were forced to inbreed, which eventually caused what biologists refer to as a lack of genetic diversity.

Evolutionary biologists believe the purpose of sexual reproduction is not simply to ensure propagation of the species, but to allow a healthy mixing of genetic material throughout the population. This is why the taboo of incest was created in primitive societies: Even then it was recognized that mating with close relatives was likely to end in a high incidence of natural abortions or deformity. In a broader sense, lack of genetic diversity weakens a species, and at the very least, genetic flaws begin to crop up, some of which can be serious. Unless fresh genetic material is introduced from time to time, genetic faults become increasingly common and can grow more dangerous with each generation.

According to a theory proposed by David Dolphin in 1985, among many defects, an hereditary disease called "porphyria" could have developed through lack of genetic diversity in some isolated villages of Bohemia. This disease prevents the production of a protein that is responsible for binding a chemical called a porphyrin ring to iron found in the hemoglobin of the blood.

The consequences of this are wide-ranging and could help explain many of the characteristics usually associated with vampires. First, those suffering from porphyria look severely anemic because their hemoglobin is not being utilized efficiently and their blood is not as oxygenated as it should be. Second, the porphyrin rings that cannot do their job are deposited in the subcutaneous fat beneath the skin. The chemical is photosensitive, and in sunlight it can release electrons that damage the skin and may cause severe blistering—hence the vampire's fear of sunlight.

Legend tells us that vampires cannot stand garlic, and there may be a sound biochemical reason for this too. Garlic contains enzymes that might, under the correct conditions, replace the function of the missing protein in porphyria sufferers. If this is consumed by a vampire, they receive a sudden rush of the necessary biochemical, which could kill them.

A further complication caused by the malfunction of the porphyrin ring is that gums recede. This has the effect of making the sufferer's teeth look larger than normal and could be the source of the image of vampire's fangs. Furthermore, the inability to bind the porphyrin rings could create a craving for blood in the genetically defective individual, which might account for the blood lust at the heart of the vampire legend.

Comparisons and links between fact and myth could extend even further. According to accounts in popular literature, vampires are often associated with aristocrats or noblemen. This may have its roots in the fact that the deficiency of the protein causing the blood disorder often only appears after the sufferer reaches sexual maturity, most usually around the age of sixteen or seventeen. This also happens to be the age at which females in such communities were married. The link arises from the fact that in the Bohemian community where the myth could have originated, the chief or leader would have taken a young bride, possibly a virgin, and soon after she was lodged in his castle, she would grow pale and suffer the effects of porphyria. To the simple villagers this would be clear proof the feudal lord was a bloodsucking creature, and the tale became elaborated and increasingly farfetched as it was passed on from generation to generation. In fact it was merely a coincidence.

Similar genetic anomalies could also explain other forms of human variance. On some Pacific islands the local diet consists largely of shark and dugongs—a herbivorous marine creature sometimes called a sea cow. Both of these creatures have livers extremely rich in vitamin A. This vitamin is most commonly found in a form called retinol, which has been found to be the cause of certain rare birth defects. It is quite possible that at some point in the past children with webbed feet were born on these islands. If some of these cases were bad enough, the tale could have crept into folk legends and eventually entered western mythology via tales communicated to early explorers—hence, the mermaid.

Genetic defects might also explain the legend of the werewolf. A boy called Jean Grenier who lived in the early part of the seventeenth century was born with a deformed jaw that endowed him with a canine appearance. He also appears to have been mentally retarded and was caught attacking a shepherdess. He later confessed to a series of murders, and from records, appears to have believed he was some sort of human beast or werewolf.

Another famous case is that of Peter Stubb, who lived in Germany in the sixteenth century. He was said to have made a pact with the devil, who gave him a special wolf fur belt that transformed him into a werewolf. During the course of a decade, he killed several people, including young children and pregnant women, and was finally caught after he savagely murdered his own son. He was tortured on the wheel and confessed to a multitude of crimes before being decapitated and burned.

Tales of man-beasts, men who were part wolf or part bear were common in the seventeenth and eighteenth centuries. Many of these seem to be linked with mental illness, and it is interesting that central to the legend of the werewolf is the notion that transformation usually occurs at the time of a full moon. For a long time, the moon has been associated with mood swings in humans. There has never been any form of empirical link made between the movement of the moon and the brain state of people, but many enthusiasts insist there is one. It has been speculated that the gravitational field of the Earth's closest neighbor can interact with the electronic currents within the brain and affect mood in the way the close approach of a thunderstorm is known to cause depression in rare, sensitive individuals.

In an alternative theory, occultists point to the fact that the moon exerts a powerful gravitational influence upon the Earth, resulting in the production of tides. Perhaps, they speculate, the moon could also alter the fluids in our own bodies. This, they argue, could in turn produce biochemical effects upon our brains, which then alter our emotions and our moods.

This may sound plausible at first glance, but it is in fact impossible. The gravitational effect of the moon on such a tiny amount of water as that in a human brain is infinitesimally small. It could possibly be argued that the difference between a stable and unstable brain might arise from minute chemical differences and that even the tiny effect of the gravitational interaction between the Earth and the moon could be enough. But if this were the case, *any* object would do the same job. In fact, any object close by—a table, a chair, another human being—would have a far *greater* effect upon the "tides" in the brain, so this theory makes little sense. (See Chapter 18: We Are Made of Stars)

In the case of someone so physically deformed as Jean Grenier, it would seem more likely that reports of his behavior were exaggerated through the gullibility and cruelty of people around him, people who viewed him as some form of dangerous freak, part-man, part-beast.

The legend of the werewolf could simply be a concoction based upon fear and exaggeration. All primitive races feared wild animals, and the wolf was probably the most ferocious animal to be found in Europe in modern times. It is not surprising that legends and myths grew up around them.

An alternative theory is that werewolves were nothing more supernatural than rabies sufferers. Rabies was a common disease until this century and was responsible for the agonizing death of many unfortunates. The biologist Louis Pasteur witnessed the suffering of victims and regularly sanctioned euthanasia. The notion that humans could transform into wild beasts could have derived from the very real and horrifying sight of someone who had been bitten by a wild dog suffering the final agonies of rabies.

During the twentieth century, stories of human variants have become increasingly confined to the pages of Gothic horror novels and Hammer horror films—unless, that is, you visit the clubs of Los Angeles! But the sightings of creatures that fit under the subheading of evolutionary cul-de-sacs or ECS's, have increased rather than diminished. Today, ECS's ranging from sea beasts to yeti are reported from all parts of the world. And along with these sightings has come a growing collection of photographic and video evidence.

Probably the most famous example of what believers claim to be an ECS is the Loch Ness monster, or "Nessie."

Loch Ness is the largest and deepest freshwater lake in Britain and is some 37 kilometers (23 miles) long and up to 230 meters (750 feet) deep. It was formed around 250 million years ago, but as a result of the most recent Ice Age, it was only cut off from the sea in relatively recent times—a little under 7,500 years ago. Peat on the bed of the loch makes the water unusually dark, and with three-meter-high waves not uncommon, many see it as a small inland ocean rather than a lake.

The myth of the monster began in the sixth century, when St. Columbia is said to have seen the beast of the loch, and there have been reports of sightings ever since. Nessie first hit the headlines in 1934 when a gynecologist, Robert Wilson, claimed to have seen a large marine creature rising its head and neck out of the water and took a photograph that was soon splashed across the front pages of newspapers around the world. Since then, tens of thousands of visitors flock to the loch every year, hoping for a glimpse of the monster, and many claim to have seen something odd in the murky waters.

Robert Wilson's photograph, which became known as the surgeon's picture, was eventually revealed as a fake when the perpetrators owned up that the "monster" had actually been part of an elaborate prank. Apparently, the beast in the picture had been made from a clockwork toy submarine and some canvas sheeting. Nevertheless, such revelations have done nothing to deter increasingly sophisticated searches of the loch.

During the 1970s there were two major searches sponsored by the Academy of Applied Science from Boston, one in 1972 and the other in 1975. These involved taking underwater photographs and sonar scans of the loch. The photographs from one of the searches revealed the upper torso and the "gargoyle-shaped" head of a monster. Sadly, later analysis showed the team had photographed a giant plastic model of a monster used in 1969 during the filming of *The Private Lives of Sherlock Holmes*.

Undeterred, another team returned to Loch Ness to conduct Operation Deepscan in 1987. This ambitious project involved a fleet of twenty-two boats creating a sonar sweep the entire length of the loch. Again the search provided no definitive results or clear photographs, merely a few blurred frames that were later revealed to show nothing more exotic than tree trunks and assorted detritus. The other problem with this search was that the sonar did not reach to the bottom of the loch and could cover no more than eighty percent of the volume of water. It is easy to imagine any shy creature living there heading for the bottom as soon as a flotilla of boats passed across the surface, so it may have been missed anyway.

But just because we have not yet seen the monster close up or found evidence it exists does not mean it is definitely not there. If Nessie does exist, what could it be and how has it survived?

The most popular suggestion is that the Loch Ness monster is not one creature but a community of plesiosaurs. The plesiosaur was an aquatic reptile that ranged through the oceans of the world from the late Triassic until the end of the Cretaceous periods, about 195 to 65 million years ago. The theory is that a group of plesiosaurs survived the extinction of the dinosaurs at the end of the Cretaceous period. Then, much later, a community became trapped in the loch when it was sealed off from the sea and have lived there ever since.

Although there are significant problems with this theory, it is by no means an impossibility. The central idea that small groups of creatures could survive species extinction is actually not so farfetched as we might imagine. In 1938, African fishermen found an armor-plated fish two meters long in the Indian Ocean. After careful analysis, it was found to be a coelacanth, a creature thought to have become extinct seventy million years ago. This specimen had evolved into a much larger fish than its prehistoric forebears and is thought to be part of a community living deep in the ocean.

Evolution is the process by which species develop utilizing two factors—natural selection and random chance. If a characteristic appears in an individual, it may be passed on to future generations if that individual survives long enough to breed. Both good and bad characteristics are passed on to future generations, but species evolve into improved or better-adapted forms because "good" characteristics make individuals better able to survive, stronger and dominant, and therefore more likely to pass on their design improvement. Luck comes into the equation because changes in genetic characteristics, or mutations, appear randomly.

All species evolve, but the environment can affect the rate of evolution. If the environmental conditions are stable for a long period and the community can survive within narrow limitations, evolution will probably be slower than it would in a more challenging, competitive environment. It is possible that a small community of creatures such as a group of plesiosaurs would have evolved far slower than other species throughout the world during the same time period—say 65 million years.

But the problem with isolation, and a major limitation to the survival chances of small communities, is the damaging effect of restricting the genetic diversity of the species. A small community of plesiosaurs is no different to an isolated Bohemian community of humans in that genetic faults could develop and threaten the health and eventual survival of the group. If we presume that the original community consisted of no more than a few dozen individuals, the likelihood of degeneration of the gene pool would become large fairly early in the life of the group.

A further problem is the need for resources. In order to survive,

Pyramid of biomass for a simple ecosystem

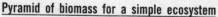

Figure 9.1

any living creature needs to draw upon what is called a *biomass*. On average, this biomass has to be about ten times its own weight, and from this comes the concept of the *pyramid of biomass*.

Imagine a simple system containing foxes, rabbits, and grass. This is a pyramid of biomass with foxes at the top and grass at the bottom. The grass is called a *producer* or an *autotroph* because it acquires its food directly from photosynthesis using sunlight. The rabbits exist on what is known as the next *trophic* level and are called *primary consumers*, because they are the first layer above green plants in this particular food chain. In order for the fox community to survive, they need ten times their biomass of rabbits. In their turn the rabbits require ten times their total biomass in grass. This is because for the primary consumers (rabbits) and secondary consumers (foxes), conversion of resources into energy is only around ten percent efficient. Without the grass, the rabbits cannot survive, and without the grass and the rabbits, the foxes cannot survive.

A group of plesiosaurs in Loch Ness would be at the top of the biomass pyramid for their own *closed system*, and in order to survive they would need a substantial mass of smaller creatures beneath them in the pyramid. They may have adapted to an omnivorous diet that would give them a more varied food supply, but even so, a group of creatures their size would stretch the resources of the loch to its limit, perhaps beyond. But if we argue that the community is very small, we then come up against the problem of lack of genetic diversity. If the Loch Ness monster really is a group of dinosaurs, they would represent a very delicate community. The group could not grow too large because of the lack of resources, but could not become so small that inbreeding creates intolerable genetic defects. One of these factors may well have wiped them out long ago.

Those who believe Nessie is a long-lost group of plesiosaurs suggest that the modern creature can live in both seawater and freshwater, so the fact that the loch gradually changed from a saltwater lake to a freshwater one during the past 7,500 years would not present a serious problem. But the plesiosaur was a marine reptile and it would have spent some time on land; it is surprising there have not been many more witnessed encounters between inquisitive humans actively looking for monsters and the creatures themselves.

To counter this argument, some believe Nessie is a strange, very large fish, and again this is feasible in evolutionary terms. Seawater fish could evolve into freshwater creatures as the loch itself slowly changed its chemical composition once sealed off from the ocean.

So, the Loch Ness monster could be a group of survivalist dinosaurs that have somehow overcome the problems of the biomass pyramid and lack of genetic diversity. They could be marine reptiles that have adapted to freshwater existence, or they may be large fish that have performed the same biochemical and anatomical trick. A further alternative is that all sightings of the monster may be accounted for by resorting to logs, tree trunks that have become entangled near the banks of the loch, or even mirages.

This last suggestion is actually more feasible than the dinosaur enthusiasts would like to accept. The environmental conditions at Loch Ness are surprisingly conducive to mirage formation, and these could be produced by objects in the water such as surfacing fish or large branches thrust skyward by freak currents.

Mirages are caused when rays of light are refracted to different extents by layers of air at different temperatures. The eye always interprets light as traveling in a straight line, so if the beam is bent or refracted on its way to the eye, the original image appears to have moved or grown large.

If an object is seen on a cold surface such as the loch in winter, light rays from the top of the object are bent in one direction by the slightly warmer air and light from the bottom is bent in the opposite direction. This effect is called *temperature inversion*, and although it was only explained during the last century, sailors have reported the illusion it produces since ancient times, calling it "looming"; hence, "looming large."

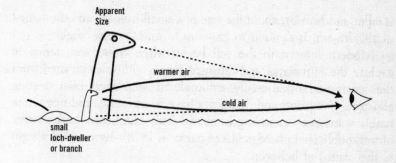

Layers of air at different temperatures cause light rays to be refracted and produce illusions of size.

Figure 9.2

Monster enthusiasts dismiss this explanation as being too limited and point to cases in which bow waves have been seen along with a long-necked creature. But recent meteorological research could also have an explanation for this. The effect of a bow wave or any other disturbance on the surface of the loch could be produced by what are called "water devils." These are thought to be smaller cousins of water spouts, distant relatives of tornadoes, and derive from vortices produced by pressure variations in the water.

These explanations may account for many of the frequent sightings of Nessie and other lake and sea beasts from around the world. Fakes and tricks could account for many more, but there is still no definite evidence either way. Loch Ness is vast and impossible to search inch by inch at present. Our best hope lies with future development of supersensitive sonar equipment or sophisticated thermal tracking and imagining devices that could show what strange creatures may live there, if any.

Almost as famous as Nessie are land-based ECS's. These include the yeti or the Abominable Snowman of the Himalayas, and the bigfoot of northern America.

The yeti appears in the legends of the local people of Nepal who worship the beast and will not trespass into its lands. They believe there are three types, the smallest, *yeh-teh* (from which the popular name derives), is about the size of a monkey. The *meh-teh* is taller and heavier, about the size of a small human; and the largest of all, *dzu-teh*, is thought to grow up to nine feet in height.

Modern interest in the yeti began when westerners started to explore the Himalayas and returned home with tales gleaned from the locals and occasionally embroidered with their own fleeting glimpses. Footprints and droppings have been found, and many reputable witnesses—including Lord and Lady Hunt, who took a series of well-publicized photographs of tracks in 1978—have added weight to the claims of believers.

The American version of the yeti is the equally shy, equally massive bigfoot which is believed to live in wooded areas but has been reported in almost every state. Other similar creatures have been documented in China, in the Urals—where they are known as *yag-mort*—in Siberia, where they are given the name *chuchunaa*, and in Australia, taking the name of *yowie*. Mulder and Scully had some dealings with a land-based ECS in the *X-Files* episode "The Jersey Devil," in which they investigated sightings of a humanoid creature who lived in the woods and ventured into the suburbs under the cover of dark to forage for food.

It is quite possible that there are small communities of people living in the wildest regions who have shunned contact with civilization for generations, and there have been documented cases of children found living with animal communities, including the famous Wolfboy of France who was found near Avetron in the early 1800s. His discoverer and mentor, a doctor named Jean Marc Itard, tried to introduce the boy to civilization, but with only limited success. The wolfboy died at the age of forty, still unable to adapt completely to human existence and viewed by many as a freak.

More challenging for science is the idea that there may be groups of creatures living in isolated spots that have evolved in seclusion and about which we have almost no knowledge—true evolutionary cul-de-sacs.

The same rules of biology apply to land-based ECS's as they do to lake monsters or giant sea creatures. For a community to survive for so long, it originally had to consist of a sufficient number of individuals to allow for genetic diversity, but at the same time not so large that the resources of the closed environment in which the animals live would be unable to support it.

The Himalayas is one of the most hostile regions on the planet, but it sustains an ecosystem containing a large variety of hardy plants at the bottom of the food chain which support the yak population and other relatively small mammals and birds. A yeti would be at the top of the food chain, but like the Loch Ness monster, with such a delicate infrastructure, it would live very close to extinction.

In order to maximize resources, the yeti would certainly be omnivorous, supplementing its diet of yak and other animals with local vegetation. They may have a thick layer of protective fat and a very thick coat, enabling them to live at high altitudes away from prying humans, only traveling to lower altitudes to hunt. They would almost certainly be solitary, highly territorial creatures that occupy a specific area and meet only to mate (which itself would be a rare event).

The American version of the yeti, the bigfoot is at once easier to accept and more problematic for the scientist. Resources, especially food, would be far more plentiful in the woods of the United States, and the climate is significantly better. But it is a highly populated part of the world, and it would seem unlikely that a creature as large as the bigfoot of legend would be spotted as infrequently as it has been. In the case of the yeti, it is easy to explain how no bones or remains have ever been located and that footprints are seen only rarely, but this could hardly be applied to a mysterious creature living in Kentucky or California.

With any of these large, humanoid ECS's, we also have to address the question of how they could have developed in the first place. Judging by the size of the yeti or the bigfoot, it would seem they followed an evolutionary path separate from Homo sapiens at a far distant point in our evolution. Where the separation could have occurred is almost impossible to speculate. The yeti and the bigfoot both appear to have more in common with the gorilla than Homo sapiens. The gorilla is the largest of the primates and can grow to a height of almost two meters and weigh 180 kilograms (400 pounds). They are also very territorial, but usually live in small groups.

The idea that "humans descended from apes" is a very common

misconception. We did no such thing, and despite popular myth stemming from the 1860s, Darwin never said we did. Homo sapiens and all other primates have a *common ancestor.* Biologists agree that the evolutionary line that ultimately led to the human form diverged from the ape line during the Tertiary period. But this is a very broad time span, ranging from the end of the dinosaur age (the Jurassic and Cretaceous periods), some 65 million years ago, to a time only two million years in our past.

Theorists are in some disagreement over when the split between the lineages occurred. Recent paleontological and anatomical evidence from fossil remains suggests that the pongid (ape) and hominid (human) lineages diverged during the Miocene epoch, which is placed at around twenty million years ago, but evidence based upon comparative immunology experiments point to a divergence as recently as four million years ago.

One staggering result from genetic research conducted during the past two decades reveals that the genetic makeup of a human is only very slightly different to that of a gorilla or an orangutan. In fact, it is now clear there is only a one percent difference between the human genome (the entire genetic composition of a human) and that of a chimpanzee.

All of this would imply that if a type of ape ancestor did diverge from the other branches of the evolutionary tree at some point millions of years ago, its genetic makeup may be very similar to modern primates, including Homo sapiens. Of course, we have no way of telling how these creatures might have adapted over thousands of generations. To determine the evolutionary path the creature had followed, biologists would need a sample of blood or some remains from which DNA could be extracted. But if one day such a find is made, it would not surprise most researchers if they discovered the Abominable Snowman is in fact a very close cousin of the Abominable Businessman.

Planet Earth is a very large place. We know more about the far side of the moon than the bottom of some of our oceans, and it is certain that humans will live on Mars before they build homes at the bottom of the Marianas Trench, six miles (almost eleven kilometers)

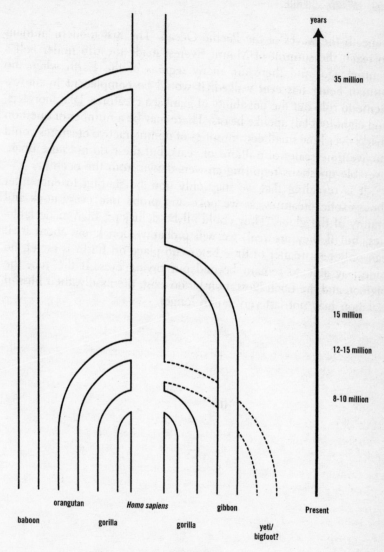

Figure 9.3

There is considerable disagreement about the time scales for the evolutionary chart of primates. Fossil records give older figures than immunological studies based upon biochemical research. It is impossible to place even an approximate date at which the evolution of the yeti diverged, but based upon its apparent physical characteristics, it could have been close to the point at which gorillas diverged from the branch that led eventually to *Homo sapiens*.

beneath the waves of the Pacific Ocean. The first modern humans to reach the summit of Mount Everest made the trip under half a century ago, and there are many regions of the Earth where no human being has ever walked. It would be complacent in the extreme to rule out the possibility of giant sea creatures, lake monsters, and eight-foot-tall apelike beasts. There may be a number of question marks over how small communities of noninteractive creatures could survive time scales of millions of years, but these do not raise unanswerable questions requiring answers drawn from the occult.

It is revealing that we may only now be starting to encounter these exotic creatures, as we poke and probe into every nook and cranny of the globe. They could all be delusions, man-made fantasies, but if they are real, we will probably soon know about it. It can only be a matter of time before no place on Earth is secret, no hideaway able to remain beyond our prying eyes. If the yeti, the bigfoot, and the Loch Ness monster do exist, then sadly, their blissful isolation may not last very much longer.

⑩ Time and Again

There was a young lady named Bright,
Whose speed was far faster than light,
She traveled one day,
In a relative way,
And returned on the previous night.

A.H.R. BULLER[1]

Almost everyone at one time or another has fantasized about time travel. One of the most profoundly frustrating aspects of living in a certain era and limited by our allotted three score years and ten is that we cannot see what will happen in the future; perhaps this is one of the reasons science fiction and fantasy are such popular genres.

Equally, how many times have you wished you could travel back into the past and change something, perhaps some pivotal event in your life? Where would you go in order to correct a problem, unsay something you wished you had never said, undo a wrong perpetrated by or against you? And if you wanted to travel in time simply as entertainment, what point in history would you choose—the Battle of Waterloo or a grandstand seat at Wembley to relive the 1966 World Cup final?

For the past three hundred years, ever since modern science stepped beyond the bounds of guesswork, physicists have believed time travel to be impossible. Yet there have been many documented cases in which people claim to have stepped backward or forward in time by seemingly supernatural means. These have been scoffed at, and perhaps for good reason, but since the late 1980s scientists have begun to consider seriously ways in which time travel could actually be possible. Although all the cases of apparent time-slips could be explained away quite easily, it is also conceivable that such things are possible in the natural world and may one day be understood thoroughly or even made practical.

One of the most famous cases of a time-slip occurred in 1901. Two highly respectable elderly spinsters, Charlotte "Annie" Moberly, who was principal of an Oxford college, and Eleanor Jourdain, the headmistress of a girls' school, claimed that during a walk around the grounds of Versailles outside Paris they inadvertently traveled back in time.

It was a hot summer afternoon and the two women were trying to reach the Petit Trianon in the great park of Versailles when they began to see some rather incongruous things. First they spotted a woman dressed in eighteenth-century clothes shaking a white cloth out of a window of a building. Next they passed a couple of officials in grayish-green coats and three-cornered hats. Later they crossed the path of a group of children all dressed in old-fashioned clothes, then encountered a man in a black cloak who had a face badly scarred by small-pox. Finally they came across a woman dressed in fine eighteenth-century dress sitting at an easel and painting. The whole experience lasted about half an hour, and the two women claimed they had reached the Petit Trianon and were given directions out of the park before returning to the world of 1901.

Ten years later Charlotte Moberly and Eleanor Jourdain published their account pseudonymously under the title *An Adventure*. During the decade between the experience and the appearance of the book they had conducted extensive research into the layout of the grounds at Versailles, the fashions of the period, and the history of the gardens. They reached the conclusion they had somehow traveled back 120 years to the time of the French Revolution. They further supported this dubious claim by showing that the landscaping of the grounds had changed since the 1780s and that on their walk they had followed a route that no longer existed. Finally, they believed the "painting woman" was none other than Marie Antoinette herself, a conclusion based upon a picture they had discovered in which the French queen was wearing an identical outfit to their phantom woman.

Miss Moberly and Miss Jourdain were respectable establishment figures not likely to deliberately lie or to fabricate such a story, even if it meant they would write a best-selling book based upon their experiences, yet there are many problems with their story.

Although they appear to have researched their account thoroughly, they failed to mention some key facts. First, during the time of their visit, a French nobleman, Comte Robert de Montesquieu, spent a great deal of effort and money putting on what he referred to as *tableaux vivants*—what we would call "happenings" or interactive art events—set in the grounds of Versailles. These performances involved large groups of the comte's friends dressed up in the costumes of the eighteenth century acting out scenes from history.

It may have been that Miss Moberly and Miss Jourdain had actually stumbled upon nothing more supernatural than a pageant or role-playing game. They may not have been aware of this, but considering how much research they had done for their book, it seems a strange oversight.

A second factor that casts doubt upon the story is the fact that they saw a woman they believed to be Marie Antoinette. Despite matching a drawing of the queen with the woman they saw, it seems improbable that if they had traveled back in time they should inevitably encounter so famous a person. But further discredit came from a letter that appeared in *The Times* in 1965. The author of the letter was a T.G.S Combe, who had delivered a lecture on the subject of *An Adventure*. At the end of the talk a member of the audience told him that as a child he was told of an eccentric woman who lived in the same district near Versailles who, on summer afternoons around the turn of the century, took to dressing up as Marie Antoinette and sat in the garden of the Petit Trianon.

Most other cases of time-slips are usually less dramatic. Often they involve individuals, or sometimes small groups of people, thinking they saw a building or a road, and when they return to the spot they find the topology of the area is actually quite different from how they remembered it.

One striking case involved four English tourists who stayed in "a quaint and old-fashioned" guest house in Montélimar in France in 1979. They described the rooms, the other guests, and the patron, recounted how they had only been charged less than the equivalent of two pounds for the four of them, and even took photographs of each other in the hotel rooms. Returning later to see if they could stay there on their return journey, they found a garage on the site of the hotel.

When they had their holiday photographs developed, the mystery deepened. They discovered that those they believed had been taken in the mystery guest house were missing, but the sequence of the negatives had not been disturbed; it was as if the pictures had never been taken. When asked how it was that the other "guests" did not seem to been disturbed by their presence in modern dress and stepping out of a modern car, or how they had been able to pay the bill in present-day currency, the two couples were at a loss to explain.

Of course, this case and many others like it could simply be a contrivance, a story deliberately created for publicity or financial reward, but an alternative explanation is that these people had experienced a group hallucination. As we saw in Chapter 8, such things are well-documented in cases of hauntings and poltergeist activity. Perhaps the tale covered in *An Adventure* was akin to a collective or group hallucination, sparked off by the women seeing the Comte Robert de Montesquieu's *tableaux vivants*. In the case of the two couples on vacation in France, it could be that their genuine hallucination was initiated by too many glasses of the local wine.

An alternative explanation may be due to the playback system described in Chapter 8. Perhaps events in the past had been imprinted or recorded in the environment and for some reason triggered to replay at the moment these people entered the area.

However, there are some cases in which the idea of a played-back image could not be used as an explanation. This is when witnesses claim to have seen the future. A frequent tale involves people observing long cigar-shaped flying craft or ground vehicles passing by at high speed. Sometimes the occupants are looking out of the windows of their craft and pointing excitedly toward the witness; in other cases, the "future people" are unaware they are being watched.

Some have suggested that these are visions from a future time projected back into the past, which may occasionally involve a two-way correspondence through which the parties can see one another. Enthusiasts of this idea go on to speculate that such experiences might even account for many unexplainable UFO sightings. This would imply that UFO witnesses do not see alien spacecraft, but vehicles from our own far future, perhaps occupied by humans who have evolved into slightly different physical forms.

Much of this is pure speculation. The fact is, we have no idea how a time machine could be constructed, and physicists working at the very edge of science are only now beginning to piece together theories that may explain how time travel could be possible. But all of these ideas exist only as mathematical concepts. We are no nearer building a time machine than we are in reaching the center of the galaxy in our own time frame.

But the initial step toward building such a device is understanding the mathematics behind it. And before we can develop such a theory of time travel, we have to come to grips with the meaning of time itself.

We all experience the passing of time, but no one seems able to explain what "time" is. Some even suggest it is nothing more than a construct of our own minds, that we piece together events in a logical, linear order because that is the only way our brains can operate and make sense of the universe.

There is no material evidence to support this concept, and although we have seen already that common sense and cutting-edge physics are often estranged, we all seem to have an in-built awareness of the direction of time, a concept dubbed the "arrow of time."

Curiously, at the fundamental level, almost all processes in the universe, whether they are interpreted using classical physics or the QM of Schrödinger and Dirac, can be conducted in either temporal direction. This means that if two subatomic particles come together and interact to form two other particles, the reverse process is equally viable—the two product particles could just as well interact to create the starting particles.

Yet, we don't experience this reversibility in the "real" world, within the macrocosmos of everyday existence. We don't see shattered glasses reform, light does not leave our eyes and travel to distant objects, and the dead do not rise from their graves. So, as we saw with other aspects of QM, it might appear that the principles governing the behavior of "simple" systems—those that operate on a quantum level—are in conflict with more complex systems such as those that encroach upon our everyday existence on a macrocosmic level. But even this seems paradoxical because we and every material thing in the universe are made up of fundamental particles. If they behave reversibly in simple scenarios, what is it about complex systems that seems to make them act differently?

The answer lies in the difference between something being *impossible* and just very unlikely. Physicists believe that it is not impossible the dead could be made to rise again (ignoring spiritual considerations), or for a broken glass to reform by chance. It is just that these events require so many improbable steps to juxtapose perfectly, at least compared to the interaction of two subatomic particles, that we would almost certainly have to wait for a period longer than the lifetime of the universe to see them happen *naturally*. This means that although they are not impossible, they are highly improbable.

In order to relate this to the arrow of time, we have to consider one of the most fundamental rules of the universe, a principle called the second law of thermodynamics.

This law lies at the very heart of physics. In his book, *The Nature of the Physical World*, the physicist, Arthur Eddington said:

> The second law of thermodynamics holds, I think, the supreme position among the laws of Nature. If someone points out to you that your pet theory of the universe is in disagreement with Maxwell's equations — then so much the worse for Maxwell's equations. If it is found to be contradicted by observation — well these experimentalists do bungle things sometimes. But if your theory is found to be against the second law of thermodynamics I can give you no hope; there is nothing for it but to collapse in deepest humiliation.[2]

Unlike some of the exotic aspects of quantum theory and relativity, the second law of thermodynamics is actually a law based entirely upon common sense. Put simply, it says that: everything wears out. In more formal terms; the entropy of a *closed system* always increases.

Entropy is the technical term describing the "level of disorder in a system." So, by this law, a cup of tea exhibits a higher level of entropy than the individual tea leaves, water, and milk, because they have been mixed together. It would take more energy to separate them out again than was used to mix them in the first place.*

Returning to the example of the broken glass. If we tried to bring together the pieces, like running a film backward to recreate the glass perfectly, we would need to lower the entropy of the system. This is possible—in fact, living creatures spend most of their lives attempting to produce a local lowering of entropy—but it requires energy, and for this to happen naturally by chance is incredible unlikely.

In a similar way, a garden left to overgrow will gradually increase its entropy level quite naturally. In order to restore the tangled weeds and vines to their former order, work would have to be done, or energy expended. It is extremely unlikely this will happen naturally without the interference of intelligence (and muscles).

Because the natural processes of the universe are all ones in which disorder or entropy is seen to be increasing, it gives us an indicator, a way to view the progress of the universe, or in other words, the direction in which time flows.†

So, if there is a definite direction to time, can intelligent beings or even information move in a nonlinear way from one time frame to another? Would it ever be possible to build a time machine?

Currently, physicists are giving serious consideration to two possible mechanisms through which a genuine time link could be produced. The first of these is to employ the ever-useful wormhole.

We saw in Chapter 6 how using wormholes to transport information may be within the bounds of acceptable physics utilizing Einstein's theory of relativity. In that discussion, I used them to explain how precognition could be a genuine natural process, but this system might also be used to create a two-way time-travel system.

---

*Our universe is a closed system, so entropy will always rise in the universe.

†We might even say that although it would seem that intelligent life is constantly attempting to decrease entropy locally, life itself could be the very reason the universe will one day end. This is because every time a process occurs there is a loss of *useful* energy from the system (energy that cannot be harnessed to do work). As the physicist Barry Chapman has said: "It may be that the purpose of life is simply to facilitate the heat death of the universe."³

In the example, the observer in "our time" viewed the other end of the wormhole as the future and could witness events that had not yet happened; but of course, to those at the other end of the wormhole, we exist in their past. So, a wormhole could produce a time loop—travel one way through it and you emerge in the past, take the opposite direction and you arrive in the future.

Although the evidence for the existence of black holes is considered very strong, wormholes could prove to be nothing more than imaginary entities. It would be a far less interesting universe if wormholes and time loops did not exist, but sadly, the laws of physics are not malleable to human desires. Conversely, if in the future black holes are shown to be common, then it is very likely that wormholes could exist, and if we advance sufficiently to first transport information and then objects through these temporal and spatial highways, time travel could become a real possibility. Perhaps others are already doing it.

Stephen Hawking and others have speculated that mini-wormholes exist naturally in our universe, and it may be that these are responsible for the accounts of time-slips. It could be that the information content of an incident at the antiquated hotel in Montélimar was somehow transported to a future time quite naturally and was observed by the four British tourists who claimed they had visited the place. We saw in Chapter 6 that some enthusiasts have suggested that particularly traumatic events may somehow trigger wormholes, or the energy from such happenings could utilize naturally occurring wormhole links between different times. Perhaps these amazingly handy devices are more common than we think. It is even possible that our universe is interlaced with a vast network of interlinking wormholes just waiting to be employed by a sufficiently advanced technology.

There is another possibility that involves black holes but does not require the existence of wormholes. This is the idea of "skimming" the intense gravity well of a black hole.

Again, this idea is based upon the equations of Einstein's theory of relativity. The first mathematician to seriously speculate upon the possibility of time travel using the equations of relativity theory was a friend and colleague of Einstein's at the Princeton Institute for

Advanced Study, Kurt Gödel in 1949. Fourteen years later a New Zealander, Dr. Roy Kerr, published a paper speculating upon the idea of a time machine using the theory of relativity as applied to black holes. At the time, black holes were still unchristened (that honor fell to John Wheeler in 1967), but Kerr knew such objects were feasible in principle, and employed the fact that time is affected by velocity and gravitational fields to demonstrate a theory of time travel. By an amusing coincidence, his paper was published on the eve of the first episode of *Dr. Who* in November 1963.

Kerr's theory suggested that if a time machine was fired at a black hole and made to skim the edge of the gravitational well without being sucked in, time would travel far slower for the occupants of the machine. Meanwhile, the events in the world outside would be whizzing by. If the machine then traveled back to a point beyond the black hole, they would find themselves in the future.

A decade later the concept of using black holes as time machines was extended by Frank Tipler from the University of Maryland, and in 1974 he detailed his ideas in a paper published in the highly respectable journal, *Physical Review*.[4]

Tipler took things much further than Roy Kerr. In his scheme, a very advanced civilization could produce a special type of black hole called a *naked singularity*. To make this, the singularity (found at the heart of a black hole) would have to be rotating. The effect of the rotation is to twist space-time so much in the region near the singularity that time itself becomes another dimension of space through which a carefully piloted craft could be maneuvered.

Tipler then went on to detail the design spec for the artificial naked singularity. According to his calculations, you would need a cylinder 100 kilometers long and about ten across made of superdense material, similar to that found in a neutron star where all the electrons of the atoms of the substance had been fused with the protons in the nucleus. Finally, the object would have to spin precisely twice every millisecond.

For all the ingenuity of this scheme, on its own it might be considered nothing more than a fantasy, but amazingly, there are naturally occurring objects in the universe that almost fit the bill.

In Chapter 2, I described the story of how the British astronomer Jocelyn Bell discovered the first pulsar in 1967 by tracing a regular radio signal that was believed initially to be a message from an alien civilization. Although this signal turned out to be a natural pulse, the consequences of its discovery may be every bit as exciting as if it had been a beacon placed there by an alien race. This is because special objects called *millisecond pulsars* have since been discovered that are so close to being Nature's time machines they may only need slight adjustment by an advance technology to be usable. Millisecond pulsars are made of material with almost the right density, and they spin once every 1.5 milliseconds, one-third the speed needed for Tipler's design.

Even though we may be thousands of years away from having the technology to utilize such objects, the discovery of millisecond pulsars combined with the innovative ideas of Frank Tipler and others is now generating great excitement within the physics community. But there remains another delicate matter—the problem of temporal paradoxes.

The concept of time travel has created agonizingly complex paradoxes for centuries. Long before any serious thought had gone into how a time machine could be designed, it was believed that these paradox problems could actually be so severe they alone would prohibit practical time travel.

H. G. Wells set the tone with his classic novel *The Time Machine*, published just over a hundred years ago, in 1895.[5] Unless Wells had a time machine himself, he would have known nothing of relativity because the creator of the theory, the sixteen-year-old Albert Einstein, had just squeezed his way into a technical college in Zurich at the time and was struggling with elementary math. Not surprisingly, the author offered little by way of explanation for his time travel system, but he was careful not to send his hero into the past, almost certainly because of the problematic paradoxes such a journey could entail.

Happily, the problem may not be as bad as Wells and others feared. By utilizing some intriguing aspects of quantum theory, physicists are now concluding that temporal paradoxes could be completely fictitious.

In his short story, "All You Zombies," written in 1959, Robert Heinlein offered what must be one of the most confusing examples of a time travel paradox ever imagined.

The story centers around a character, Jane, who is mysteriously abandoned at an orphanage in 1945. The child grows up with no idea who her parents are, but in 1963, at the age of eighteen, she falls in love with a drifter who visits the orphanage. For a while things go well, but then the drifter leaves her and Jane finds she is pregnant. The delivery of the child is difficult and she has to undergo a caesarean section. During the operation, surgeons discover Jane has both sets of sex organs, and in order to save her life they have to convert "her" to a "him."

Subsequently, the baby is mysteriously snatched from the hospital, and Jane drops out of society and finally ends up a vagrant. Seven years later, in 1970, he stumbles into a bar and becomes friendly with the bartender, who offers Jane a chance to avenge the drifter who had ruined her life, on the condition that she joins the "time travelers corps." The pair then go back in time to 1963, and the vagrant "Jane" seduces the eighteen-year-old female Jane at the orphanage, making her pregnant before disappearing. The bartender then travels forward in time nine months, snatches the baby from the hospital and deposits it at the orphanage in 1945 before dropping off Jane in 1985, where he joins the time travelers corps which has been created after the recent invention of time travel.

Jane, the time traveler, distinguishes himself in the corps and eventually becomes a highly successful bartender, opens his own place and returns to 1970 to persuade a young vagrant to join the time travelers corp.

So, in this tale, Jane is her own mother, father, and daughter. She is also the drifter and the bartender. But who are Jane's grandparents? She seems to be a creature out of time, self-created and totally independent of the universe; in other words, a paradox.

There are other simpler examples of this twisting of events. Imagine a time traveler journeying a hundred years into his past to the studio of a struggling artist. There, he tells the artist that in the future, he is world-famous, recognized for a distinctive style very different from the one he is currently using, and then proceeds to show him a catalogue of his future work. Distracting the visitor, the artist photocopies the artwork, and the time traveler returns to the future. The artist then starts to copy the paintings he has photocopied.

The disturbing thing about this paradox is that it seems to offer a free lunch, and taken on face value, it breaks the laws of physics. Which came first, the paintings or the artist's fame? It also seems to cancel out the principle of free will. If beings from the future are able to manipulate the past and change our lives, where is the element of self-determination?

Fortunately, there is a solution to this set of possible paradoxes, utilizing a concept physicists call the many universes interpretation.

I touched on this theory in Chapter 6 when discussing various interpretations of Schrödinger's cat experiment, but the implication of the idea may have even stronger relevance to the subject of time travel.

The simplest interpretation of the many universes theory is that whenever any fundamental event occurs, the future splits into two possible outcomes or separate universes. It is easy to understand this when we refer it to our own lives. Suppose we have an important job interview. Perhaps along one route we make the interview, get the job, and eventually become the chairman of the company. In the other, we miss the train, fail to make the interview, and lose the chance of a perfect job.

But this example is one on a macrocosmic scale. According to the many universes interpretation, every time any subatomic change occurs anywhere in our space-time continuum, the path splits, creating two different universes. These universes may be so similar that any difference may be completely imperceptible to us. Perhaps the only variation is the position of one electron situated on the other side of the universe. Even so, they will be different. And it is because of this that the troublesome time travel paradoxes could be written out of the equation.

Consider another example. If in Universe A, "our" universe, we go back in time and persuade our grandfather not to go on a crucial date with our future grandmother, a paradox will be avoided be-

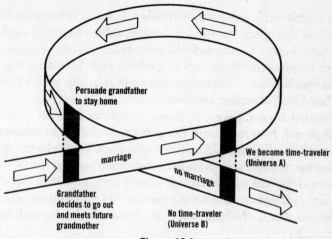

Figure 10.1

cause, at the instant we arrive in the past, two possible futures or universes are created simultaneously, Universe A and Universe B. In Universe A, our grandfather goes out to dinner and starts a lifelong relationship completely unaware of our arrival. This leads to a future in which we are born and become a time traveler who returns to the past. In Universe B, the grandfather misses the date and we are never born. But because we have come from Universe A, we do not suddenly cease to exist and there is no paradox.

Stephen Hawking used to say that if time travel was really possible, we would be visited by time tourists; but as we are clearly not, it is impossible. This philosophy is wrong for at least three reasons. First, time travelers would almost certainly be sophisticated enough to cover their tracks. Second, our space-time constitutes only a vanishingly small part of the entire life and volume of the universe, so it is highly probable time travelers have not yet visited this time or place. Third, if the many universes interpretation is correct, only versions of ourselves in certain universes would ever be aware of the visitors. Hawking has since changed his mind and now believes time travel is theoretically possible.

It seems probable that one day we will be able to utilize natural aspects of the universe such as the properties of a black hole, or exotic objects such as pulsars, to develop a device to travel backward in time. If this system is found to be impossible, then we might still have the chance to use suitable wormholes. Either of these will require dramatic developments in physics, and in particular a successful combination of quantum theory and relativity, which remains the Holy Grail of modern physics.

Whether or not fluke conditions sometimes arise on Earth enabling people to wander into the past or the future by accident is still open to conjecture. Given the tremendous forces at work within pulsar time machines or wormhole temporal highways, this would seem unlikely, but then there are almost certainly other exotic objects like pulsars and neutron stars just waiting for curious scientists to discover them. Perhaps some of these will be located nearer home.

Finally, the ideas of quantum mechanics seem capable of solving the problem of paradoxes. If the many universes interpretation is proven one day to be the way in which the multiverses of creation really behave, and the technology is in place, then the last barrier to time travel really could be overcome. Perhaps it is premature to book a ticket on the Temporal Express, but we can dream.

Into the Light

> They were seated in the boat. Nick in the stern, his father rowing. The sun was coming up over the hills. A bass jumped, making a circle in the water. Nick trailed his hand in the water. It felt warm in the sharp chill of the morning.
>   In the early morning on the lake sitting in the stern of the boat with his father rowing, he felt that he would never die.
>
> ERNEST HEMINGWAY[1]

We have countless euphemisms to deal with it—passing on, passing away, kicking the bucket; but whatever we call it, one day we will each face the Big D, or as the geneticist Steve Jones has said recently while discussing the odds of dying from various causes: "Not every gambler will win the lottery, but all of them will die."[2]

Morbid? Yes, and that is why since members of our race first became sentient, we humans have tried to figure out ways to convince ourselves that death is not the end, that there is an afterlife, be it in heaven, hell, or right back here again. But what evidence is there? Is it all wishful thinking, as materialists would claim, or is there, after all, something special about us as individuals, something immortal?

Until the 1970s there was little attempt to make a psychological study of the process of dying; naturally, those who had died could tell us nothing. But gradually, stories were documented that involved individuals traveling to the brink of death, even being declared dead, only to return to life. Some of these people were able to recall what they had felt, and the term "near death experience," or NDE, was coined to describe the process.

The first person to popularize the idea of the NDE was a doctor working in Georgia, Raymond Moody, who wrote a book on the subject called *Life After Life*,[3] which has since sold over three million copies. Although Moody made no attempt to analyze or judge the experiences he described, through a collection of real-life cases he introduced into the language many notions synonymous with NDE—seeing one's own body from above, traveling along a dark tunnel to the light beyond, and meeting a spiritual guide. The first serious analysis of these experiences and an initial attempt to quantify claims came with the work of another American doctor, Kenneth Ring, at the University of Connecticut.

In 1980, Ring categorized five separate stages of the near death experience. These he labeled: peace, body separation, entering the darkness (or tunnel), seeing the light, and entering the light. The majority of individuals who have reported an NDE only recall the initial stages, and the full five stages have been experienced by fewer than ten percent.[4] Yet, strikingly, most analysts believe there is a certain uniformity in many of the descriptions, and a typical full experience might consist of the following details:

> I was suddenly aware of a calmness, a still, peaceful feeling. Then, all at once, I felt myself floating above my body and looking down at the proceedings, totally detached, as though I was watching a movie. I could see myself from above, I could see the tubes and pipes and doctors around me, the flatline readout on the screen.
>
> Gradually I became aware of a darkening around me, as though the scene was narrowing, and then I could see a tunnel with a bright light at the end of it. I felt compelled to move toward the light and felt myself accelerating that way. As I traveled I could see scenes from my past flashing before me, both good and bad moments. Then, as I approached the brightness, I could see someone; they were surrounded by light. Suddenly, I recognized the figure as my father. He had died ten years earlier, but there he was, just as I remembered him. He was asking me if I was ready for this journey. Did I want to go on with him, or go back? Then, a moment later, before I even had time to answer, I felt myself being dragged, sucked back into my body. I snapped awake and the agony returned in a great rush. I felt terribly angry and then gradually calmed down.

To many, descriptions like this are clear evidence that we live on in some form after physical death. But sadly, upon close scrutiny, such claims may be quite unfounded. Let us take each stage of this amalgamated description and consider the possibilities.

The sense of relaxation, or impression of peace, is perhaps the easiest aspect to explain using accepted medical knowledge, and almost certainly comes about because the body is being flooded with "calming chemicals" such as endorphins. These are naturally occurring agents produced by the brain that serve to relieve pain during times of extreme physical stress. Fitness enthusiasts make use of them in the gym—it is the endorphin rush that keeps them going, and by controlling it, athletes can push themselves through the pain barrier during training. It is no surprise these chemicals are released; it is another example of how the body switches into survival mode during physically demanding moments.

The next stage of an NDE involves the patient seeing themselves and their surroundings. This is almost without exception from a bird's-eye perspective, looking down upon the proceedings with feelings of detachment. This stage of the NDE has been given the name Out of Body Experience, or OBE.

Claims of OBEs are not restricted to traumatic situations such as a near-death experience. There are many people who claim they have become dislocated from their bodies and have been able to wander around in an astral world. In many cases this behavior seems to be indistinguishable from dreaming. One astral traveler, Robert Monroe, has described vivid tales of his trips to the planets of the solar system and beyond to the distant stars. He claims to be able to travel across intergalactic space outside of time. Unhindered by any of the tiresome restrictions of relativity, he travels instantly to any location and supports his stories by suggesting that he has seen "markings" on planets before NASA spotted them using satellites.[5]

Naturally, there is little to back up these farfetched stories, but they do make good material for best-selling accounts. More interesting, but a little less dramatic, are reports from individuals who claim to have had OBEs while in a traumatized situation and as part of a near-death experience. These are intriguing because they are part of a process and not apocryphal recollections or one-of-a-kind experiences that offer little chance of rational explanation.

Much has been made of the OBE aspect of the near-death experience. This is perhaps because it is the only stage that can be related to the living world and other individuals. Often, victims report snatches of conversation heard from close by or can quote what doctors have said as they were making attempts to resuscitate them. They are also able to describe bits of machinery surrounding them and the sounds they made. Very occasionally a patient will claim to have seen something that has not occurred in their immediate vicinity or to report an incident in an adjoining room, and it is these apparently supernatural abilities that have most excited believers in life after death.

It is actually not surprising that people suffering trauma can recall some of the things they have seen and heard around them under these conditions. It has been found that people are quite capable of recalling things that have been said nearby while unconscious or asleep, and events that occur in the "real world" close by often appear in the individual's dreams.[6]

They do not need to be astrally projecting to achieve this. Probably the most common example is the behavior of new parents toward their baby. The slightest sound can wake them from deep sleep if it is associated with the child, but extraneous, unconnected auditory signals are ignored. Again, this is a survival mechanism dating from primitive times, a deeply rooted instinctive reflex.

There have also been extensive studies of anesthetized patients who register external stimuli, although often many of them do not recall the details afterward.[7]

But, all of these rational arguments could be overridden if there were proof that an individual in the midst of an NDE could perceive information they could not possibly have gained through normal physical means. Unfortunately for the enthusiasts, when studied closely, any "evidence" melts away. The psychologist Susan Blackmore has made a thorough study of OBE claims and has analyzed several hundred cases, yet not one stands up to close scrutiny.

One of the best examples comes from a cardiac patient, a woman named Maria who recovered from a heart attack and later described how she had floated over her body and traveled to another room in the hospital, where she saw a tennis shoe left on the window ledge. Later, a nurse went to look for the shoe and found it exactly as described. However, Maria is now dead, the shoe was not seen by anyone else, and the only support for the claim comes from the singular testimony of the nurse.

In another story, a woman claimed she had left her body and saw her son, Graham, arguing with a nurse before he was told to leave the room. The woman then traveled astrally above her son as he walked along a corridor. She saw him enter a room where his partner, Chris, was waiting. Chris asked Graham what had happened, and he replied that they would not let him see his mother. He then kicked a chair before sitting down and lighting a cigarette.

The problem with this is that every aspect of the story is predictable. Despite being in a state of trauma, the woman's brain could have registered her son arguing with the nurse. She then imagined the logical series of steps that would follow such an incident. She would have known who was most likely to be at the hospital, she would have been able to predict her son's likely reactions (kicking the chair), and she may have further assumed he would light a cigarette. There is no proof here, nothing to pin down or to demonstrate anything that could not be explained by quite natural means.

Yet, there are some intriguing aspects coming out of experiments conducted by doctors who have treated patients who believe they have had a near-death experience. One interesting piece of research comes from cardiologist Dr. Michael Sabom. He has conducted experiments to see just how much can be remembered of their environment by those who claim to have recovered from an OBE. He asked these individuals what they could recall of their own anatomy and the function of various pieces of surgical apparatus in the operating theater. To his surprise, he discovered that those who claimed to have had an OBE were considerably more knowledgeable than a test group who had undergone the same surgery without reporting an OBE.

At face value, these could be viewed as circumstantial evidence, but so little is known of the brain state during what patients believe to be an OBE that we cannot jump to conclusions. It could merely be that patients who do have what they perceive as a mystical experience actually experience a mental state that makes their short-term memory far more receptive. In this way, they could absorb more

information about their environment than people in a less traumatized state.

One of the most intriguing elements of reported cases of NDE
is that in the overwhelming majority of them, the patients "see"
themselves from above. The almost universal character of this aspect
of the near-death experience has lead enthusiasts to use it as evidence that an astral body really does detach itself from the physical.
But in fact this universality is quite suspicious. Why, if the spirit or
the astral body is leaving the physical body, should they only see
their physical body from above?

Psychologists believe that this "floating above the body" perspective supports the conclusion that all OBEs are self-induced hallucinations or visualizations originating from a traumatized brain. The
reason for this is that as living, breathing individuals, we build models of the world as we see it through our eyes at the time we experience it. So the world is always visualized at eye level. But in our
memories we almost invariably visualize a scene either from a bird's-
eye view or from the viewpoint of a third party. We very rarely
imagine or remember a scene in which we played a part from our
perspective at the time, despite the fact that almost all the audiovisual information gathered by our brains comes from this angle.

When our brain is dying and the audiovisual impulses are becoming confused or are largely shut down, we turn to our memories
and dreams to construct a model of the world, and these images are
almost always seen "from above." This is the viewpoint of the
dreamer, those in a trancelike or soporific state and those suffering
from severe brain trauma surviving at the limits of consciousness.

Susan Blackmore, at the University of Bristol, and Harvey Irwin,
from the University of New South Wales, Australia, have conducted
experiments on sets of people who claim to have had OBEs, and
have found that in many cases these individuals are particularly good
at switching viewpoints. Most significantly, they report their dreams
to be largely perceived from a bird's-eye view and can more easily
imagine a waking scene from this angle.[8]

After this sensation of floating above the body, most of those who
remain in a near-death state report that their vision becomes re-

stricted and their entire environment appears to be taken over by a tunnel, at the end of which they can see a bright light.

Believers in life after death perceive this to be a "passage to the afterlife," at the end of which is a heavenly light, a place where we will meet those most important to us and decide either to travel on to an existence beyond our physical lives or return to recover here on Earth. But again, work conducted in recent years by neurophysiologists and psychologists is casting doubt on this very attractive idea.

One suggestion for the origin of this tunnel came from Carl Sagan, who proposed that the experience was a "birth memory," but this idea has fallen out of favor recently.[9] Psychologists point out that birth memory is very rarely a fully formed notion, and that being born is actually nothing like going through a tunnel. It was then proposed that the tunnel was a *visualization* of birth—an image the brain created in order to picture what birth was really like. The idea was finally disposed of with a survey in which 254 people were asked if they had ever had a near-death experience. Thirty-six of them had been born by caesarean section, but almost equal numbers responded with a conviction that they had indeed endured an NDE.[10]

A more likely explanation is that the brain is suffering what is called *cerebral anoxia*, or oxygen deprivation. It is this purely physical phenomenon, psychologists now believe, that is responsible for the tunnel and the light. It is also thought to lead to the frequently reported encounters with long-dead relatives or even a godlike figure.

As discussed in Chapter 3, the brain operates by electrical impulses passing along neurones. These impulses cross gaps called synaptic gaps, between the ends of neurones. The messages that pass across these gaps have to be carefully controlled. Too many impulses too rapidly, and the signals cannot be processed properly; too few too slowly, and brain function is impaired. This means the inhibition of signals through the brain is every bit as important as excitation.

It was found in the 1950s that hallucinogenic drugs like LSD and mescaline operated by quashing the inhibitors in the brain so that signals passed across the synapses with increased speed and intensity. LSD suppresses the action of the *Raphe* cells which regulate activity in the visual cortex, a part of the cerebral cortex, or the thought-processing center of the brain.

Experiments conducted on rats have shown that when a brain is deprived of oxygen, the inhibitor signals are shut down before the excitor signals, so the nerve cells pass on their impulses more rapidly and efficiently. This will have two dramatic effects. First, just as happens with LSD, reception of impulses will be altered in the visual cortex. Second, cerebral anoxia produces hallucinations.

The visualizing of a tunnel had been well documented long before the NDE was popularized, and was known to be part of an hallucinatory experience created with drugs such as psilocybin (magic mushrooms) and LSD. In the 1930s, Heinrich Kluver, a psychologist at the University of Chicago, noted that there were four basic images seen during drug-induced hallucination. These are the lattice, the spiral, the cobweb, and the tunnel, but it has only been during recent years that the reason for this limited collection has been explained.

Neurobiologist Jack Cowan worked on this problem during the early 1980s and came up with a link between signals registered by the retina and the images visualized in the brain.[11] He concluded that a traumatic disturbance in the brain, such as oxygen deprivation causing loss of inhibition signals, would create what he called stripes of activity in the visual cortex. When these stripes were translated into a perceived image, they appeared as concentric circles.

More important still is the fact that when we register information on the retina, processing the visual data from the center of the visual field requires the most neurones. If the brain has been starved of oxygen and the inhibitor signals are suppressed, this will affect all neurones equally, but because the center of the visual field requires more neurones, this will suffer the most dramatic disruption—hence the bright light at the end of the tunnel.

Fortunately, this theory can be verified using computers. If a processor is programmed with the information needed to recreate a simple model of the link between the retina and the visual cortex, and then a simulation of increased excitation is added, a tunnel-like image is produced on the screen. This image is bright at the center and darkens at the edges. If the level of activity is increased to replicate the effect of increasing the cerebral anoxia, the "tunnel" appears to come closer, which creates the illusion of moving along it toward the bright light at the end.

Some very intense experiences have been reported by those brought back from the very edge of death. They sometimes report that as they enter the light (the fifth and final stage of any reported NDE), the tunnel image appears to break up and occasionally transforms into fields, trees, and other distorted images. Amazingly, this is exactly what happens in the computer simulation. If the stimulus is increased to represent further oxygen loss and greatly increased inhibition of signals, the tunnel transforms into a set of spirals and complex rings.

But what of the overpowering impression that as the patient emerges from the tunnel and enters the light, they often encounter lost loved ones, or even, in some cases, the Creator?

It seems likely that these images are also created by cerebral anoxia, because as the normal brain functions become confused or overloaded, we are forced to call up dreams and memories to deal with the situation. As Susan Blackmore describes it: "The system simply takes the most stable model of the world it has at any time and calls that 'reality.' In normal life there is one 'model of reality' that is overwhelmingly stable, coherent, and complex. It is the one built up from sensory input. It is the model of 'me, here, now.' I am suggesting that this seems real only because it is the best model that the system has at the time."[12]

People severely injured or suffering a heart attack or other such medical emergency would naturally cling to the most secure of their fantasies or dreams. The most likely image to appear in these hallucinatory states would be lost parents, partners, or in some cases, an image of their God.

One of the most striking aspects of NDE, and one that stimulated the hopes and imaginations of believers very early on, is the claim that the experience is uniform, that the images are universal and cut across all religions and cultures. But according to some critics, this is not entirely true.

First, they point to the fact that many people do not report their experiences from fear of ridicule; this lowers the sample size considerably. Even then NDE seems to be a very rare phenomenon. In this book, *Pseudoscience and the Paranormal*, the skeptic Terence Hines claims that there are a number of cases in which people report very unpleasant near-death experiences analogous to the occasional "bad trip" recalled by users of hallucinogenic drugs.

This may be caused by a variety of factors, including the brain chemistry of the patient at the time of the NDE, and obviously the reason for them being there in the first place. It could also be influenced by the patient's mood, life situation, or the medication they are receiving as the doctors try to bring them back from the dead.

Most psychologists agree that the reason why the majority of NDEs share a set of common elements, irrespective of class, race, sex, or religion, is simply because similar brains create similar images. This would imply that many aspects of personality—a person's tastes, interests, their job, or their preoccupations—are actually "non core," or peripheral activities that have little effect on the way their brain behaves during shutdown. It would appear that the deeper drives, strong emotions—their loves and hates—play a far greater role in the creation of hallucinatory images, both drug-induced and trauma-associated.

No one who has endured a genuine near-death experience has been left unaffected by it. Many people believe it to have been a pivotal event in their lives. Some have been changed beyond recognition by the experience; for most, it has radically altered their perception of life and their place in the scheme of things. Almost everyone (excluding those unfortunate enough to have had a bad NDE) has come away from it no longer sharing the fear of death so natural to most others. Some even claimed to be looking forward to dying.

Science can show that the claims of those who have lived through an NDE are not based upon the supernatural. These people are not treated to a glimpse of an afterlife, but are nevertheless very lucky for two other reasons: they survived, and also, they experienced a very rare phenomenon, one most people could only hope to experience just before they die. There is no doubt that NDE is a genuine and fascinating phenomenon. Those who claim to have seen a tunnel, even those who believe they met God for a moment, are not lying, they believe genuinely they experienced the real thing. What they did perceive was a representation of powerful, personal images associated with life and death, all within their own cerebral cortex.

Perhaps the ancients knew about the NDE. Perhaps the legend of heaven and hell has grown out of the "good trip" and the "bad," each stimulated in part by our deepest drives, desires, and fantasies. Maybe this is the real reason why believers have always declared that the good of heart go to heaven and the wicked go to hell.

The NDE is a very reassuring discovery. It is comforting to think that we can at least hope to have this naturally induced solace when our brains close down forever. Many die instantly, by a wide variety of means, but perhaps the greater number of all the humans who have ever lived slipped away from physical existence. And if comforting chemicals and cushioning brain processes create dreams and hallucinations, it may be thought of as a neat biological trick to aid our passing.

For many, life may be cruel, but perhaps it's not entirely malicious.

12 The Healing Touch

There was a faith-healer of Deal
Who said, "Although pain isn't real,
If I sit on this pin,
And it punctures my skin,
I dislike what I fancy I feel."
ANONYMOUS[1]

"Rise up my child, you're cured," the evangelist shouts, and the huge audience cries out in harmony, "Hallelujah." Then on the other side of the stage a man crumpled into his wheelchair suddenly shouts out, "Praise the Lord," and eases himself from the chair he has not left for over a year, takes a few tentative steps toward the preacher, and the crowd roars.

At first glance such dramatic shows seem very impressive; the evangelical healer would claim that God had given him or her the power to repair damaged limbs, to eradicate cancer, and to restore sight to the blind. But how truthful are these claims, and is there any scientific substance to any form of faith or spiritual healing?

Faith healing is one of the oldest mystical arts, and has a place too in orthodox religions. Jesus was supposed to have returned Lazarus to the living, and according to the Bible, the sick could be instantly cured by simply touching the hem of his robe. Throughout the Dark Ages and into the modern era, countless quack doctors and miracle workers have claimed to possess the healing touch, whether it is a form of internal power or because they act merely as conduits for their all-seeing, all-knowing, all-caring gods. The healer is another example of an image that has become so ingrained that it is an archetype. Perhaps this is one of the reasons why even discredited evangelists maintain such huge devoted followings.

Healers, whether they enwrap their practices in religion or use it in its purest form, fall into several categories. The two most important types are those who claim they channel energy from beyond themselves to the patient and those who believe the energy they use comes from within them.

The typical evangelical healer is of the first type and engenders the belief in his followers that his powers come from God. And, in the majority of cases, followers believe the healer because they desperately need to. Naturally, the preacher/healer (who is actually only interested in making money from the hapless members of his audience) emphasizes this need and exploits it to the very limit.

Massive shows such as those put on by the evangelist Morris Cerullo at auditoriums the size of Earls Court or Madison Square Garden are such gargantuan events they generate the near hysteria usually associated with football matches or pop concerts. In this environment of religious fervor and the deep-rooted need to believe, it is easy to see how people become even more open to suggestion, more readily manipulated.

Shows like Cerullo's are carefully choreographed and stage managed. The healer pretends that he knows nothing about the people he "cures," but they have actually been selected before the show by assistants. These helpers pump the chosen ones for information soon after they arrive, and this is passed on to the healer. More often than not, it is those with the least disturbing or grotesque deformities or illnesses who are selected to participate in the show. The truly disabled—the quadriplegics and those with cerebral palsy, are positioned at the back of the auditorium and play no role in the dramatic, photogenic happenings on stage.

Yet, the most insidious aspect of these shows, and a practice that has caused many fatalities, is how some very ill people manage to convince themselves they are no longer ill and injure themselves or make their condition far worse. Crucially for the preacher, these symptoms do not reappear until sometime after the show has ended and everyone has gone home.

In his book, *Healing: A Doctor in Search of a Miracle*, investigator and M.D. William Nolen recounted the case of Helen Sullivan, a fifty-year-old woman with metastasized cancer who was swept up by the excitement of a healing service held by the evangelist Kathryn Kuhlman. To show she had been cured, Mrs. Sullivan took off the brace that had supported her back for months and walked on stage in front of thousands of hysterical supporters.

"At the service, as soon as she [Kathryn Kuhlman] said, 'Someone with cancer is being cured,' I knew she meant me," Mrs. Sullivan later explained. "I could just feel this burning sensation all over my body, and I was convinced the Holy Spirit was at work. I went right up on the stage, and when she asked me about the brace, I just took it right off, though I hadn't had it off for over four months, I had so much back pain. I was sure I was cured. That night I said a prayer of thanksgiving to the Lord and Kathryn Kuhlman and went to bed happier than I'd been in a long time. At four o'clock the next morning I woke up with a horrible pain in my back."[2]

X rays later showed that Mrs. Sullivan had compacted a vertebra which was already weakened by cancer. She died two months later.

The charismatic or evangelical healing circus is the most dramatic and disreputable face of faith healing, and to many serious well-intentioned practitioners it is itself a cancer that generates bad press for them. Healers such as Cerullo, Pat Robertson, and others have legions of critics but still they thrive. After one televangelist, Oral Roberts, demanded that his followers should send him cash lest he perish, car bumper stickers began to appear brandishing the message "LORD," standing for: "Let Oral Roberts Die."

Healing, in all its forms, is very much part of the 1990s ethos, and not surprisingly, it has become increasingly popular in recent years. According to the Federation of Spiritual Healers, there are more than twenty thousand practitioners of some form of faith or alternative healing in Britain, and a recent survey showed that fifty percent of people believe that spiritual or faith healing works.[3] Many celebrities subscribe to a wide-ranging selection of treatments and gurus with very good public relations. One recent newspaper reported that: "At smart dinner parties or girly gatherings, the talk is all of toxins and unblocking, of quests and negativity. Bulging address books reveal aromatherapists and reflexologists, the best place for colonic irrigation or pendulum swinging. Tracking down the latest healer has become a kind of spiritual shopping."[4]

Bill Clinton is said to be a follower of holistic medicine, actresses such as Demi Moore and musicians like ex-Beatle George Harrison have their favorite therapies and treatments, but in the late nineties what was once the preserve of the rich and celebrated has filtered into middle-class circles.

There are a number of reasons why such regimens are becoming so successful. First, people who either have every material benefit given to them or who have struggled to achieve the comforts of modern life still feel that something is missing—an element that will satisfy their spiritual needs. Second, aromatherapy, rebirthing, crystal therapy, or any one of the dozens of modern variants of traditional therapies are easy. They require little effort and can be bought almost off the shelf—so much easier than working out, dieting, or following the teachings of an old-style guru such as George Gurdjieff, who preached that it was "impossible to achieve the aim without suffering." Finally, following the teachings of a health guru absolves the individual from responsibility for their failings. Anything but total fitness and perfect health can be blamed on environmental toxins, the air, the water, stress, but never anything lacking in their own psychological makeup.

When viewed alongside the fads, crazes, and, more insidiously, the hundreds of cases of false cures and actual harm that many evangelical healers are responsible for, it is easy to see how the whole issue provokes extreme emotions both for and against. It is also easy to see how legitimate healers who believe they have a genuine gift (and make far less money than the big-time showmen) would feel aggrieved.

The showmen exploit simple biology and psychology to achieve their temporary wonders. The reason Mrs. Sullivan was able to take off her back brace and felt no pain until much later was because her body was flooded with endorphins. As we saw in Chapter 11, these are produced by the body during times of extreme physical stress and act as pain-blockers. Experiments conducted during the early 1980s showed that the endorphin level of the subject is directly related to what has been dubbed the "placebo effect." This describes the situation where patients feel better merely because they believe in the treatment. A simple experiment shows how it works.

A group of patients are divided into two groups, A and B. Group A are treated with a recognized drug and are told so, group B are given vitamin tablets but told that they have been given the same regimen as group A. It has been found that there is almost no difference in the response to the treatment. But if the patients in group B are told that they are only being given vitamin pills, their recovery rates are far lower.

A recent study has shown the placebo effect reduces pain because the subject releases endorphins and that these can be inhibited by a drug called naloxone.[5] Further experiments involving rats have demonstrated that this release of endorphins is a conditioned response that can be learned and controlled.[6]

If we take Mrs. Sullivan as an example of how the placebo can work, the excitement generated by the healer and the audience triggered the release of high levels of endorphins into the blood. This, coupled with a strong desire for the healing ability to be genuine, was enough to swamp any pain she would have otherwise felt when she removed the brace and left her wheelchair. It was only several hours later, when the endorphin levels had dropped sufficiently, that she would have begun to feel the intense pain produced by a collapsed vertebra. Naturally, few others in the audience of thousands that night ever heard the full story and the agonies the poor woman suffered; for them, Kathryn Kuhlman was a miracle worker.

The energy produced by evangelist shows is real and undeniable, but it is not what the evangelical preachers claim it to be. After attending a performance given by Pastor John Arnott of the Toronto Blessing, a deeply skeptical and cynical journalist wrote recently that such a show was "a profoundly unpleasant experience. I felt I had wandered inside a bizarre mass psychosis, which to resist required an astonishing degree of emotional energy. I felt drained, used up, exhausted."[7]

There are other factors at work in these situations. Fraud and individual conviction play a major role, but timing is also important. All serious diseases follow a nonlinear pattern. That is, within what may be a general decline, the patient suffers bad periods and good periods. Even those suffering fatal illnesses and who die from the affliction eventually go through times of better or worse health. Logically, an individual is most likely to feel the need to visit a faith healer when they are at a particularly low point. Often, faith healers are used as a last resort, when a patient feels they cannot possibly get any worse. If an individual then attends an evangelical performance or visits a private healer, the chances are they will soon enter a less severe spell of illness purely as a consequence of the natural waxing and waning of the disease (or else they will die). But not realizing that this is a natural pattern within the course of the disease, the patient will feel justified in attributing the sudden improvement to the healer, and the healer will of course be quite happy to take the credit.

A different kind of healing is generated by a belief in the holiness or the curative powers of a particular location. Usually these special places are identified with mythical or religious events and hold a powerful attraction for the desperate believer. Such places play the same role as the charismatic healer—they trick the faithful into believing they are cured because the sick acquire a short-term pain inhibition from the excitement of being there or by taking part in a ritual. This effect is particularly powerful when the healing site has strong religious connotations.

The most famous example of a Christian shrine is Lourdes in France. Lourdes became a center for pilgrims after a girl claimed a visitation of the Virgin Mary there in 1858. Within twenty years a shrine had been constructed on the site, and today it has over five million visitors a year. Although tens of thousands claim to have experienced "miracles" at Lourdes each year, there have only been sixty-four cases accepted by the Catholic Church over the past 130 years. These have been passed and accepted as miracles by a specially appointed medical group formed in 1947 called the Medical Bureau. To qualify as a miracle, an incident must conform to a set of tough criteria established by Cardinal Lambertini (later Pope Benedict XIV) in 1758. These state that the disease must be incurable and unresponsive to treatment, it must be at an advanced stage but not at the point where it has resolved itself, and any medication the patient has received must be shown to have already failed. Finally, the cure must be instantaneous and total.

Even the sixty-four approved cases are open to criticism and some claim that as medical knowledge has improved, the cases passed in the early days of the bureau would not now be allowed. It is also telling that there has never been a clear-cut, obvious miracle that was undeniable and without controversy. As the writer Anatole France commented when visiting Lourdes at the end of the nine-teenth century and seeing the discarded crutches of the "cured": "What, what, no wooden legs?"

Lourdes and other shrines energize the wishes of the desperate. They focus the needs and desires of the very sick and can, for a short time perhaps, create the outward signs of a cure—hence the discarded crutches. But just as with the televangelist con show, follow-up investigations into what later happens to pilgrims are rare.

One investigator, a doctor from Southampton, England, Peter May, has spent twenty years carrying out detailed investigations of those who claimed to have been cured at Lourdes. He has come to the conclusion that there is no substance to big-time faith healing, to the televangelists and the world-renowned shrines, and sees the entire phenomenon as a mish-mash of misunderstandings and misinterpretation.

"The fact is that a large proportion of medical conditions get better anyway, treated or not," he said in a recent magazine article. "I've seen supposed cures of sarcoidosis"—a disorder of the lymph nodes—"describing it as a 'rare and potential killer disease'—when we know that eighty percent of cases get better with no treatment. Migraines, backaches, nausea, phobias, and eczema regularly feature on lists of cures, but they all come and go, apparently independently of treatment. In usually irreversible conditions, such as sensorineural deafness, for which healing cures have been claimed, there are cases of spontaneous remission in medical literature."[8]

So, if this is the case for large scale phenomena such as Lourdes and the preacher-healers who are supported by millions around the world, what of the individual healer working alone and dealing with individual patients? Is there any chance that some individuals are truly gifted, or that there are special techniques that science cannot explain that do facilitate genuine cures?

Almost all evangelical healers claim their spurious gifts come from God, that they are mere conduits who channel the energy given to them, but there are also a number of well-intentioned, nonevangelical healers who believe they possess this power.

One of the most famous is Matthew Manning, who as a schoolboy started to become the center of apparent poltergeist activity at his parents' home and at his boarding school. He soon developed an ability for automatic writing and for reproducing works of art in the style of the great masters. Manning has no inherent artistic ability, but the paintings he reproduced were of the highest quality and in a range of styles.

Today he is a professional healer, and claims that he channels the energy he used for automatic writing and painting into curing the ills of others. "It is not me," he has said, "I simply switch on the energy."[9]

Although his abilities are open to the same questions and doubts that confront all faith-healers, his inexplicable artistic skills and his laboratory tested psychokinetic demonstrations seem to show that something incredible (but not necessarily supernatural) is happening within his mind.

The problem with the notion that Matthew Manning or any other claimant is a genuine conduit for a supernatural force is the question of relevance: Why should a spirit, a God, or an alien entity want to assist in these processes? Although science strains to explain how ghosts could be playbacks of past events or how poltergeist activity is probably collective hysteria, the acceptance of immortal spirits, anthropocentric gods, or concerned aliens is simply illogical and based entirely upon an exaggerated sense of self-importance. Even if for a moment we accept that there may be a divine being, even if we allow ourselves to believe in life after death, why should an all-powerful universal being care about the fate of a single organism on Earth?

The Earth is a tiny planet, one among perhaps millions of inhabited worlds in a near-infinite universe. Does it make any sense to suppose that God would be interested in the chilblains of Mrs. Smith at No. 46?

Yet, Matthew Manning has demonstrated amazing powers as an artist and now claims to be applying the same gift to a different end. It is perhaps significant that many skeptics refuse to even mention him, knowing they are unable to dismiss him using present-day scientific understanding. Is ability like Manning's truly a gift from the gods, as some claim, or is it rather an example of the enormous potential of the human brain? We use little more than ten percent of our brain capacity. Could there be hidden regions where exceptional powers of memory, of visualization, of dexterity, would allow an ordinary individual to reproduce a Leonardo perfectly? And could similar abilities explain the claims of faith-healers?

It is because of this question that those healers who believe they are able to focus energies within themselves to cure the ills of others are far more convincing than the evangelical healers. Yet, interestingly, those who practice healing often have no idea where the energy they use comes from.

In many cultures there is a long tradition of employing "internal healing energies." The shaman of African tribes have practiced healing techniques for thousands of years, and the alternative medical practices of acupuncture, crystal therapy, and systems such as Shiatsu all rely upon balancing the internal "energy fields" of the body.

But what are these energy fields? Orthodox medicine has failed totally to locate them, to detect or observe them, yet techniques such as acupuncture are being absorbed into the body of conventional medicine because they have proven so successful.* Is acupuncture really another example of a placebo, because, if we accept that acupuncture works—which is, after all, a technique dependent upon the notion of an undetectable energy system—then where does science and medicine draw the line? If acupuncture, why not crystal therapy, which suggests that certain minerals can be made to resonate with the "human energy field"? And if crystal therapy is acceptable, what should we make of healing by the laying on of hands?

The closest we have come to any form of observing an energy field around animate objects is through the technique of Kirlian photography.

The method of photographing what enthusiasts believe to be a representation of an aura surrounding living things is credited to a

---

*According to a report in The Mail on Sunday in December 1991 there are upward of forty NHS doctors who are also faith-healers.

Russian couple, Semyon and Valentina Kirlian, who produced their first image in 1939. The Kirlian photograph is created by passing a small electric current through living tissue, which produces a brightly colored pattern that appears to be affected by the emotional condition and the general health of the subject.

Enthusiasts suggest that this image represents the energy field of the person, animal, or plant subjected to the electric current. Their originator, Semyon Kirlian, said of the images: "The inner life activities of the human being are written in these light hieroglyphs."[10] The skeptics, however, have another explanation and claim that the excitement Kirlian photographs have generated should be seen as nothing more than a simple misinterpretation of the scientific facts.

The image is created by the electric current ionizing the air surrounding the object, and the shape is determined by its outline. Scientists believe the bright colors and the often dramatic images are greatly enhanced by the presence of moisture. This, they say, is why living objects create strong Kirlian images but inanimate objects do not. They further suggest that the reason the image appears to be influenced by mood is due to the fact that moisture levels close to the skin are subject to emotional changes—clammy hands being a sign of anxiety, for example. Further doubt comes from the fact that if the Kirlian photograph is taken in a vacuum, no image is visible, which implies the brightly colored fringe is an effect produced entirely by the ionization of particles in the air and has nothing to do with the living object itself.

Supporters of the aura theory have countered this with an impressive body of research conducted since the 1930s which has revealed a growing collection of odd effects. The most striking result, and one the skeptics have so far failed to explain fully, came from an experiment first carried out on a leaf. The leaf was photographed using the Kirlian technique and showed the usual glowing outline. Then a section of the leaf was cut away and the remaining section rephotographed. To the astonishment of the experimenters, the Kirlian image mirrored the full outline of the uncut leaf as if the missing part were still attached.

This and other experiments have led researchers of the technique to suggest the existence of what they have dubbed a *bioplasmic field* and the notion of *bioenergy*—new names for the traditional ideas of the "energy field of the body" and the energy that many healers believe to be responsible for their abilities.

As we saw in Chapter 7, the mind seems capable of controlling the body to a surprising degree, and this ability can be honed and developed in appropriate individuals. In the same way, it may be possible that a talented, highly trained healer could instruct the body to repair itself or defend itself against disease by suggestion—amplifying the wishes of the sick individual. This may take the form of hypnosis in which the patient and the healer each play a role in creating a barrier against anything from an allergy to a cancerous tumor.

In one sense this is not a supernatural phenomenon but one utilizing extreme powers of the mind and body and exploiting metabolic self-control. It might seem supernatural because it is far from commonplace, but it does not require occult agents or even religious observance or devotion.

During the 1950s psychiatrist Stephen Black conducted an impressive series of experiments to demonstrate the power of hypnosis in healing and disease prevention. He took a group of subjects who were susceptible to hypnotic suggestion and also suffered common allergies. In traditional fashion, he divided his subjects into two groups. The first were hypnotized and the material to which they were allergic was injected just beneath the skin. This allergen very quickly produced a characteristic red "weal and flare" which increased in intensity over a period of twenty minutes.

Black then repeated the test with the second group but induced hypnosis, with the suggestion that they would not react to the allergen when it was introduced. The material was injected under the skin, and none of the hypnotized subjects showed any sign of an allergic response. Black then conducted biopsies on both sets of subjects and discovered that in the second group, the chemicals responsible for the lesions had been blocked.[11]

The placebo effect upon which most skeptics rely could not have played a role in this set of experiments because the hypnotized subjects were aware of the process. However, the fact that they were told they would not respond to the allergen could have been the trigger for the release of the all-important endorphins that seem to be the principle agent behind the placebo effect.

The growing field of psychoneuroimmunology (PNI) is beginning to throw up new and exciting variants on this theme. PNI attempts to explain how mood and emotions can influence the body's defenses and responses to disease. Until recently it was believed that the body defended itself against viruses and bacteria independently of the brain—that specific regions under attack responded autonomously by triggering the appropriate biochemical agents. Now, neurologists are changing their minds and coming to accept a role for peripheral nerves that link the thymus gland, the lymph nodes, and bone marrow, which produces white blood cells (lymphocytes)—the body's army, navy, and air force.

During times of stress or depression, the immune system is suppressed by a group of biochemicals called *cortiocosteroid hormones*; by suppressing the action of the lymphocytes, these chemicals weaken our natural defenses. Conversely, at times of extreme happiness or euphoria we produce a chemical called *interferon* (which is now gaining use by doctors as a major cancer-suppressing material), which assists the immune system to defend against attack.

The release of these chemicals can be triggered by a variety of sources. If we consider only the "positive" biochemicals (those that enhance health), endorphins are released in a rush over a short time frame by transitory excitement—a response to strenuous physical activity or the buzz created at an evangelical meeting. These chemicals get us through emergencies and give us an extra boost, and it is because of this that the televangelist is a danger, as the unfortunate Mrs. Sullivan discovered.

More useful in the long term are those agents released slowly and over longer time periods. These could be encouraged by a caring healer, by a belief in a technique such as acupuncture, aromatherapy, or colonic irrigation. Such techniques provide a more controlled, balanced, and easily assimilated neurological response. Surprisingly, perhaps, pets have been shown to be another source of comfort for the ill—they relieve stress and offer what one psychologist calls "a form of unconditional love that assists the immune system."

So it would seem that almost all aspects of faith healing may either be dismissed as dangerous exploitation or explained by modern neurophysiology. The placebo effect is a term that is overused and often inappropriately employed, but there is a growing body of evidence to show that the brain does play a very active role in bodily health, and with training, this ability could be gainfully exploited. However, there remains one form of faith healing that has so far defied scientific explanation and has become another topic conveniently ignored by skeptics.

One day in 1993, when racehorse breeder Jan Piper discovered a lump over the eye of her favorite breeding mare, Jessica, she was immediately concerned and feared the worst. The vet arrived later that morning and soon found several more lumps in the horse's throat and groin. A week later Jan Piper had the diagnosis. The horse was suffering from *equine lymphosarcoma*, a rare form of cancer that effects the white blood cells before attacking the internal organs. It is fatal and quite incurable.

In despair, and on the eve of Jessica being put down, the breeder heard about Charles Siddle, an animal healer. Despite the skepticism of her husband, Jan Piper called Siddle, and the next evening he arrived at the stable. He was there for less than half an hour. After placing a large crystal on the floor of the stall, he laid his hands on Jessica and ran his palms over the horse's coat, along her mane, and across her eyes. "She'll probably be tired tomorrow," he said quietly, "but after that she'll be fine." Then he left.

Next morning, Jessica started eating for the first time in weeks, and within days she was completely cured. She now has a foal and is still in the best of health.

Animal faith healing is a true mystery. Because it involves animals and not free-thinking humans, skeptics cannot accuse believers of wishful thinking or even the placebo effect. Some doctors suggest that apparent miracle cures such as Jessica's are an example of spontaneous remission, and indeed, cases of sudden natural cures in humans and other animals are well documented, if rare.

The British Veterinary Association refuses to acknowledge the work of animal faith-healers, but the prestigious Royal College of Veterinary Surgery has recently allowed them to work with animals in the presence of a qualified vet. But, frustratingly for the scientific investigator, the healers themselves talk in the same vague terms as those who deal with human patients. Like them, they seem to have little idea how they do what they do. "I don't really know what happens when I heal," one said during a recent television documentary on the subject. "The energy comes through my hands, so I guess I act like a channel."

It has been suggested that the healing process for any animal is the same—a matter of "regaining the aura" or readjusting the energy field. The problem with this attempt at an explanation is that enthusiasts are again trying to solve a mystery using a phenomenon that itself has no scientific support. It is as bad as suggesting that ghosts are definitely the spirits of the dead—faith healing is a mystery, and so is the concept of the energy field or the aura.

Yet, Charles Siddle continues to have remarkable success with the animals he treats. One prize racehorse he attended had a ligament so badly damaged the horse could hardly walk and the lower leg was severely swollen. After one visit, the swelling had almost gone and the horse was cantering around the paddock as if it had never been injured. This certainly cannot be explained by spontaneous and natural remission.

There is other evidence to suggest the placebo effect is not the only mechanism by which an individual can control symptoms or keep disease at bay. In Canada, a healer named Oskar Estebany and a researcher, Dr. Bernard Grad, working at McGill University in Montreal have found that mice are susceptible to what appears to be Estebany's healing powers. Three groups of sixteen mice each had a small patch of skin removed. The first group were subjected to Estebany's healing touch, the second group were warmed (to rule out the possibility that it was heat from the healers hand that affected them), and the third group were ignored. The first group showed significantly faster recovery rates than the other two.

In a similar experiment, this time using human guinea pigs, a team based at the JFK University of California appear to refute completely the idea that all unorthodox healing is simply placebo. A piece of skin was removed from the shoulders of forty-four subjects who were told they were involved in an orthodox medical experiment. Each day the subjects were asked to place their wound in a window in the belief they were being photographed. Some of the volunteers were actually being treated by a healer, others were not. The wounds of those singled out for psychic treatment recovered far faster than those who were not.

Some scientists have taken the investigation of healing powers even further and conducted experiments on plants. Dr. Tony Scofield and Dr. David Hodges, working at London University, collaborated with healer Geoff Boltwood. The experimenters took a batch of watercress seeds and treated them with saltwater to "make them sick." Then, following the usual pattern, the sample was divided into two groups. Boltwood held half the seeds in each hand. The first group he tried to heal by concentrating his abilities and directing his energies toward them, the second group he held but ignored. After six repeats of the experiment conducted under scrupulously controlled conditions, the team were surprised to discover that in five of the tests the seeds subjected to Boltwood's efforts grew at almost double the rate of the others.

It could be that the researchers missed some crucial element in these experiments, and their results have been vilified in the scientific press, with some skeptics just stopping short of denouncing them as frauds. But theirs is not an isolated experiment. Teams in the United States and Europe have shown similar exciting results.

How any of these nonplacebo effects could work remains a mystery. Enthusiasts of the phenomenon of psychic healing see them as clear evidence that the healers are either able to channel external energies or are in some way capable of using their own energy fields to correct those of sick people, animals, or even plants. The skeptic points to the valid fact that these are tiny trials, and despite the best efforts of the experimenters, errors could have been made, or unknown extraneous but quite natural factors could lie at the root of the effect.

Like telepathy, precognition, and many of the other topics covered in this book, healing is another borderline activity. And so it seems to be with almost all aspects of the paranormal. Alien vehicles have not landed in Parliament Square, the sick do not grow severed limbs spontaneously, and no one, not even self-publicists like Uri Geller, has managed to give a clear undeniable demonstration of paranormal abilities. This is very disappointing but as we will see in some of the following chapters, it should not lead us directly to the conclusion that all occult matters are bunkum. Much of it certainly is, but not all. As the astronomer Sir Martin Rees has said: "Absence of evidence is not evidence of absence."

⑬ Swept Off Their Feet

> When a man no longer believes in God, he'll believe in anything.
>
> CHESTERTON

One warm evening in October 1957, a twenty-three-year-old farm laborer named Antonio Villas-Boas, living near the small town of São Francisco de Sales in Brazil, was, he claimed, snatched by a group of strange beings and dragged kicking and screaming into their craft hovering nearby. The hapless young man was then laid on a table in a sealed room, smeared with a thick jellylike substance, and had a blood sample taken. But then it seems his luck changed. Minutes later he was visited by a nude female humanoid alien who forced him to have sex with her twice. During this process she apparently yelped like a dog, and bit his chin before being led away by her colleagues. As she left the room, she pointed to her belly and then at the stars, and was gone.

Villas-Boas was examined by a doctor a few months later, when the young man had finally plucked up the courage to tell his tale and the full account came out. The doctor found small scars under the man's chin and on other parts of his body and concluded that these marks were similar to radiation burns.

Although it sounds ridiculously farfetched, this account, the first widely publicized abduction case in the world, is actually quite typical and has been repeated with slight variations ever since. In fact, reports of alien abductions are on the increase, and one poll has stated that during the past fifty years in the United States alone, five million people could have been abducted.

It might sound flippant to call this case "quite typical," but it is not really, because claims of abduction fit into a very limited pattern. In fact, investigators have established five distinct criteria for abduction reports. These are: reporting seeing lights in the sky, missing time, flying through the air without any apparent means of support, paralysis in the presence of strange beings, and later, after the physical ordeal has ended, finding strange marks on the body or small, seemingly alien objects located inside the body.

Another famous case will serve to illustrate many of these classic elements.

One night in late September 1961, an interracial couple, Betty and Barney Hill, were driving through the White Mountains of New Hampshire when they spotted bright lights in the sky above the car. They stopped, and an object—"pancake in shape, ringed with windows in the front through which we could see bright blue lights"—hovered over the road. The next thing they knew, they were traveling along the road again and noticed that two hours were unaccounted for. Traveling at forty to fifty mph, the 190-mile journey to their home should have taken five hours, but it had taken them seven.

The couple said nothing, but each started to experience weird dreams in which they were captive aboard a strange craft. It was only after strains in their marriage led them to get psychiatric help and they underwent hypnosis that they discovered what has since been claimed to be repressed memories of the experience. According to their accounts, they were taken aboard a craft and medically examined, poked and prodded by bald-headed beings with huge, lidless eyes. A large needle was pushed into Betty's abdomen, a process she later took to be a pregnancy test, and she found herself discussing "great matters of the universe" with the captain of the craft. Barney had a similar account, which came out during individual hypnosis sessions during 1963.

To this day, Betty Hill (now a widow) sticks by her story, and it has become the most famous account in UFO lore, probably because it was also the first publicized case originating in America.

It is interesting to note the marked similarities between this case and that of Antonio Villas-Boas and to link it with the countless thousands of documented cases over the years. UFO enthusiasts would point to this consistency as some form of verification that the experiences of these people are examples of genuine abductions by aliens from another world (what are called "close encounters of the fourth kind"). Skeptics would say the uncanny similarity is actually a point against the possibility, that rather than verifying these extraordinary claims, they demonstrate the extreme gullibility and straightforward lack of imagination of those who report them.

Perhaps one of the most intriguing aspects of the Hills' case is that it was a "multiple-abduction scenario." In other words, more than one person was abducted and their stories corroborated. There have been many such cases where either several people were abducted or the abduction was seen by a group of witnesses.

A case almost as famous was reported in 1975 by one Travis Walton, who was snatched by a UFO while driving home to Snow-flake, Arizona, with six colleagues, all of whom had been working in a nearby forest. They too saw lights in the sky. Walton got out of the car to investigate, disappeared, and was found five days later delirious and half naked in the forest close to where the abduction occurred. His account has a familiar ring to it: "I was lying on a table . . ." I saw several strange creatures standing over me. I became completely hysterical and flipped out. I knocked them away, but I felt so weak I collapsed. They forced me back on the table, placed a mask over my face, and I blacked out."

So, what should be made of these accounts and thousands of similar stories?

For the followers of paranormal phenomena, alien abduction really is the "Big One." As we approach the end of the millennium, there are no half measures when it comes to alien abduction: either you believe that the experiences of many people are delusions, or explicable using ideas from psychology and neurophysiology—or else it is all terrifyingly real and part of a massive conspiracy.

The enthusiasts claim that advanced alien beings are visiting our world in large numbers and with the full knowledge of many western governments. Furthermore, they believe state agencies are working with some groups of aliens and against others. In short, we are engaged in some form of galactic intrigue that, to the nonbeliever, sounds reminiscent of a badly written and very dated science fiction novel. The purpose of many of the alien abductions, it is claimed, is to conduct ongoing genetic experiments. Some of the aliens are keen to breed alien-human hybrids and to this end they are "tagging" humans in order to find the most fruitful genetic crossovers. Apparently they have no qualms about what they are doing.

Of course, this theory "explains" the similarity in accounts. People are invariably snatched at night before being medically examined, and there is often a "sexual aspect" to their stories. In order for this scheme to work, say the believers, the governments of the industrialized nations, in particular—the U.S. and Russia, as well as British and other European states—are all involved with the conspiracy. In other words, at some senior level, government agencies are aware of what is happening and are party to the political network connecting what are thought to be at least three or four different alien races.

This is an entertaining and interesting story—the backbone of *The X-Files* television series. One of the recurring threads and the original intrigue of the program is that Mulder and Scully are caught up in a web of information and disinformation, and an alien intelligence is intimately linked with human agencies to suppress knowledge of breeding experiments.

Is there any way in which this tale could be true? Is it possible that we are not only being visited by alien beings, but that the earth is a key interchange for all the experimenters and political schemers of the galaxy?

First, we have to accept that there is almost certainly life on other planets. In fact, the galaxy is likely to be teeming with life of all types, and there are probably alien civilizations more advanced than us living happily and unhappily all over this galaxy and beyond. Second, it is extremely difficult to travel interstellar distances, to cover the vast expanse of space between stars (see Chapter 1). But that does not mean it is impossible. It is currently impossible for *us* and will remain so for the foreseeable future unless some stunning revolutionary set of ideas is stumbled upon and used properly. However, we live on a tiny speck of dust in an almost infinite universe. This universe began some fifteen billion years ago, and life on this planet started about four billion years ago. The series of steps that led from the first bacteria to you reading these words was staggeringly complex and dependent on an almost infinite number of events going right. One of these steps is: When did life begin? Another is: How quickly did we evolve? If either of just these two has a different answer for another world, it would make a huge difference to their present level of advancement.

On Earth, the dominant species is just about capable of reaching nearby planets. As we saw in the first chapter, on another world where evolution acted faster or was set in motion earlier, there could live alien life-forms we would view as "superbeings," aliens who could travel between the stars using one of the loopholes in the theory of relativity or perhaps some process we cannot even begin to imagine.

So, the idea that we are being visited by aliens is not beyond the realms of reason. In fact, it is quite possible that at some time in our history a craft built by an advanced race could have passed this way. It is even possible that we are still being visited from time to time. But as I will try to show in this chapter, the idea that we are being visited by hoards of aliens, that millions of humans have been snatched for genetic experiments, and that there is some vast cover-up, is pure science fiction, the result of overactive imaginations. It does not make logical sense, and rather than being the start of an alien invasion, it is actually, I believe, a result of very human and, sadly, rather mundane events.

Alien abduction is a huge phenomenon in the United States. Although America cannot lay claim to the first publicized case (that of Villas-Boas), UFO lore has a rich and fertile past in the United States. The term "flying saucer" was first coined by a USAF pilot, Kenneth Arnold, flying over Washington State near the Canadian border in 1947. And a few days later several thousand miles south, the *Beowulf* of UFO legend took place—the Roswell incident, a case I will discuss in detail later in this chapter.

American culture has always encouraged open exchange of ideas and views—freedom of speech lies at the heart of the American dream, the moral structure of the nation; so it is perhaps inevitable that the wilder limits of human imagination find fertile ground there. The other two countries with high numbers of cases is England and Brazil. The fact that England has bought into the phenomenon will also surprise few. Enthusiasts would argue that the U.S. and U.K. are obvious sites of alien activity because, along with Russia and China, they are the most important military powers in the world.

Until recently, the former Soviet bloc has been as secretive about interest in the paranormal as it has about military matters, and the authorities in China are so secretive about anything that may be linked to military matters that we can hope to learn little from there just yet. Skeptics would also argue that the U.K. often follows trends started in America, and that Brazilians are particularly keen to report imagined or deliberately made-up cases because Villas-Boas has set a precedent.

America has also produced the vast majority of experts in the field of alien abduction, and not all of these are fringe figures. They include university professors and established intellectuals. Best known of these is the abduction investigator, John Mack, a Pulitzer prizewinner and a professor of psychiatry at Harvard University.

Mack believes that the human race is the subject of a huge breeding program, and that its perpetrators intend to produce a race of alien-human hybrids. "A huge, strange interspecies or interbeing breeding program has invaded our physical reality and is affecting the lives of hundreds of thousands if not millions of people," he has written.[1]

However, he seems a little confused about the precise nature of these alien experimenters and has said that they "penetrate and enter the physical world, and to that extent they're a little different from spirit entities."[2]

Instead of being a fully paid-up member of what has been called a Hidden College of UFO researchers (important academics and intellectuals who believe but keep their heads down), Mack is an evangelical, proselytizing believer who goes out of his way to publicize his beliefs and appears to take the wrath of his superiors and the anxieties of his friends in his stride. He has written widely on the subject of UFOs and lectures around the world. His most important contribution so far is a book called *Abduction: Human Encounters with Aliens*, in which he models his elaborate theory about abduction based upon the testimony of over a hundred abductees (or "experiencers," as they prefer to be called) who have found their way to his couch during his many years of research in the field.

John Mack's academic credentials are impeccable, but his past reveals him to be a man who desperately wants to find alternative paths through life rather than following any form of orthodoxy. Now in his late sixties, he has been in and out of every possible form of alternative research and experimentation. He has conducted psychology experiments with psychedelic drugs, became involved with EST—a form of alternative "therapy" widely discredited in the late 1970s—and traveled to India to find enlightenment in eastern religion and philosophy.

There is nothing wrong with any of this, but to some skeptics it does imply that Mack is not merely open-minded, but has left the door to his mind flapping wildly in the breeze—a state which many scientists believe to be the wrong approach to a subject like alien abduction. Mack and many other enthusiastic believers around the globe would of course argue the exact opposite—that the problem with scientific investigators is that they approach fields like this with closed minds and are therefore unqualified to deal with something so far outside their normal mode of thinking.

To study those who claim to be experiencers, Mack uses the technique of *regressive hypnosis* or *hypnotic regression*. For many researchers this is the most invaluable tool in the armory of the investigator. It involves relaxing the subject and putting them into what some psychologists call a *nonordinary state* in which they can remember experiences they have suppressed and which they cannot recall in an "ordinary state." It is used by some researchers of the paranormal to open up what they claim to be experiences individuals have had in previous lives.

Regressive hypnosis may be a useful tool for the investigator, but it has its pitfalls. It is highly susceptible to deliberate fakery, but most important, it can be used to plant ideas in the minds of subjects and, in extreme cases, to create what has recently been dubbed "false memory syndrome."

False memory syndrome has made headlines recently because of the suggestion that some claims of child abuse investigated by psychologists are actually fictional ideas planted by the investigator and expanded upon by the distraught subject. In some accounts these false memories are incredibly vivid, and the result of these investigations has meant shattered families and even police investigations. It has been suggested, not without reason, that the use of regressive hypnosis could also explain many cases of so-called alien abduction.

Naturally, researchers like John Mack strongly deny this claim. They may not be planting the ideas deliberately, and from all accounts, Mack genuinely believes in what he is doing and is deeply committed to the idea that alien abduction is not a human invention. But the similarity between alien abduction experiences as described by investigators using this technique are uncannily similar to the overview created by adults who have been regressed to their childhood and unearthed what they believed to be real events but were later proven to be false memories.

At one point in his book about alien abduction, Mack writes: "The undoing of denial is effected by having the abductee stare into the 'engulfing, searching eyes' of the alien. This will make them real and remove once and for all the denial that has operated as a psychological defense." With this, Mack found there is "a shift in their relationship to the alien being."[3]

Recently there appears to have been a reversal in the interest shown in regressive hypnosis, and within the orthodox psychiatric community it is viewed as questionable, and possibly damaging. Although it continues to be used enthusiastically by psychologists like Mack, who work on the fringe, it is losing credibility in the mainstream. Psychiatrist in chief at Boston's Beth Israel Hospital, Professor Fred Frankel, has said of the technique: "Hypnosis helps you regain memories that you would not have otherwise recalled . . . But some will be true, and some will be false. The expectation of the hypnotist and the expectation of the person who is going to be hypnotized can influence the result."[4]

A huge dent in the credibility of Mack's experiments with experiencers as well as the entire practice of regressive hypnosis was made by a forty-one-year-old Boston-based writer, Donna Bassett. Hearing that Mack was "strip-mining the emotional lives of distraught people and failing to help them with follow-up therapy,"[5] Bassett decided to pose as an abductee to test out Mack's techniques and to question the entire phenomenon.

She read everything she could find on the subject of UFOs took method-acting lessons to produce convincing testimony in a faked hypnotic state, and gradually ingratiated herself with Mack. She told him she had been visited by "little people" whom she called "angels from God," and described how she had been experiencing weird events since childhood—balls of fire hovering above her parents' house at night, the healing touch of an alien who visited her after she had burned her hands. Mack spent hours with Bassett and became so convinced of her story he even used her account of an abduction in his book.

Other American investigators are just as enthusiastic. David Jacobs, an historian of modern American history at Temple University, is almost as well known as Mack and suggests ideas every bit as radical. In a recent magazine article, he was asked why it was that during over fifty years of investigation there has never been a clear-cut abduction incident in daylight with reliable witnesses, complete corroboration, and without underlying psychological abnormalities linked to any of the participants. He retorted: "The secret aspect of the phenomenon is remarkably efficient and extraordinarily effective. The way in which the alien program is instituted mitigates against having a lot of cases from the same day. And so does the way in which we find out about the cases. Most people who have had abduction experiences don't really know what has happened to them. They might know that an odd thing has happened here and there, but linking it to a UFO abduction is not something most of them would probably do. So of all the abductees out there, we only hear from about 0.001 percent of them. But every once in a while we'll have a case where somebody who is an abductee will come up to another person and say, 'I know you. I've seen you before.' And they will trace it back to an abduction event they have shared."[6]

Jacobs is particularly interested in multiple abductions. There are a surprising number of these; one estimate suggests that about one-quarter of all cases involve more than one individual, and there have been reported incidents involving up to eight people being abducted simultaneously. But these run into all the same pitfalls as individual cases—questionable techniques applied to extract the story, and clear inconsistencies that are surprisingly overlooked by investigators.

A good example is the testimony of an anonymous man who was supposedly abducted along with his entire family. One of the things he remembered about the incident was lining up in a hall with dozens of others who had been abducted around the same time before being led into an investigation or interrogation room by his alien hosts. While standing in line, he noticed that a man in front of him had a mole on his left shoulder. To Jacobs, this was a convincing detail that added weight to the man's story. To a dispassionate outsider, it seems ridiculous that anyone would notice such a thing as they were about to be examined by an alien, let alone in what experiencers happily relate as being an otherworldly or "trippy" state.

Another aspect of reports often ignored by believers is what independent-thinking investigators call "contamination." Frequently in multiple abduction cases, those involved are all interested in UFOs, have read the literature, and invariably behave in a highly competitive manner with one another. Psychologists investigating many of these cases believe that quite often one of the individuals will suggest to the others that something has happened when it hasn't, and so the idea spreads among them. Peer pressure and simple wishful thinking eventually leads them to actually believe a false memory themselves and to let this come out in whichever way is most suitable—usually by having the story teased out under hypnosis.

So, if in fact humans are not being abducted by aliens, then what do we make of the swelling flood of reports coming in not just from the United States and Europe, but from all over the world? By applying ideas from psychology, neuroscience, as well as a sprinkling of common sense and logic, there may be some answers.

First, the commonsense part—some statistics. I mentioned earlier in this chapter a survey finding that as many as five million Americans might have been abducted by aliens. This survey was actually carried out on 6,000 people during 1991. Of these, 119 people fitted the stereotype of a classic abduction scenario. When this figure is extrapolated up to the population of America, we arrive at a figure of five million abductions. Strictly speaking, this in itself is not really a reliable figure because extrapolating from 6,000 to some 250 million (the population of the United States) is not statistically valid. But let us ignore this and look at a more significant absurdity offered by these figures.

Taking the figure of five million American abductees, investigator Robert Durant reached an interesting conclusion. He assumed that the average abductee has been taken ten times—some have claimed many more abductions, some have had only one or two experiences. If five million individuals have been abducted an average of ten times during the past fifty years, that constitutes one million abductions per year, or 2,740 per day in the United States. Durant went on to assume that each abduction required an average of six aliens, based upon the growing abduction lore and data gathered by enthusiasts, and that this team can carry out twelve abductions per day. This means that each day there are 288 teams of aliens operating, constituting a total of 1,370 alien visitors.

Perhaps more startling than these figures is Durant's conclusion, which inadvertently reveals more about the sort of sloppy thinking of UFOlogists than any statistics could alone. Starting his investigation, he claimed, "very skeptically," Durant concluded that 1,370 alien visitors per day was actually not too absurd a number. He concluded that even with backup and support, the figure of 1,370 aliens would only need to be increased to, say, 5,000; a figure, he pointed out, that is about equal to a single aircraft carrier crew. But where he fell down in his argument was with the simple fact that he had considered only American statistics. In a breathtaking display of parochialism, he ignored the other 95 percent of the world's population. This would mean that to account for the figures presented by the community of UFO enthusiasts themselves, there needs to be in excess of one hundred thousand aliens on Earth engaged upon missions to abduct humans, all undercover, all going about their business without a single convincing verifiable piece of evidence for a period of at least fifty years.

But it gets worse. If there have been five million abductees during the past fifty years in America, there must have been 200 million abductees during that time throughout the world. This is about one in twenty people.

These statistics offer just one argument against the entire notion of alien abduction. A counter argument could be that the vast majority of incidents are accountable by some other means—hallucination, hysterical reaction, hypnotic suggestion—so that perhaps only one percent of reported cases are genuine alien abductions. This admittedly lowers the alien population on earth to a believable 1,000 and the number of abductees worldwide to just two million. But there are other logical arguments used by skeptics against the entire abduction lore.

It is impossible to convincingly explain each and every case of abduction, but, the skeptics argue, this is actually quite unnecessary. The concept of alien abduction as it is reported by thousands of experiencers is not logical—the overall story simply does not work, it is internally inconsistent.

Imagine human explorers visiting a remote part of the Amazon jungle. During their travels they discover a new, as-yet-unheard-of species of animal or plant. Within six months of the discovery, the BBC would have made a documentary about the creature, complete with David Attenborough wandering through the undergrowth. Within a year, books about the new species would be in the shops and there would be CD-ROMs available detailing every aspect of the creature's anatomy, behavior, and habits.

Now, when we are talking about alien visitors, it is easy to forget just how technologically advanced these beings must be to have traveled here in the first place. Any scientist will admit that we cannot hope to travel interstellar distances at practical speeds in the foreseeable future, and until we can crack the light-speed problem, a journey even to the nearest star system would take thousands of years. With this in mind, it is clear that any alien visitors must possess technology thousands of years in advance of ours. In which case, why would they possibly need to spend at least fifty years snatching a total of perhaps millions of humans in order to learn about our species?

Considered from this perspective, it is clear that the very notion of alien abduction is an example of human vanity at its worst. We imprint upon an alien race our own technological limitations and lack of imagination, and at the same time imply that we are so special and important as a race that, first, they are drawn here, and second, we are so endlessly fascinating and complex that they can't stop snatching us to find out how we tick.

Enthusiasts would argue that this misses the point, that alien abductors are not merely finding out about our anatomy and behavior, as we would a new species of animal from the Amazon jungle, but are engaging in complex genetic research.

This is again indicative of human egocentricity and a lamentable lack of imagination. Genetic engineering is an exciting new field of research and has sparked the public's imagination, but to an advanced race that could travel across the galaxy, it would be ancient knowledge. If they had advanced in the areas of physics needed to accomplish interstellar travel, then they would almost certainly be equally advanced in the field of genetics. We might need to abduct creatures, dissect them, stick metal probes into them, and conduct simple genetic experiments using sperm and eggs, but is it reasonable to assume that such an advanced people would need to do this?

Even assuming they would be interested in experimenting with less advanced life-forms, aliens who can travel across interstellar or even intergalactic distances would be able to extract materials from other beings without the need to resort to anything so crude as physical abduction and internal examinations.

Part of abduction lore insists that there are many cases in which women have been abducted, impregnated, and then found that the fetus they were carrying has disappeared. None of these cases have been proven, and it is a very simple matter to explain them as phantom pregnancies, or to show a mitigating psychological disorder in the experiencer. But even if we assume for the moment that this really happens, that aliens somehow dematerialize a fetus and transport it to their craft in *Star Trek* fashion, would not those same aliens be capable of creating the hybrid pregnancy without recourse to abduction and impregnation?

Some skeptics have also pointed out that the entire idea of abduction is nonsense because advanced civilizations would be advanced morally as well as technologically and so would consider it unethical to interfere with other less advanced civilizations and abduct innocent, unwilling individuals. This may be true, but it is by no means a certainty. Although we are far more advanced technologically than people of, say, the fifteenth century, are we any more developed ethically or morally? We may kill each other in different ways today, but we still kill. Skilled individuals may steal using the Internet rather than pickpocketing, but it is still theft; the rich western nations may exploit their poorer neighbors in a clandestine manner, but it is still a form of imperialism.

However, the matter of whether or not an advanced alien race is morally any better than us and would or would not stoop to abduction is actually irrelevant. Far more significant is our total insignificance as a species and our unremarkable position in the galaxy. The Earth is a tiny planet, lost in an ocean of stars and other worlds. Why would an alien race want to come here? Furthermore, why would they want to bother with us? What could an advanced race possibly gain by interbreeding with humankind? Perhaps there is a plentiful supply of dilithium crystals somewhere on earth.

Putting even these logical arguments aside, the fact that large numbers of seemingly sane individuals are experiencing something strange cannot be ignored easily. If not alien abduction, what then could they be experiencing? If they are not part of some massive global conspiracy, and millions of individuals who appear to be caught up in the web are not in fact smuggled aboard alien vehicles, what are they living through?

As we have seen, stories related by experiencers follow a very clear template, but there are also some common threads which, to dispassionate psychologists, indicate obvious psychological and neurological patterns.

First there is the question of the emotional and sociological profile of abduction victims. According to the enthusiasts, cases encompass all socioeconomic classes, professions, and intellects. But they have to admit that there are a significantly greater number of female abductees. This is an interesting aberration because, in general, men are more likely to be out at night alone and therefore more vulnerable than women (it is important to remember that both male and female abductees are equally powerless in the face of alien interference, so physical strength is irrelevant). This must mean that statistically there is an even greater slant toward women believing they have been abducted.

Another interesting aspect of abductions in relation to gender is that reports often have a strong sexual element, with abductees examined and sexually abused. In many cases the reports of experiences have a marked similarity to the accounts of rape victims, and even the most enthusiastic researchers are keen to highlight the fact that abductees often feel abused and violated, and that they did not want to recall the experience under hypnosis, but feel compelled to pour out what they had lived through.

The fact that experiencers are resistant to telling their stories even under hypnosis should not be surprising, but it certainly does not validate their supposed encounters with what they believe to be alien beings. First, there is a social resistance to tell such tales. It may be that a few abductees have deliberately concocted their stories to make money or to gain their fifteen minutes of fame, but even they have to create a pretense of being resistant to the fact that they were snatched and manipulated by extraterrestrials. Taking this further, if experiencers show resistance to their memories beyond the fear of ridicule, this is quite understandable because, if the experience was created in their own minds, they will have produced a barrier against the memory as a form of self-protection. If the false memory they have nurtured is painful, which almost all abductions appear to be, then they will try to submerge this false memory and will have to be encouraged to speak about it.

Another possible explanation for the abduction phenomenon is that the individuals who come forward are merely satisfying a deep-rooted egomania or overcoming a powerful sense of inferiority. What more exciting and special experience could anyone ever have? To be abducted, no matter how unpleasant the experience, indicates that an individual is important, special, a notch above everyone else. Often, experiencers recount how they met the commander of the interstellar vessel they were taken to and engaged in a detailed discussion with the crew about the propulsion system of the craft, about the ecological dangers facing humankind, or, on occasion, "the meaning of the universe." Remember, these are invariably ordinary citizens who have little knowledge of philosophy let alone high-powered physics. Is it reasonable that such people would be conversant with such things when they had just met an individual from another star?

There is also an element of oneupmanship within the UFO community, a clash of egos, so that experiencers are now trying to outdo each other. In *The Alien Jigsaw*, American Katherine Wilson describes her 119 abduction experiences spanning twenty-six years. She tells of encounters with several different groups of aliens, about out-of-body experiences and time travel. She describes being privy to a meeting between Bill Clinton and George Bush shortly before the 1992 U.S. election, and how an alien ambassador negotiated a smooth transfer of power and "arranged" the election, a tale which in itself weakens her claims—after all, why should aliens be that interested in the politics of a single country?

To rival these fantasies, there is the marvelous tale of Peter Gregory, who claims to have been taken on board an alien craft and given a lecture on the perils of ecological mismanagement on Earth before meeting Oona, an adult alien-human hybrid with whom he appears to have fallen madly in love. In Gregory's tale, he describes the control room of the alien craft including a "viewing screen" and a control consul with an array of flashing lights, à la *Star Trek*, circa 1966.

But these are all outstripped by the testimony of an anonymous twenty-six-year-old businessman from West Virginia who claims to have been abducted no fewer than fifteen hundred times.

A further explanation for the huge increase in the numbers of apparent abductions is linked with the social climate of the late twentieth century. It is clear that we are moving into a godless age, particularly outside the U.S. and the third world. Church attendance in Britain is less than ten percent what it was only fifty years ago, and even in the far more religiously inclined United States many people who cannot find a faith system based upon orthodox religion are turning to alternatives. The past two decades have seen a staggering rise in cult religions in the U.S. and elsewhere. These range from suicidal groups who believe they are being contacted by aliens to those who follow eastern religions and practice alternative health regimes and obscure meditative techniques. For such people, UFO lore and abduction is merely another tendril of a many-headed creature that offers an alternative universal view to orthodox religion— a different way to cope with existential fears and the corrosive angst many of us feel at one time or another.

Alien abduction offers a great deal to people who cannot accept institutionalized religion. It is fresh and different, it displaces the conventional "Godlike" figure with something tangible. It offers hope, excitement; it draws in the experiencer, drags them into a bigger world in which they have played a significant role. Finally, it is more manageable than orthodox religion—after all, Christianity has been unable to offer a spectacular show for at least two thousand years, and the faithful are expected to take their cue from a very fallible clergy who are losing credibility and offer a doctrine based upon an ancient text facing its own credibility crisis.

Journalist Tom Hodgkinson puts the matter succinctly: "The UFO offers a satisfying blend of techno-futurism, religion, and spiritual quest which is personally motivated and does not require a commitment to an externally imposed set of social rules. In a reason-based society, it is almost easier to believe in aliens poised to descend and save the earth at any moment, than it is to believe in God."[7]

In other words, "aliens are easy." It is a simple matter to transfer one's anxieties, fantasies, fears, and hopes onto an alien race and visiting flying saucers. Whatever your problem, you can always say: "Aliens made me do it." There are some who have become paid-up members of the newly created Church of Elvis the Messiah in the United States. Should it come as any surprise that others have found a new religion in the lore of UFOs and abduction?

There is also the question of hysteria. We should not underestimate the power of public obsession and the way in which shared hysteria over something felt passionately can escalate almost out of control. The recent sad death of Diana, Princess of Wales, presents a clear illustration of this phenomenon. For many people who had little time for the princess before her death, the circumstances of her passing were very sad and disturbing, but the reaction of the public verged on hysteria. The Royal family and the press were quickly made scapegoats following a public outcry over the circumstances of her death and what was viewed as a cold response from her former inlaws. The public demanded that the Royal family show remorse, and the baying mob even insisted that the young princes, William and Harry, be made to appear outside Buckingham Palace to inspect the thousands of floral tributes. Only then was the public anguish appeased.

And what was this anguish based upon? The vast majority of the mourners never even knew Diana, and most of those gathered outside Buckingham Palace were not rabid Royalists. The reason for the reaction was angst, the individual's fear of death, of the abyss—one of the key reasons for a belief in UFOs and alien abduction. The thought that we are being visited by aliens lifts us out of the parochial framework in which we live our daily lives, the limitations of being alone in the universe and restricted to this tiny planet. Abduction lore and UFO enthusiasm are every bit as potent as any religion and satisfy exactly the same needs.

Most abductees really do believe they have been taken aboard alien vehicles. This in itself is a fascinating neurological phenomenon, and a number of possible explanations for it have been offered by psychotherapists and neurophysiologists.

The idea favored by most psychologists is that abduction scenarios are created in the brain by stimulating the temporal lobes—an area of the brain in which we store memories. No one yet knows what may cause this stimulation, and it may be the result of a collection of external (and perhaps internal) stimuli, but the effect has been shown to work in laboratory experiments. If the temporal lobes are artificially stimulated by the application of a magnetic field (using a specially designed helmet), subjects recount experiencing many of the symptoms of an abduction.

The psychologist Susan Blackmore is interested in the phenomenon of abduction and other paranormal claims, and she underwent experiments using the above mentioned helmet:

I was awake throughout. Nothing seemed to happen for the first ten minutes or so. Instructed to describe aloud anything that happened, I felt under pressure to say something, anything. Then suddenly my doubts vanished. "I'm swaying. It's like being on a hammock." Then it felt for all the world as though two hands had grabbed my shoulders and were bodily yanking me upright. I knew I was still lying in the reclining chair, but someone, or something, was pulling me up. Something seemed to get hold of my leg and pull it, distort it and drag it up the wall. It felt as though I had been stretched halfway up the ceiling. Then came the emotions. Totally out of the blue, but intensely and vividly, I suddenly felt angry—not just mildly cross but that clear-minded anger out of which you act—but there was nothing and no one to act on. After perhaps ten seconds, it was gone. Later it was replaced by an equally sudden attack of fear. I was terrified of— nothing in particular. The long-term medical effects of applying strong magnetic fields to the brain are largely unknown, but I felt weak and disoriented for a couple of hours after coming out of the chamber.[8]

Linked to this is the idea that abductions are the modern equivalent of what are known as "sleep paralysis myths." During normal REM sleep (rapid eye movement), the muscles of the body are temporarily paralyzed. Because REM sleep is also the period in which most of us dream, it is believed that this muscular paralysis is a defense mechanism triggered to prevent us from hurting ourselves if we "act out" our dreams. However, sometimes people become mentally alert while their body stays paralyzed—the mind is awake, but the body is asleep. These experiences are often associated with sexual arousal.

Throughout the ages, people have reported vivid nocturnal experiences, sometimes of a sexual nature, but just as often they are frightening and vivid recollections of being chased or attacked by a monster and being powerless to do anything about it. These are the sleep paralysis myths reported by all sorts of people, from young children to horrified nuns, who believe their chastity has been sullied by a "night caller." The connection between this phenomenon, the effects of magnetic fields on mental processes, and reports of alien abduction is striking.

Another suggestion, made by Albert Budden in his book *Allergies and Aliens*, is that the illusion of abduction is precipitated by what he calls "electromagnetic pollution." This pollution is created by the increased number of electronic machines in the environment that leak electromagnetic waves. A particular culprit is high voltage power lines, and Budden points out that a great many cases of abduction occur near such power lines.

Other psychologists are more candid. For them the answer to alien abduction is a simple one. "I have not come across the phenomenon of abductions by aliens except as a delusional belief of someone suffering from schizophrenia," claims psychotherapist Sue Davidson.[9]

For the skeptic, one of the most powerful arguments against alien abduction is the way enthusiasts dig their own traps. Recently, UFOlogists have begun to attempt to explain how alien visitors arrived here, how their technology works, and even to discuss openly what they claim to be the physiology of alien races. These attempts have appeared thanks to what enthusiasts claim are leaks from government sources, and relate directly to tales of captured alien technology and the suggestion that human agencies are in possession of alien beings, both dead and alive.

The most famous example of this is the Roswell incident and the supposed goings on at Area 51—a military establishment in the Nevada desert.

According to U.S. authorities, there is no such place as Area 51, which is a flagrant lie and a misguided attempt at disinformation. The attempts of the military to deny that there is a top secret military establishment at the map reference corresponding to Area 51 is an unfortunate and typically heavy-handed attempt to disguise the truth. It has been a counterproductive move because the denials have simply attracted more and more interested investigators, who risk serious trouble in trying to probe the secrets of the establishment.

Area 51 is almost certainly a top-secret military establishment where experimental craft are tested and cutting-edge research is conducted, but the idea that a craft that crashed in nearby Roswell in 1947 and was taken to Area 51 along with a collection of dead alien occupants is far from being substantiated. In fact, the reports of those who claim to have leaked information about the aliens and their vehicle proves the very opposite and damages the UFOlogists' cause.

A few years ago what was purported to be a film of an alien autopsy was screened by television networks across the world. The film was purchased by a British film producer who claims to have acquired it from a man involved with the autopsy of the Roswell aliens in 1947. It shows what look like small humanoid figures on a slab, partly dissected and fitting the description of an alien that has become almost a cliché—the archetypal "gray" with large black eyes, an abnormally large bald head, and long arms. However, even UFO enthusiasts are beginning to suspect that the film is a fake and contrived merely to make money. Yet, the associated lore is widely accepted within the UFO community and discussed as if it were academically sound, established fact.

According to the enthusiasts, grays have a combined heart and lungs, and a combined pancreas and spleen. This they claim is a piece of information that has seeped out from places like Area 51 and is based upon close study of alien anatomy. The problem is, the idea is ridiculous, but it is easy to see the "logic" behind this phony anatomical description.

Those who make such statements obviously believe that alien visitors are more advanced than us technologically. This is perfectly reasonable and quite obvious because they can come here, but we cannot go there. However, they take this concept too far. They assume that alien bodies must have evolved further than ours, and to those without scientific training it would seem to make sense that a combined heart and lungs would be an improvement on separate organs.

But actually, it would be the opposite. It would be a great disadvantage.

We have a separate heart and lungs for a number of very good reasons, but most important is the advantage this arrangement offers as a defense against disease. If our heart becomes infected with, say, a bacteria, it may not spread to the lungs and we have a better chance of fighting off the disease. If we had a joint heart-lungs organ, then any attack would cause us far greater problems—we would have two infected organs in one.

In other accounts, UFO enthusiasts apply pseudoscience to "explain" how alien ships work or how they communicate. In one account entitled "Aliens in Our Ocean" from a magazine called *Encounters*, the author discusses how human military forces can counteract alien weapon systems. She says: "The question should be whether a rotating and oscillating high voltage, electromagnetic standing wave which changes gravitational phase, could be interrupted by cool longer waves . . . Without being too technical, it is important to have specifics about phrases like, 'The energy produced from this generator was to be concentrated so that it could be aimed and used as a weapon in order to destroy the alien craft and beam weapons.' I do know that in radio astronomy, radio frequencies are measured in a beam of hydrogen atoms and that the vanished scientist in question had been testing particular [sic] his device beneath the water."[10]

It is really anybody's guess what this is all about. It is a blend of scientific terms and contrived phrases used in an attempt to impress people who know little about modern science. For example; what is "a cool wave"? Even more confusing is the last paragraph. The author says that radio astronomers measure radio frequencies "in a beam of hydrogen atoms." This is not a very precise description, but it would seem the author is confusing the fact that astronomers measure frequencies *against the standard of the hydrogen spectrum* using a technique called "spectroscopic analysis." This has nothing to do with "beams of hydrogen atoms." And what is to be made of the link made between this and that scientist in question was "testing particular [sic] his device beneath the water"? Is the implication that the hydrogen spectrum used by radio astronomers is somehow linked with hydrogen because hydrogen is one of the atoms in the water molecule H_2O? If it is, then the author deserves an award for writing creative fiction.

A further example is the often very mundane descriptions of the control rooms of flying saucers offered by experiencers. These are almost always descriptions straight out of B-movies or old-fashioned science fiction novels. Abductees use their experiences of television and cinema as reference points to describe the interiors of craft that are supposed to be designed by creatures thousands of years more advanced than us. Is it not reasonable to assume that the interior design and gadgetry at the disposal of such aliens would be totally unrecognizable to an abductee? Would a lay person propelled from their living room in 1950 and put in the bedroom of a late-1990s teenager know what to make of a Nintendo machine or even a PC?

There are many examples of this sort of flabby thinking in the UFO literature, which has found fertile ground in this age of X-*Files* mania and what has been dubbed premillennial tension. Sadly, such material undermines those who are interested in the idea of life on other planets and the serious search for extraterrestrial life. Because of the growing crank element, those in power and those with money to help fund projects are reluctant to assist serious scientists. There are even those who suggest that some of the more extreme ideas of the UFOlogists, and particularly those who try to convince us of alien abduction, are actually dangerous. UFOlogist Jacques Vallee believes the phenomenon creates a loop of consciousness that escalates the idea far beyond its original foundations: "Conventional science," he claims, "appears more and more perplexed, befuddled, at a loss to explain. Pro-E.T. UFOlogists become more dogmatic in their propositions. More people become fascinated with space and with new frontiers of consciousness."[11]

An elevating of consciousness is a good thing, but unmanaged, it leads to greater confusion and flabby thinking. Any faith system— and UFO lore is simply that—that threatens to damage real science is a danger. The irony for the UFO enthusiasts is this: if aliens are visiting our planet, they have managed this feat by the application of science. To accomplish something as amazing as this, many alien scientists would have had to work for many years refining their ideas and learning how the universe worked. This may involve an holistic approach—incorporating what many would call "alternative" mechanisms and thought patterns, combined with empirical and intellectual rigor—but there would have been no room for the gullibility and the anti-intellectual stance epitomized by some within the UFOlogy community.

Alien abduction is a product of human minds spurred on by ulterior motives. These vary from sexual fantasy to the need to suppress a more painful experience under the guise of alien abduction.

It is fueled by a form of low-key public hysteria driven by a natural need for spiritual and emotional fulfillment in our modern age of vanishing family security and a religious orthodoxy out of touch with the needs of modern people.

I genuinely wish aliens were visiting us on a regular basis, and that there was a great network out there in the galaxy of which we are a tiny part; but not only is there no evidence, there is also no logical reason for it to be true. This does not negate the fact that there may well be a network out there operated by advanced alien cultures, thousands of years in advance of us. Maybe one day we will join that most exclusive of clubs, but I'm sorry to say we are not involved to any degree just yet, and one day we will look back upon the belief in alien abduction as an aberration symptomatic of the era during which we all now live.

14 The Cult of the Cult

> Kill a man, and you are a murderer. Kill millions of men, and
> you are a conqueror. Kill everyone, and you are a god.
>
> JEAN ROSTAND

The scene that met the officers of the San Diego Police Department as they entered Rancho Santa Fe in a suburb of San Diego on March 26, 1997, was one of clinical horror. No bloodstained walls, no gore, no mutilated bodies, for this was a death scene from the final years of the twentieth century, but inspired by the twenty-third.

Inside the house the police found thirty-nine bodies. Each was dressed in overalls and new, previously unworn Nike trainers. Each of the former members of the cult group know as Heaven's Gate had a bag beside their death bed. The bag contained a change of clothes. In the top pocket of their tunics they had their passports, and in their trouser pockets, a five dollar bill and a roll of quarters. Over their heads lay a square meter of purple silk scarf.

They had, the investigators discovered, died in waves over a period of three days. Each had enjoyed a final meal, made a personal, upbeat goodbye video message, consumed a pudding liberally laced with phenobarbital chased with vodka, and placed a plastic bag over their heads. Those in the second wave removed the bags from the heads of those in the first, and two women—the last to die—removed those from the heads of the second wave, tidied up thoroughly, deposited the bags and other detritus in garbage cans, and quietly killed themselves in the same way.

In some respects the cult called Heaven's Gate that self-immolated in March 1997 bore startling similarities to other suicidal cults, but in other ways it was very different. Although, like all other such groups, Heaven's Gate, as led by a mentally damaged charismatic, Marshall Herff Applewhite, offered a prescription for an alternative reality, it was different because the cult members seemed to be entirely happy with their exit, and optimistic about their "future"—there appeared to be no form of oppression or dictatorship within the cult.

Cult members were often seen in the neighborhood, and the women gave away bourbon pound cakes they were forbidden by their creed to eat. Both men and women had their heads shaved and wore plain tunics which disguised their gender. At least six of the men, including Applewhite, had been castrated long before their suicide.

One commentator has said that the cult members of Heaven's Gate "died of kitsch." This is actually a pretty accurate description. Fed on a constant diet of science fiction—from videos, TV, books, magazines, and the Internet—the members of Heaven's Gate were actually living out a science fiction fantasy in which they were the lead characters, and for them, their mission was not to die, but merely to enter the next stage of the plot—to "shed their containers," as they put it. Ironically, one of the thirty-nine members was Thomas Nichols, brother of Nichelle Nichols, who played Lieutenant Uhuru in the original *Star Trek* series.

The philosophy behind the group (if "philosophy" is what you want to call it) was of course a simple one—the simple script is always the most effective—and it came entirely from the mind of their leader, Marshall Applewhite.

Applewhite was a most modern guru, but as we will see, he fits neatly into the mold created by others through the ages. He was sixty-five at the time of the mass suicide and had lead a checkered life. His early years were spent with his family in Texas. His father was a Presbyterian minister, and he was brought up in a strict, censorial environment. He went to college and then the University of Alabama, dropping out in the mid-sixties. A talented musician with a good singing voice, he was physically attractive as a young man. Applewhite performed in university musicals and worked for a while in a small school in Houston, where he produced several musicals before leaving under a cloud in 1970 after a scandal involving a sexual relationship with one of the boys in his charge.

From that time on Applewhite seems to have gradually and deliberately retreated from the world. There followed bouts of alcoholism and drug abuse, he was committed to a succession of mental institutions, and was seen by some who came into contact with him as a "poor-man's Timothy Leary."

Toward the end of 1970 he met his soul mate, Betty Lu Trusdale Nettles, a nurse with a keen interest in the occult who had set herself up as an astrologer and "spiritual guide." Nettles was five years Applewhite's senior, and soon after they met, she left her husband and four young children and went into the wilderness with her new companion, never to return to her family.

Throughout the 1970s the pair traveled around the United States, spending a good deal of time in the South and forming a menagerie of obscure cults and groups who traveled with them, meanwhile collecting minor convictions for car theft and drug possession. To show that given names were "meaningless," they assumed pseudonyms such as Bo and Peep. At one point their group was called HIM (Human Individual Metamorphosis) at another they dubbed themselves Total Overcomers Anonymous, then "Undercover Jesus" followed by "E.T. presently Incarnate." Later Nettles and Applewhite named themselves simply "The Two."

When Nettles died of cancer in 1985, Applewhite began to remodel his "philosophy" and to gather a new group of devoted followers. His vision was a peculiar blend of Gnosticism and New Age/Cybertech fantasy, a doctrine in which the physical was scorned, most probably because he had been fighting inner sexual demons all his life. Applewhite was a homosexual, and both he and Nettles repeatedly emphasized that their own relationship was completely nonsexual. The cult referred to the physical realm—everyday life—as "meatspace," and preferred to spend most of their time in cyberspace. Their ultimate aim was to actually leave this environment altogether, to progress to what they called "the next level."

The members of the cult were so calm and "happy" about what they were doing because they believed they were merely going to step into a new and far better environment. They believed that an advanced extraterrestrial race was about to arrive near Earth and take them away to an alien world. Indeed, later they concluded that they were themselves alien beings who were merely occupying gross, human bodies. Their mission over, they were ready to return, to join Peep, a.k.a. Nurse Nettles, and to live happily ever after.

This was the situation in early 1997, when the cult was pushed over the edge into affirmative action, action precipitated by two things. First, Applewhite learned that he was seriously, possibly terminally, ill. This meant he had little to lose and was able to drag the other cult members to their own graves as he approached his—a form of ego-boosting sacrifice. More important perhaps, the group discovered what they adamantly believed to be a spacecraft approaching earth.

In 1995 two astronomers—a professional, Alan Hale from New Mexico, and Thomas Bopp, an Arizona amateur—spotted a comet approaching earth. It was named Hale-Bopp after the pair of them, and when news of this object made the Internet and, later, the headlines of the world, the members of Heaven's Gate were convinced it was concealing a craft in its tail and that this was the signal they had been waiting for.

Comets and other fast-moving celestial objects have for long been associated in lore with doom and destruction. At the same time, the members of Heaven's Gate believed they were to be saved and the rest of humanity was to be destroyed—"spade under," as they charmingly referred to it. Strikingly, this did not prevent one of the members joining after leaving her family and children, including a seven-month-old baby.

Then, in November 1996, came the real clincher. An amateur astronomer named Chuck Shramek took a picture of Hale-Bopp that showed a bright disk-shaped object in its tail. It looked for all the world like a classic flying saucer hiding in the wake of the comet. Shramek's photograph became an overnight sensation and was posted all over the Internet before journalists around the world picked up on it, propelling it into UFO legend.

The object was actually a star, SAO 141894, that lies many light years from earth, but when viewed by terrestrial telescopes, it appeared in the field of vision perfectly within the tail of the comet. It was actually nothing more than the poor quality of Shramek's optical equipment and imaging devices that created the illusion that the object was in our solar system, and shaped almost too perfectly like a 1950s image of a UFO.

The scientific explanation for this fell on deaf ears within Heaven's Gate. They knew what they wanted to believe—the original story as proclaimed by Shramek. It spurred them on and reaffirmed their ideas—it was irresistible, and initiated what the cult called "closure"—Heaven's Gate was about to close.

The cue was the comet's closest approach to earth, the night of the first deaths at Rancho Santa Fe. The members of the group believed they were to be transported aboard the craft, and Peep (or Nettles) would be there waiting for them to take them home. Despite a thorough investigation of this entire story, to date no one knows what lay behind the need for purple silk scarves, rolls of quarters, and five-dollar bills. One wit suggested the coins were needed for the cult members to "phone home" from the spacecraft; others have suggested that it was done as a joke, a touch of black irony from the members, a final gesture of rejection and contempt. We will probably never know.

Heaven's Gate is just one example of the many hundreds of cults that exist around the world today, a figure that is growing steadily as we approach the millennium. Many of these cults are benign and merely act as a safety valve for people who find it difficult to integrate within society. Some cults and groups are "isolationist," that is, they seal themselves off from the world and retreat into their own way of life. Some of these can become dangerous, some become self-destructive. Other groups are outward-looking and seen more as religious movements than alternative, isolated communities of severely misanthropic individuals.

Cults have many shared traits. They usually consist of young, well-educated people; they have a leader, often a figure perceived as almost godlike; they have an idiosyncratic language—deliberately created to isolate themselves and exclude outsiders; and most important, they offer a totally alternative and hermetically sealed lifestyle.

It is this last factor that has drawn so many people to cults in recent decades. As we have seen, today there is a pervading sense of dissatisfaction in western postindustrial society, a feeling that material gain and a standard of living vastly improved since WWII is not the be-all and end-all of a happy life. Many people are seeking spiritual fulfillment, and they are not finding answers in orthodox religion. Cults and strange belief systems are filling this vacuum and drawing a growing number of people to an ever-increasing array of ideologies.

The group calling themselves Heaven's Gate shared many of these characteristics. Its leader was as crazed as any other leader of such inward-looking and ultimately self-immolating groups, even though their ideology was a little different than the run of obsessive "fire and brimstone" megalomaniacs the world has witnessed in recent decades. Others have been far more violent.

In November 1978 a cult comprising over nine hundred people, including 260 children, drank a specially prepared soft drink laced with cyanide. Their bodies were found scattered around a commune that had grown out of the jungle of Guyana, a place called Jonestown. The group had been settled there for almost four years and had grown steadily since the guru of the site, an American lay minister named Jim Jones, arrived there in 1977.

Jones was mentally ill, and, as we will see, he fitted precisely the template of the insane guru. Originally, life in Jonestown was considered by some a rewarding experience, but gradually it turned into hell on Earth. The cult members who saw Jones as a god were completely under his control. Jones gradually became more and more despotic, leading a double life as an absolute religious leader and a pseudo-head of a tiny, isolated state. His illusions of grandeur developed into a conviction that he could wield absolute power, that he was immortal and divine. He tortured and maimed children and murdered anyone who showed any signs of dissent. Almost unbelievably, he separated families and punished young children for minor offenses in front of their mothers.

It is perhaps significant that when the authorities discovered the mass suicide in Jonestown, a significant number of the victims had died of gunshot wounds, including Jones himself. No one knows what happened there during the last few hours of Jones's mad "regime." It could have been that the community finally turned against their leader, but this would seem unlikely. It may have been that a small faction could not go along with the final stages of the plan and had to be "helped."

A man suffering a comparable psychosis, and every bit as destructive as Jones, was David Koresh, leader of a cult that called itself the Branch Davidians. In April 1993 he was the absolute head of a community of eighty-five people who went on to die in the fire that engulfed Apocalypse Ranch in Waco, Texas. Spurred on by delusions of divinity, he called himself "Yahweh Koresh."

David Koresh was born Vernon Howell in 1959. He had lived through an unfortunate childhood. His mother was fourteen when he was born, and his father had disappeared. He was raised by his grandmother and aunt, until Howell's mother married when he was five. He hated his stepfather and left home as soon as he was able. He had been bullied at school and seen as retarded, even though he had an average IQ and was an able sportsman. In the late seventies he joined the Seventh Day Adventists, but was expelled from the church when he was caught in a compromising situation with the sixteen-year-old daughter of his local pastor.

Koresh was obsessed with sex. Of the twenty-two children who perished in the Waco fire, seventeen were fathered by him. He married the fourteen-year-old daughter of an Adventist Church official in 1983 and was soon sleeping with her twelve-year-old sister. In the doomed Waco Ranch, Koresh split up families and had men and women sleeping on different floors of the building in order to make it easier for him to seduce almost all the women who lived there.

Mirroring the events in Jonestown a decade earlier, Koresh's deteriorating mental state meant that some of the cult members were desperate to leave before the final siege. Koresh had specially trained members of the cult posted as guards on the inside of the building to keep anyone from escaping, and he too tortured and maimed, murdered and raped his followers.

There are many other examples of such cults, and sadly, the list will grow still longer as the millennium is upon us. Some gurus, the likes of Koresh and Jones, are totally destructive, akin to figures like Hitler and Stalin but without the political intelligence, the opportunities, and the advantage of good timing. They are, however, motivated by different variations of the same mental force. By controlling the minds of susceptible individuals, they succeed only in destroying, in snuffing out life.

Other gurus are less obvious. It has been claimed that both Jung and Freud were gurus. They had a body of followers, they created an alternative explanation for fundamental processes—in their case, the functioning of the human mind—and they also shared some of the mental characteristics and even a few of what has been identified as the influences and hallmarks of a guru's life.

In most cases the cult of the guru dies with them. As far as we know there are no serious followers of Koresh or Jones left on the planet, and it would be surprising to find any existing believers in the cult of Heaven's Gate after the body bags were removed from Rancho Santa Fe and the Earth continued on its merry way, untouched by the cataclysmic interference of aliens. But there are some gurus whose ideas continue after their death.

One example is the founder of the religion called Scientology.

Scientology is a fascinating example of a religion that has grown out of its original cult status. Some point to the fact that Christianity followed the same course—that Christ, who incidentally also fits perfectly the guru template—was originally a cult leader and is now perceived as one of the great religious icons of Planet Earth.

Scientology is now officially a "religion" or a "church"—it has been granted charitable status in Britain and claims to have eight million followers worldwide, 100,000 of these in Britain alone. But to many, Scientology is nothing more than an elaborate con based upon fantasy and illusion. Indeed, the British Home Office declared Scientology to be "socially harmful" as recently as the 1970s. Yet to others, Scientology is their religion, and the church can point to several high-profile members—including Hollywood stars John Travolta, Tom Cruise, and Kirstie Alley.

Scientology was created by a man named Ron Hubbard during the 1950s. Broke and drifting from one job to another, Hubbard wrote a book called *Dianetics*, which explained his theory of life. It became an instant best-seller and has remained so ever since, selling principally to the ever-growing numbers of new church members worldwide.

According to one version of Hubbard's life, he was an all-action superhero, a great writer who worked on many Hollywood classics, was honored and decorated as a war hero, was an accomplished scientist and charismatic leader. In the alternative version he was a con man par excellence, a hack who hit on an idea at the right time and followed it through to its logical conclusion.

It is interesting to note that Scientology may be seen as another influence in Marshall Applewhite's doctrine. Hubbard created the idea that we are all inhabited by the spirits of beings called "Thetans." Most of us, he claimed, are unaware of this, but we can become "Operating Thetans" by a process of mental auditing or cleansing. This is done by exorcising painful memories. He devised a system for doing this—the basis of his famous book—and created a machine, an "electropychometer," which helps to clear the mind of unwanted material, to enable followers to attain the enlightened state of being Operating Thetans.

Hubbard was certainly a guru. He fitted into the usual profile, but like Christ, Jung, and others who are loosely clumped together in this study, he did not enforce his will upon a small community, he did not turn inward and create a self-destructive ghetto. Instead he projected his ideas, and in doing so, captured a far wider audience, which flourishes today.

So, given that gurus can encompass such diverse figures as Christ, Freud, and Jim Jones, what is it that unites them, what lies behind the "template of a guru"?

In his excellent study of gurus, *Feet of Clay,* the eminent British psychologist Anthony Storr has pinpointed the attributes common to almost all gurus. First, they must preach and found a sect. Their style of preaching may vary, depending upon the nature of their core belief—Applewhite traipsed around the southern U.S. preaching self-enlightenment through suppression of natural drives, a doctrine spiced up enormously by his conviction that he was an alien trapped in a human body. Koresh preached fire and brimstone self-restraint (while indulging himself with unfettered enthusiasm). Bhagwan Shree Rajneesh, the guru from India who led a community of over two thousand into the 1980s, extolled the belief that enlightenment could be attained through unlimited amounts of sex. Clearly, all gurus need a "message" and a medium through which to preach and to gather a devoted following.

Gurus also need to be highly charismatic individuals. The word "charisma" comes from the Greek and originally meant "the gift of grace." The psychologist Max Weber initially coined the modern usage of the word to describe someone who possessed special powers of attraction and who then acquired the belief that they were better than the run of people and in some way superior. The sociologist Eileen Barker has pointed out that "almost by definition charismatic leaders are unpredictable, for they are bound by neither tradition nor rules; they are not answerable to other human beings."[1]

Gurus are almost invariably self-taught. Each creates their doctrine, formulates an approach, a method to attain enlightenment which they then feel the desperate urge to pass on to others. If we are to include figures such as Freud and Jung in the collection of gurus, the definition needs to be stretched a little because both men received a high level of training in their discipline. But each then rebelled against the establishment and produced their own framework, their own philosophy and methods. So in this respect they too were self-taught. Jesus is another example. There is no record of Jesus being taught anything. He is always perceived as the teacher, an autodidact.

A third trait is intolerance of criticism. Almost all of us face criticism every day, for both minor and major errors. "To err is human." We take it as a basic aspect of all our lives.

Freud commented that the man who is predominantly erotic will give first preference to his emotional relationships to other people; the narcissistic man, who inclines to be self-sufficient, will seek his main satisfactions in his internal mental processes."[2]

All gurus are of the latter type. They are narcissistic in the extreme, turned inward and totally incapable of accepting criticism, often reacting violently against it. Koresh killed people who opposed him, as did Jim Jones.

The Armenian guru and mystic, Georgei Ivanovitch Gurdjieff, was explicitly intolerant of any form of criticism and lashed out at anyone who offered even measured critiques of his views. Jung and Freud were highly sensitive to criticism and indeed fell out irretrievably because they could not agree over their different approaches to their profession. When Jim Jones's wife Marceline tried to persuade their son Stephan to coerce his father into reducing his intake of narcotics, he is reported to have replied: "You're talking about going to God and telling him he's a drug addict?"[3]

At the other extreme, there is no account of Jesus shunning criticism in the Bible, but then that should come as little surprise, given the reverential nature of the written history of his life and times, and the sanitizing of those accounts during the past two millennia.

Applewhite somehow managed to keep a certain balance within his group, but nobody really knows what went on in Rancho Santa Fe before death stalked the community. It would be safe to assume that Applewhite was as intolerant as all the other gurus we have ever seen. One clue lies in the way the cult responded to the reaction they received when they began to emerge on the Internet some months before their mass suicide. Expecting to be received warmly by the cyber community, they were instead "flamed," that is, insulted and treated with contempt. When they received this welcome, the members of Heaven's Gate realized that their time had drawn near. As one former member, a man calling himself "Jwnody" (who was left behind, apparently to spread the word) has put it: "This was the signal to us to begin preparations to return home. The weeds had overtaken the garden and truly disturbed its usefulness beyond repair. It is time for the civilization to be recycled—spaded under."[4]

It follows from this that all gurus are elitist and totally undemocratic. Within all the enclosed communities known, there has only ever been one absolute leader in each; anyone who came close, anyone who posed the slightest threat to the leader's absolute position, was discredited and expelled immediately, or more often, slaughtered.

The final pair of traits, also linked together through character and circumstance, is that all gurus are isolated and friendless. These two characteristics are common to what could be called "positive" and "negative" gurus—those who destroy and kill and those who merely enlighten.

Jesus had no friends. He had disciples who proclaimed they

would lay down their lives for their savior, but disciples are never friends. Friendship is a two-way process, a symbiosis. Gurus have to be above this arena, they cannot truly share their burden, nor do they wish to, beyond the preaching of their doctrine.

For a guru, the greatest achievement is to engender devotion, which leads us to perhaps the most succinct and clear way of seeing what being a guru really means. A guru has sublimated love and sacrificed tender normal human relationships—for power. They have, for their own, various reasons, and in their own different ways, nullified the instinctive drive to share, to love and be loved. Instead they crave and sometimes find a different mechanism to get through life—they use power over others to gain satisfaction. The further they can push their disciples, the greater their sense of self-importance. By ultimately driving them to make the final sacrifice, to take their own lives for a doctrine the guru has created entirely themselves (or so they wish to believe), gives them the greatest of all emotional rewards.

But why? Why do some individuals need this? And how do gurus arrive at this stage in their sad lives?

The prevailing theory is that people who become gurus are trying to resolve an inner conflict. Many of us suffer in this way, and it may be said that each of us has his or her own way of dealing with it. For some, immersion in the world of commerce, jobs, marriage, domestic commitment, is a way of coping with this internal schism. Others, those with creative powers and enough drive, express their inner turmoil in various forms of art. Musicians, writers, painters, creative scientists, share some of the characteristics of gurus. They have deep-rooted inner conflicts that need to find resolution. But an artist never truly believes that inner conflict can be resolved completely. An artist is constantly searching for resolution knowing that with each piece of work they only reach a partial conclusion, resolve only an element of their psychic disturbance.

A guru is different form an artist because he believes that he can find a resolution to his inner conflicts. A guru almost always passes through a period of mental breakdown, or schizophrenic episode. They find their way out of the maze of this severe mental or emotional eruption by settling on a "solution." This may take any number of forms, but it always comes from what psychologists refer to as the "Eureka pattern"—a moment of great inspiration, of realization—what Plato called "divine fury," and what was understood in Renaissance times to be akin to the spirit of religious or mystical revelation. From this Eureka pattern comes the realization, the plot, if you like, for the new path, the new teaching, the new revelation.

For Jim Jones, a doctrine of severe antiracism was the key. He saw prejudice in everyone and in every aspect for society, and created Jonestown on extreme antiracial principles. For Koresh it was the gaining of enlightenment through fundamentalist biblical principles, which he twisted and perverted at will. For Applewhite it was a late-1990s countercultural idealism based upon science fiction, wish fulfillment, and emulation of cheap cable TV. For Gurdjieff it was the ideal that enlightenment could be attained through self-denial, intense physical discomfort, and labor. For Rajneesh, Nirvana was hidden in the heart of total sexual liberation and complete immersion in carnal desires.

Another link lies with a recognized psychical condition called *folie à deux*. It has been known for some time that if two people live together and one of them is mad, the other will begin to accept and even adopt at least some of the other's delusions. In this way, Nettles and Applewhite would have reinforced one another's strange concepts, and within a community of believers, this effect becomes marked and self-reinforcing—adding power to the guru and helping to perpetuate the ideology they expound.

Gurus are also insecure. In many ways they are even more insecure than their disciples and followers. The true difference between them is that the guru has the charisma or force of personality to impose their beliefs on others, beliefs that followers are able to accept wholeheartedly. For the guru, the follower is their salvation, their vindication, and they actually need the follower more than the follower needs the guru.

In some ways all gurus of the isolationist type—the Koreshes, Joneses, and Applewhites—are on a road to self-destruction from the moment of the Eureka pattern. By forming a group and acquiring a self-supporting community around them, the guru is actually buying time. In almost all cases of isolationist communities, the leader's mental health is seen to go into rapid decline from the beginning of the group's formation and the point where they seal themselves off from the outside world. It is almost as if the community is the guru's last ditch attempt to save themselves, to find solace. When that support system is found to be insufficient for their egos and their insecurities, they have little room left for maneuver and the inevitable follows.

So what of the disciples? It is tempting to believe that followers of gurus are simpleminded, of low IQ, and excessively gullible; but this is almost invariably not the case. The members of cults are usually from a range of socioeconomic backgrounds, education, IQ, and age, and are approximately divided evenly between the sexes. The composition of the Heaven's Gate community contained an unusually large number of older people. The average age was about fifty, the oldest member seventy-two.

Of course, cult members are unhappy with the "normal" world. Many are gullible and insecure, but the common thread is not stupidity or lack of intelligence, it is simply that some people are searching for fulfillment, seeking an alternative to what they see as the run-of-the-mill life of the average citizen. Many have gone through the standard course, followed the path of convention, and for one reason or another found it lacking.

What is astonishing about followers of extreme cults is not so much why they should join a group or a community, but the degree of devotion they then apply to this new platform. Mothers have left families, young babies, and loved ones; fathers have walked out on wives; parents have allowed their children to be abused by the guru; some have watched passively as their loved ones have been butchered.

This obsessive devotion says a great deal about the power a guru wields, the level of corruption and the total abuse of this power. It also demonstrates how minds can be molded and controlled, how some people are able to let go to the same degree that others can take command. History has demonstrated the horrors this symbiosis can produce—how else do we begin to explain why thousands came under such evil control in wartime Germany and participated in the horrors of the death camps? How else could U.S. marines be persuaded to slaughter innocent children in Vietnam?

The study of gurus and cults raises so many questions concerning the inner darkness of humanity and the points at which creativity and madness meet that we cannot yet hope to begin to understand many of the subtleties of this extraordinary power that crackles between the followers and the leaders, the sheep and the shepherds. But psychology is beginning to unravel many of the motivations, the hidden powers and the shrouded complexities of this fascinating psychic world. We can only hope that by studying the margins, the borders where insanity and art, sexual obsession and emotional desiccation, meet, we may find solutions to problems of mental illness, schizophrenia, and emotion imbalance. The study of gurus and cults is fascinating in itself, but such extremes of human character and behavior may also provide answers to age-old questions such as: What is self? How do nature and nurture function in our mental development, and how does desire so often override rationality and reason?

15 Mojo Rising

In 1930 a French anthropologist, Dr. Georges de Rouquct, went on a field trip to the West Indian island of Haiti. He could speak fluent Creole and soon ingratiated himself so well with a local landowner that he was allowed to see what the Haitian told de Rouquct were genuine zombies. De Rouquct was not allowed to touch the zombies, but could observe them closely. He described the experience in his journal:

Toward evening, we encountered a group of four male figures coming from the nearby cotton field where they had been toiling. I was struck by their peculiar gait, most unlike the lithe walk of other natives. The overseer with them stopped their progress, enabling me to observe them closely for some minutes.

They were clothed in rags made from sacking. Their arms hung down by their sides, dangling in curiously lifeless fashion. Their face and hands appeared devoid of flesh, the skin adhered to the bones like wrinkled brown parchment. I also noticed that they did not sweat, although they had been working and the sun was still very hot. I was unable to judge even their approximate ages. They may have been young men or quite elderly.

The most arresting feature about them was their gaze. They all stared straight ahead, their eyes dull and unfocused as if blind. They did not show a spark of awareness of my presence, even when I approached them closely. To test the reflexes of one I

made a stabbing gesture toward the eyes with my pointed fingers.
He did not blink or shrink back.

My immediate impression was that these creatures were imbe-
ciles made to work for their keep. Baptiste [the land owner] how-
ever assured me that they were indeed the zombies; that is, dead
people resurrected by sorcery and employed as unpaid laborers.[1]

The Haitian religion of Voodoo is an ancient one. It has its roots
in the tribes of central and western Africa and was transplanted to
Haiti with the slave trade of the seventeenth and early eighteenth
centuries. The reason Voodoo is not better known on the mainland
of the United States is that here the blacks were gradually assimilated
into the population, whereas Haiti gained independence very early
on—at the end of the eighteenth century—and its population of
seven million is today made up almost entirely of the descendants
of the slaves who arrived there centuries ago.

The fundamentals of Voodoo as a religion appear to be quite
confused—a hodgepodge of different creeds and beliefs. There is
even a name for this blending of religious ideology— *syncretism* (the
same word used to describe how the language of Creole is derived
from bits of other languages)—and an old Haitian saying declares
that Haitians are 95 percent Catholic and 110 percent followers
of Voodoo.

Those who practice Voodoo believe in a God, a deity they call
Djo or *Mawu*; but this God differs fundamentally from the Christian
God because they think that Djo or Mawu is far too important to
interfere in earthly matters. Followers of Voodoo believe that when
a human is born, they are mere animals but become infused with a
spirit called the *lao*, placed inside you as a guide during an initiation
ceremony. Another aspect of your being is the *ti bon ange*, or "little
good angel," which is roughly equivalent to the "will." According
to the Voodoo religion, at the initiation where the spirit imbues the
body, the *ti bon ange* may be extracted and stored in a jar in the
inner sanctum of the temple. When the owner of this life force dies,
the jar is opened to allow their *ti bon ange* to hover over the dead
person's grave for seven days. The purpose of leading a good life is

to enrich the *ti bon ange* so that after sixteen incarnations a human may return to God.

As well as these ancient beliefs, the Voodoo religion has incorporated random aspects of Catholicism. These ideas were picked up from the whites who transported the slaves to Haiti, and from the Europeans who were already living in Haiti and the mainland of the United States in the sixteenth and seventeenth centuries. There seems to be no real pattern to these acquisitions from European religion, and visitors to Voodoo shrines in Haiti should not be surprised to see images of Catholic saints such as St. James the Greater, or statuettes of the Virgin Mary among images of Erzuli, the female serpent/rainbow spirit or *loa* who symbolizes love.

The term "Voodoo," which derives from the African Fon language, comes from *voo*, meaning "introspection," and *doo*, which means "the unknown," and it means slightly different things to every practitioner of the religion. But aside from the basic tenets of the Voodoo deity and hierarchy of saints, there are other more exotic elements to the religion, many of which have been misinterpreted and misunderstood by nonbelievers.

First, Europeans have an image of Voodoo as simply a rather nasty, even evil, cultural phenomenon. They link it solely with zombies and sticking pins into dolls. These are indeed aspects of Voodoo, which I will explore later in the chapter, but Voodoo believers also insist that like all religions, there is a healing element to their faith. Not only is it a conduit for emotional and spiritual feelings among those who believe, it is claimed to be a mutual support system for what is a largely impoverished people. It should be remembered that the vast majority of Voodoo believers live on Haiti, a very poor island with limited resources. It is particularly important to the poorest of this poor community—those who live in the most remote parts. They have little else but their religion.

It should also be recalled that another, far more widespread religion, Catholicism, is also followed most closely and faithfully by those living in third world countries. Many put this down to the fact that those living in poorer nations rely more heavily upon religious faith, particularly fundamentalist, or what some would call divisive, doctrines. Others have suggested that those living in industrialized nations, who are better educated and more sophisticated, are less likely to believe in what some see as an outmoded faith system. It is perhaps significant that in these same industrialized countries, religion once played a far more significant role.

Equally, we should not lose sight of the fact although there are undoubtedly cruel and barbaric aspects to Voodoo—in particular, the creation of zombies and the infatuation with the power of *mojo*—what we would consider more orthodox religions also have their primitive overtones. Before we condemn outright the Voodoo priest, the *bokor*, it might be prudent to remember the Inquisition, regular sectarian killings in Northern Ireland, and the pain and anguish caused by continued backward thinking on contraception and abortion in countries that consider themselves modern states, even countries in the European community.

Putting aside the purely religious aspects of Voodoo and its comparisons with other faith systems, let us look at the controversial elements of its practice and claims for the creation of zombies, and the power to influence matter at a distance purely by the power of the mind. There are two fundamental and supposedly powerful "supernatural" angles to explore—the spirit world of the Haitian priests and their ability to raise the dead, and the power of mojo—the placing of curses, hexes, and death wishes by the use of ritualistic magic.

The most important person in some Haitian communities, especially remote settlements that have little contact with the major cities of the island, is the *bokor*—a Voodoo priest or black magician. Many Haitians live in fear of the *bokor*, who is considered a very powerful man never to be crossed. Whether or not you believe in Voodoo, there are some very good reasons to be afraid of the *bokor* because, in one form or another, zombies do exist, and to become one fits perfectly the cliché of "a fate worse than death."

According to believers, a zombie is the resurrected body of a dead person—someone who has been reconstituted by *a bokor* in order to perform a specific task. Practitioners of Voodoo claim that the *bokor* is able to capture the soul or spirit of a dead person and separate it from the body. The spirit is then contained in a special jar, and the body of the deceased is used for whatever purpose the priest chooses. To the believers, this is a purely supernatural process heavily dependent upon the skill of the much-feared priest. A particularly famous case will serve to illustrate.

On May 2, 1962, a young man by the name of Clairvius Narcisse "died" in the Albert Schweitzer Hospital in the small town of Deschapelles. The cause of his "death" was a mystery, but he had developed a fever a few days earlier and then went into respiratory failure. Pronounced dead, he was buried within a few days. Then, eighteen years later, his younger sister Angelina was shopping in the market of her home village, l'Estère, when from behind her she heard a voice she immediately recognized. She turned, and to her utter amazement saw a man she thought was long dead, her brother, Clairvius.

When she had calmed down and taken the man—who seemed confused and incoherent—back to his family, the story of his past eighteen years gradually came out.

His memory was hazy, but he recalled the scenes in the hospital almost two decades earlier. He remembered being short of breath and then slipping into a trancelike state. He could hear people talking and a doctor proclaiming him dead, but he had been unable to move or to say anything. It had been literally a living nightmare.

Checked by two doctors, one an American, the exact cause of Clairvius's condition was never ascertained, but because his skin was chalk-white and his heartbeat had slowed to an imperceptible rate, he had been pronounced dead and subsequently prepared for burial.

Clairvius remembered the sound of the lid closing on his coffin. Thinking that he really was going to die this time, one of the nails banged into the coffin pierced his cheek, and he could hear his sister weeping. After that he lost all track of time until he saw a light shining down on his face. He felt himself being dragged out of his coffin, and then several men set upon him, beating him almost to death before he was dragged away.

For the next two years he had been kept as a slave in the wild northern region of the country. He had been drugged by the *bokor* and regularly abused by him and the landowner who had enslaved him. Then, one day, one of his zombie companions suddenly awoke from his trance and turned on the *bokor*, killing him. With the priest

and controller dead, the effects of whatever was keeping them se-
dated began to wear off and the entire group of slaves escaped.
Clairvius wandered the island for another sixteen years until one
day he heard that his brother had died, and this prompted him to
return home.

At first the family were mystified by his account, and especially
the reason for Clairvius's decision to return when he had. But he
explained that it had been his brother who paid the local *bokor* to
turn him into a zombie in the first place. The reason? Clairvius had
crossed him.

Clairvius was eventually integrated back into his home village
and accepted by the people he had known so many years before.
He even became something of a national celebrity, appearing on
television shows, the first person to publicly claim they had once
been a zombie.

To rational nonbelievers, this entire story clearly has little to do
with spirits being captured or "undead" beings roaming the countryside of Haiti. But if Clairvius was not one of the "undead" what
really happened to him? Who were the zombies he encountered
and how had the *bokor* exerted such power?

In the 1980s an ethnobiologist, Dr. Wade Davis, then working
at the Harvard Botanical Museum in Boston, traveled to Haiti to
make a detailed study of Voodoo and the practices of the *bokors*
who live in remote parts of the country. He summarized his findings
in two books, *The Serpent and the Rainbow* and *Passage of Darkness*,
which came up with scientific explanations for many of the phenomena we see in the ritualistic practices of Voodoo. One of Davis's key
findings was that the *bokor* were in fact using a sequence of drugs
to control people before, during, and after what others perceived to
be their death.

The first stage is motivation for the action. Often, those who
become zombies are victims of a hate campaign or are undesirable
individuals that someone wants to get rid of. Clairvius Narcisse was
a classic example. He had apparently become involved with a
woman in whom his brother was interested, and had made pregnant
a number of other women in the village without any intention of
marrying or supporting them. The brother had finally called upon

the services of the local *bokor* to remove his wayward sibling from the scene.

The chemical route to becoming a zombie is rather delicate and requires skilled use of very particular drugs. The *bokor* makes a powder called a *coup poudre*, the preparation of which is surrounded by ritual and hocus-pocus to add status to what is in essence a simple blend of a few potentially lethal chemicals.

According to ritual, the *coup poudre* must be produced in June. To prepare it, the priest needs one "thunderstone" or *pierre tonnerre*, which is a piece of rock that has been buried underground for one year before it is unearthed by the *bokor*. To this, add one human skull and assorted bones; two puffer fish (preferably female), one of which must be *crapaud de mer*, the "sea toad;" one sea snake (the *polychaete worm*); vegetable oil; a sprig of a plant called a *tcha-tcha*; half a dozen pods of *pois gratter*, otherwise known as "itching pea;" two blue agamont lizards; one big toad, *Bufo marinus*; and finally an assortment of tarantulas, white tree frogs, and various insects according to taste.

These ingredients are to be used in the following way: the sea snake is tied to the leg of the toad, the *Bufo marinus*. The two are then placed in a jar and buried. The toad is said to "die of rage," which, according to Voodoo lore, increases the power of the poison it secretes into the jar. At no time must the *bokor* touch any of the ingredients because some of the most potent elements can be carried through the skin and are deadly in concentrated form.

As the toad and the sea snake are doing their work, the *bokor* places the human skull in a fire with the thunderstone and a collection of other ingredients until the skull turns black. While this is being prepared, he grinds the vegetable and insect ingredients together and adds some shavings of the skull, taken before it was placed in the fire. The mixture should then be ground to a fine powder along with the skull and thunderstone and the poison exuded by the toad. The entire mixture is then placed in a coffin underground for three days. This is even more potent if the coffin is the one containing the body from which the skull was removed to begin the process.

After three days, the *coup poudre* is ready. Traditionally, this is sprinkled in the shape of a cross on the doorstep of the targeted victim, but for a better chance of success it is more usually poured down the back or surreptitiously placed in a sock or shoe. The poison is then absorbed through the skin. Within hours the victim will have problems breathing, and will appear to "die" soon after.

The active ingredients in this ceremony boil down to just two components. First, the mix contains a chemical called *tetrodotoxin*, which comes from the female puffer fish. This is both an anesthetic and a poison. As an anesthetic, it is estimated to be almost 200,000 times more powerful than cocaine, and as a poison 500 times more deadly than cyanide. The other essential ingredient is another powerful anesthetic and hallucinogenic drug contained in the poisonous excretions of the *Bufo marinus*, a chemical called *bufotenine*.

Combined, these two compounds make a potentially lethal cocktail. In precisely the right doses, the preparation of which requires great skill, it can create the onset symptoms that make the victim appear to be dying. They go into a trancelike state, their breathing becomes so shallow it is almost undetectable, and they take on a deathly white pallor. This effect appears even more pronounced in hospitals such as those on Haiti (especially at the time Clairvius Narcisse was zombified in 1962), where highly sophisticated heart monitors are comparatively rare.

The rest of the components of the preparation are largely for ritualistic purposes, and their use has been refined over centuries primarily to add an element of the macabre and to instill greater fear into the minds of naïve country Haitians and those who want to believe.

So, with the first stage over, we now have someone presumed dead but actually clinging onto life with barely perceptible life signs. They are duly buried, grieved over, and left in peace. It is then that the *bokor* and his helpers return to their work.

The next stage of the process requires the reanimation of the "dead" victim. Again, this is in part a delicate operation requiring precise timing. If the victim is left buried alive for too long, he really will die, but if he's dug up at precisely the right time, he will be usable. Which, incidentally, implies an almost unimaginable degree of cruelty on the part of those who pay for the *bokor*'s services.

Not content with having someone killed, they employ the priest to make their enemy one of the "living dead."

Returning to the graveyard, the *bokor* and his team remove the victim's coffin and lift out the limp body. Next, they viciously beat them up. This might seem like unnecessary cruelty, and it is, but there are two reasons for it. The occult reason given by the *bokor* is that they have to make sure that the zombie's "will," the *ti bon ange*, is trapped and cannot return to the body, so that the victim is under the complete control of the priest. Sometimes, a *bokor* will add an extra nasty twist to this scenario. If he and his helpers are feeling particularly cruel (or they have been paid extra), instead of trapping the *ti bon ange* in a jar, they will endeavor to transpose it into the body of an insect. According to the Voodoo faith, in this way the *bokor* is almost certain to destroy the victim's chance of resurrection and eventual union with God. In effect, he is not only keeping a "dead body" alive, but destroying the very soul of the poor victim.

In purely biological terms, the beating is necessary because the bufotenine that has put the victim into an hallucinogenic trance can have unpredictable side effects, and sometimes the intoxicated zombie can go literally berserk. The beating has the effect of sedating them just as the effects of the tetrodotoxin may be wearing off.

In this state, the zombie is of little use to anyone. However, the job of removing the person from the community has been fulfilled and the *bokor* has already earned his fee. But he has more work to do if he wants to earn a further fee for turning the zombie into a compliant slave via the use of another cocktail of chemicals.

After the beating, the victim is led to a cross and baptized with a new name. This is usually something insulting and humiliating. He is then force-fed a paste made from sweet potato, cane syrup, and a drug called *datura stramonium*, commonly known as zombie's cucumber. Datura is another hallucinogen that causes psychotic delirium. It also contains a drug called *atrophine*, which is the antidote for tetrodotoxin. So, this new chemical mix brings the zombie out of the catatonic state induced by the *coup poudre* and places him in a state of constant psychotic delirium, a bit like a permanent LSD trip in which the victim is able to walk, to carry out simple tasks such as working in a field, to eat and to drink. But zombies are unable to focus on reality, they cannot speak, and only understand vaguely what is happening to them.

As reported by Clairvius Narcisse, there have been cases where a zombie has snapped out of this condition, usually because the landowner to whom the zombies are enslaved forgets to administer the drugs at the appropriate time or erroneously allows the zombies certain forbidden foods. Most important of these is salt. At no time should zombies be allowed salt, because, according to tradition, this will allow them to return to the real world. This probably has something to do with the metabolic breakdown of the datura. Large quantities of salt will metabolize the drug into a less potent form, and the delirium will gradually slip away as a result. Understandably enraged, the zombie, who is more often than not a disreputable individual or criminal, will then turn on their keepers and murder them before escaping.

This scenario accounts for the way in which the vast majority of zombies have been produced by *bokors* in Haiti. No one knows how many people have been zombified, but over the centuries it may run into many thousands. It is thought that people in this state do not usually live very long; they are worked hard and the constant application of hallucinogenic drugs severely damages the brain, leading eventually to cerebral hemorrhaging and true death.

Recently, other explanations for the existence of zombies have been suggested.[2] Professor Roland Littlewood of the department of anthropology and psychiatry at University College, London, has studied a number of Haitian zombies and reached an altogether different explanation of the biochemical process described above. He claims that many zombies are actually mentally ill members of the community. He speculates that in remote communities some people suffering from mental illness—usually paranoid schizophrenia—are dealt with by being sold into slavery. Whether or not these poor souls are also fed a constant diet of hallucinogenic drugs is still unknown.

It is most likely that the zombie community of Haiti is made up of a mixture of victims. Some *bokors* no doubt conduct their rituals on the mentally ill and in this way lay claim to creating the undead.

Other zombies are probably criminals and ne'er-do-wells who have fallen prey to what amounts to a long-drawn-out assassination.

However zombies are formed, the threat of zombification is very real in the minds of many Haitians. Sometimes families deliberately mutilate the bodies of their dead relatives just before burial to render them useless as slaves even if the *bokor* resurrects them. And, throughout their time as leaders of Haiti, the Duvaliers, "Papa Doc" Francois and "Baby Doc" Jean-Claude, regularly employed Voodoo imagery and occult ideas while officially outlawing its practice. Papa Doc even went so far as to let it be known that he was a powerful *bokor*, and he named his personal guards, his private mafia, *tontons macoutes* after the most powerful of Voodoo sorcerers. Although Baby Doc was deposed in 1986, a belief in Voodoo was never dented by the Duvalier's strictures and lives on after Haiti's former rulers have disappeared utterly from the political scene.

The other major aspect of Voodoo to be considered is the black art of harming or even killing victims by suggestion, by action at a distance using ritualistic black magic. This is the art of mojo.

Mojo is a curse or magic spell to bring harm to others. The traditional idea is of a Voodoo priest sticking pins in waxen images of the victim, which produce excruciating pain or even death. But there are other methods employed by various practitioners of black magic around the world. The aborigines use a technique called "bone-pointing" which involves no physical contact with the victim. Instead, during a carefully contrived ritualistic ceremony, a priest merely points a special bone at the victim and they are treated by the others as though they are already dead. Soon they are ostracized by the community, and often despair leads to suicide. Other priests use wooden effigies, the power of the magic enhanced by applying curses over cuttings of hair or nail clippings taken from the victim.

Again, to many Haitians this is a very real and potent aspect of their faith. When American marines landed on Haiti during a brief occupation of the island in 1994, it was widely believed that there were groups of *bokors* working to thwart the military effort by the use of curses and hexes. The same year, an American judge actually jailed a man on Haiti whom he believed was preparing a curse against him because this well-known voodoo priest had been found mixing a lock of the judge's hair into a specially prepared elixir.

Quite naturally, emotive rituals such as those performed by Voodoo priests are easily exaggerated by believers and practitioners alike. The believers want to believe and are fearful, and the priests are manipulating hysteria and anxiety for their own ends. As a result, there is no shortage of graphic tales that appear superficially to support fantastic claims that there is real magic at work when the priest points a bone or impales an image with a needle.

During the 1930s the anthropologist Dr. Herbert Basedow was one of the first to introduce Europeans to stories of priests applying the black arts and conjuring the evil of mojo. He included what was then a startling account of such a ritual in a book about aboriginal tribes called *The Australian Aboriginal*. In it he wrote:

A man who discovers that he is being boned by an enemy is, indeed, a pitiable sight. He stands aghast, with his eyes staring at the treacherous pointer, with his hands lifted as though to ward off the lethal medium, which he imagines is pouring into his body. His cheeks blanch and his eyes become glassy, and the expression of his face becomes horribly distorted . . . he attempts to shriek, but usually the sound chokes in his throat, and all one might see is froth at his mouth. His body begins to tremble and the muscles twist involuntarily. He sways backwards and falls to the ground, and for a short time appears to be in a swoon; but soon after he begins to writhe as if in mortal agony, and covering his face with his hands, begins to moan. After a while he becomes more composed and crawls to his wurley [his hut]. From this time onward he sickens and frets, refusing to eat and keeping aloof from the daily affairs of the tribe. Unless help is forthcoming in the shape of a counter-charm, administered by the hands of the Nangarri or medicine man, his death is only the matter of a comparatively short time. If the coming of the medicine man is opportune, he might be saved.[3]

As with the zombification of victims, there are two crucial elements in this series of events that go some way to explaining how such seemingly grotesque and frightening phenomena could occur. The first is the fact that the victim is almost invariably a trouble-maker. This entire ritual can be interpreted as a court dishing out a punishment. These people live in small isolated communities in which the only law is that the strong dominate the weak. Equally, in some cases, those hurt by a troublesome member of the community can rely upon the medicine man to intervene. The "victim" in the above account is basically ostracized by everyone he knows and falls into a deep depression. This is fueled by guilt and the fear of the medicine man, a fear that has been instilled into the members of the community since early childhood.

The other factor to consider is the liberal use of hallucinogenic drugs. Although Basedow made no specific mention of this in his account, it is well-known that most tribal people around the globe use some form of drug as recreation and for ritualistic purposes, whether it be alcohol, mescaline, or other less well-known but often very powerful agents. In a situation involving a combination of drugs and the power of a community turning on an individual, there should be little surprise that the effect is very powerful indeed.

This potent blend should not be underestimated. The ethnobiologist Wade Davies has suggested that the victim of a hex actually becomes a threat to the community and that the community actively conspires in the eventual death of that victim. In some cases, they deliberately mourn in front of the victim, as though he were already dead.

This conspiracy activates and encourages what scientists have dubbed a "giving-up complex." They describe this as being similar to the frame of mind sometimes adopted by people who suffer from terminally ill disease—they "lose the will to live." It is easy to imagine how anyone could adopt the giving-up complex after a terrifying ceremony in which they are cursed and have their former friends and family treating them as if they were already dead; especially if this treatment is combined with the use of mind-altering drugs.

Wade Davis describes this as a form of "total social rejection." He says, "Although physically the victim still lives, psychologically he is dying, and socially *he is already dead*."[4]

Crucial to the success of this process is total belief in what is happening both within the mind of the victim and in the community as a whole. Professor Gottlieb Freisinger of the John Hopkins University in Baltimore has declared that, "special circumstances and beliefs in a community must exist before an individual can die by hex." And psychologist Stanford Cohen, working at Boston University, believes that "hexing can be fatal when it implants a mixture of fear and helplessness in the victim."[5]

Wade Davis rightly points out: "Even doctors of the most traditional sort admit the role psychology plays on our well being." And: "If one believes strongly enough then it is more likely to happen."[6]

To demonstrate that the power of suggestion is greatly influenced by the total commitment to a belief in what is happening, Professor William Sargent, an expert in brainwashing techniques, used electric-shock therapy on a woman who was convinced she was being cursed. He persuaded the woman that the electric shocks had removed the hex, and she duly recovered.

But is this the entire answer? Surely there are some individuals who would be resilient to the pressure of those around them? This may be true, especially if we consider that many of the so-called victims of hexes are themselves actually unpleasant individuals, including in their number thieves, murderers, and rapists, or those who have crossed others seriously enough to have become subject to the attention of the medicine man. But no matter how tough they may be, they would face serious practical problems because of the treatment of the community.

Firstly, if the community is treating the victim as though they were already dead then presumably the hexed individual would find it hard to continue daily life, to eat or drink, to find shelter or gainful employment. Their only option would then be to leave, but they would probably be prevented from doing so. Inevitably their health would suffer and they would fall into rapid decline. But even then, there may be purely biochemical reasons for hastening this fall.

During the 1920s, doctors treating soldiers who had returned shell-shocked from the trenches, discovered a new syndrome precipitated by a process called, *vagal inhibition*. This is caused by an over-stimulation of the adrenal system. If the blood is flooded with adrenaline, the blood supply to the extremities of the body is reduced so that blood can be concentrated in the muscles. The reason for this is simple: Adrenaline is pumped into the body to stimulate the muscles ready for "fight or flight." However, with the supply of blood to the extemities reduced, the cells in these parts of the body become less infused with oxygen and the tiny capillaries that carry blood (and oxygen) to these regions become more permeable to blood plasma. The blood plasma then seeps into the tissue surrounding the capillaries further reducing blood pressure.

This vicious circle of events causes a continued drop in blood pressure and if left unchecked, the victim will die. The process takes a few days and is almost certainly a major factor in the gradual "fading-away" so often described in mojo stories.

So, it appears that the ability of bokor to produce the "undead" and the power of the medicine man or the priest to cause the death of someone by suggestion are both real and powerful phenomena. However, neither are "supernatural"—each can be explained using the accepted laws of science. In the case of the zombification of victims, the bokor utilizes a complex amalgam of drugs mixed with ancient rituals to generate fear in the community over which he commands a powerful position. The use of mojo, or hexes is a blend of sociological factors, that work strongly against the victim, combined with the liberal use of drugs. In each case, fear lies at the heart of the ritual and the practice. In the case of hexing, this element of fear may be the crucial factor that sets in motion an elaborate chain of biochemical events ending in death.

In Haiti there is a proverb that sums up precisely the many-layered complexities of this ancient, elemental and highly devisive religion. It says: "The closer you get to Voodoo, the more vulnerable you are to its power." A scientific explanation for the work of the bokor does nothing to diminish the truth and the power of this aphorism.

16 Miracles and Wonder

> A cask of wine works more miracles than a church full of saints.
>
> ITALIAN PROVERB

What constitutes a miracle?

In one sense, the concept of miracles lies at the heart of the paranormal. We may think of ghosts as "miracles," or even the arrival of alien beings in shiny flying saucers as "miraculous." This is because, as Arthur C. Clarke has so cogently put it: "Any sufficiently advanced technology is indistinguishable from magic."[1] In other words, the reality and the meaning of "miracles" is really built upon the perception of the witness. If the witness to a miracle has sufficiently detailed knowledge, the event is no longer miraculous. Some would therefore argue that miracles are not just for the gullible, but only meaningful to those *without* knowledge.

Miracles, as we understand the term, are also more closely linked to orthodox religion than many other aspects of the paranormal. The eighteenth-century Scottish philosopher David Hume knew this, and was as skeptical as any latter-day scientist when he wrote: "The Christian religion not only was at first attended with miracles, but even at this day cannot be believed by any reasonable person without one. Mere reason is insufficient to convince us of its veracity: and whoever is moved by faith to assent to it, is conscious of a continued miracle in his own person, which subverts all the principles of his understanding, and gives him a determination to believe what is most contrary to custom and experience."[2]

There is little to be said about telepathy, abominable snowmen, or spontaneous human combustion in the Bible or other holy works (although some would claim there is plenty written about alien visitation and abduction in ancient religious texts), but there are many references to "miraculous happenings." In fact, such events have a category of their own in the litany of the paranormal. They are called "biblical miracles," and include such things as turning water into wine, parting the Red Sea, and raising the dead.

There are also latter-day "miracles" that impact with both the occult and conventional religious faith. Perhaps the most important of these are the phenomenon of *stigmata* and claims of *incorruptibility*. It is these, and the canon of biblical miracles, that will be considered in this chapter—paranormal phenomena that have a strong religious connection.

Both the Old and New Testaments of the Bible contain dozens of stories of miracles, and to the cynical, these stories have been used to give weight to the faith system at the foundation of Christianity. Although most religions preach the concept that faith should not be based upon evidence, that the true believer does not *need* miracles, there do seem to be a substantial number of such tales sprinkled throughout the gospels and in earlier parts of the Bible to consolidate events and teachings. The devout would argue that these need not be there and that the Bible and the basis of Christianity would be just as sound without them.

When reviewing some of the miraculous tales recanted in the Bible and other ancient holy texts, it should be remembered that these stories come from a time when humans had little knowledge of how the universe worked. To people living during these ancient times, domestic life and the observed grander sweep was a mystery; they believed that nature was controlled by a personal God, or gods, and had no concept of physical laws describing the fundamental natural forces we now understand through science. To these people, life was ruled by supernatural forces, forever beyond the control of humankind. Within such an intellectual climate, the idea that the sea could be parted, that plagues and pestilence were controlled by a wrathful deity, or that individuals could be struck by a lightning bolt or reduced to a pillar of salt was unquestionably within the power of the divine.

Old Testament miracles are grand in scale, whereas those related in the New Testament are more personal and focus on the deeds of Jesus Christ. But although these lie at the heart of Christian literature and teaching, what does the dispassionate scientist think of them?

One of the grandest of grand Old Testament miracles is the parting of the Red Sea. This tale describes how the Egyptians attempted to enslave the people of Moses, but the great prophet leads them to safety after parting the sea and allowing them to reach freedom before the waters return to engulf the pursuing Egyptian soldiers.

To the rationalist, this sort of divine favoritism has no basis in logic—why should God be on the side of one tribe of humans rather than another? It smacks of race ego, the idea that a benign God always looks after "us." (Although even this is in itself contradictory, because this God was obviously not so benign toward the Egyptian soldiers in this old tale.) But even putting this objection aside, let us consider the details of the story of Moses parting the Red Sea.

To make the trip described in the Old Testament, the fleeing slaves would have journeyed close to a region called Per Rameses, which is made up almost entirely of papyrus swamps. This is negotiable on foot, because small groups with an able guide can skirt the most treacherous areas. However, soldiers unfamiliar with the terrain and with chariots and carts laden with supplies may well have run into serious trouble. It would seem likely, then, that Moses used very human earthly skills to guide his people through dangerous terrain that became the graveyard for the soldiers in hot pursuit. This version of the story is also supported by the fact that the Hebrew name for the region through which Moses led his people to safety is *yam suph*. The true translation of this is not Red Sea but Reed Sea.

Similar clinical analysis can also help to explain the torrid tales of plague and pestilence wrought upon parts of humanity by God in order to help out other groups. The most vivid and detailed of these stories again involves Moses, and is known as the "Ten Plagues," which God is supposed to have inflicted upon the Egyptians to persuade them to release the Hebrew slaves. The Ten Plagues were a particularly nasty collection of punishments. First, the Nile was turned to blood, then the country was invaded by frogs, followed by a plague of mosquitoes, then flies. Next, the cattle perished, men fell ill with boils all over their bodies, and hail rained down. Then, the country was beset with a plague of locusts, darkness fell for three days, and finally, all the firstborn died.

This is a calamitous series of events for any one nation, but the account of these terrible misfortunes is short on specifics. It does not give an accurate time scale for these events, and what we may now see as a succession of natural, interlinked misfortunes is enwrapped in a description filled with horror and portrayed with relish by the partisan authors.

To begin with, the Nile has been known to "turn red" on many occasions. This is caused by pollution washed down from the highlands of Ethiopia. This pollution, mainly stagnant water that collects in slow-moving pools during dry spells, combined with reddish sand and silt, would cause frogs to leave the Nile. Then later, when the water receded, the land that had been flooded would have retained many of the microorganisms living in the polluted water. This would then have acted as a breeding ground for mosquitos and flies. These pollutants could also precipitate a number of serious local diseases, in particular anthrax, which would inflict animals, killing cattle and bringing boils out in humans. The most susceptible to the disease would be the old and the very young. Of these, the most important to the community would have been the babies and young children. The fact that many young children died would have been given greater emphasis in any account of events, so that from the viewpoint of those looking back, it might seem that only the firstborn had died.

Finally, such a succession of disasters would have inevitably created the effect desired by the Hebrews—they would have been freed by the Egyptians. It could be argued that, from the perspective of the Egyptian rulers, the Hebrews would have represented a further drain on a society already on the edge of destruction precipitated by what was actually a series of interconnected natural problems.

Similar cataclysmic events recounted in the Old Testament can be explained using a degree of detachment and clear reasoning based upon a greater amount of information than was available to the ancients. The fall of the walls of Jericho may be explained by the fact that the site of the city is now known to have stood in an area plagued by frequent earthquakes. Likewise, the Dead Sea lies in a rift valley ravaged by earthquakes, and it is thought to have been a particularly violent quake in 2350 B.C. that destroyed five cities that once lay on the rim of the sea. These were es-Safi, Khanazir, Numeira, Bab-edh-Dhra, and Feifah. Two of these are thought to have been the Old Testament cities of Sodom and Gomorrah, which, according to the Bible, were destroyed by fire and brimstone wrought upon them by an angry God. Again, to simple people who knew nothing of seismology, an earthquake would be seen as an act of God. Indeed, even today we call natural disasters "Acts of God" — just look on any insurance policy.

In the New Testament, the reported miracles have more to do with individuals and small scale situations. The turning of water into wine has been explained by analysts as either an illusion—the skilled application of sleight-of-hand—or else Christ was able to add something to the water that made its taste approximate wine. One suggestion is that he merely added sugar to the water to liven up the flavor and the tale became elaborated later.

Christ's documented skills raising the dead are harder to explain away, but analysts have tried. Explanations range from a similar use of suggestion or illusion to the idea that in all cases those who were apparently dead were in fact merely catatonic or unconscious, for a variety of reasons, and that Christ was sufficiently knowledgeable to know how to snap them out of this state. It has also been claimed that the greatest miracle of the New Testament—the resurrection of Christ—was again a grand deception, that Jesus did not die on the cross but was taken down before he expired, and healed by a group of his followers.

Naturally, the devout will ignore all these ideas and stick to what they believe. This should come as little surprise, for the power of faith is incredibly strong; skeptics can only offer alternatives to what they consider the unpalatable and unacceptable idea that miracles do happen. Again, David Hume has much to say on this matter. Almost two hundred years ago he pointed out that "when anyone tells me that he saw a dead man restored to life, I immediately consider with myself, whether it be more probable that this person should either deceive or be deceived, or that the fact that what he relates should really have happened. I weigh the one miracle against the other; and according to the superiority, which I discover, I pronounce my decision, and always reject the greater miracle."[3]

This is sound reasoning about which any skeptic would heartily approve, but it takes no account of what people *want* to believe. Hume is absolutely right in a purely empirical logical sense, but most people allow their feelings or their beliefs to cloud the issue so they can no longer see which is the "superior" notion.

As I said at the start of this chapter, to the atheist and the dispassionate observer, a miracle is only such to someone with a certain, rather narrow, perception of the events; someone who does not question too deeply. Even so, those who probe and conduct as thorough an investigation as possible can only offer alternatives. Because the events described in the Bible occurred so long ago and have come down to us through complex and meandering paths, it is extremely difficult to say for certain what really happened. The best the skeptic can do is to offer what should be more logical and clear-cut explanations based upon analysis rather than blind faith.

Because of this, it is in some ways more rewarding to consider modern day "miracles" and to see how they could be accounted for using fundamental scientific principles. Interestingly, orthodox religion, and most especially Catholicism, has little patience with most modern miracles and has made it very clear (quite properly) that these events should not be articles of faith and should not be presented as "evidence" of God's existence or in any way validating the teachings of the Church. Indeed, the Catholic church has rigorously investigated many of these accounts in an effort to find flaws in them, and only after intense scrutiny have they considered some to be of interest, but still not intrinsic to religious doctrine.

The cynic would argue that the Church has been forced into this approach by the need to accommodate itself to science, and that it would be dangerous to support claims of paranormal happenings such as those made by stigmatics, believers in apparitions of the Virgin Mary, or those who claim that the bodies of saints have remained physically uncorrupted after death.

Probably the most sensational modern-day miracles are what has become known as "Marian apparitions," claims that the Virgin Mary has appeared to the faithful. Such claims have been made by hundreds of witnesses and may be traced back centuries. Interestingly, a high proportion of the witnesses, especially in recent times, have been children.

In some cases the vision interacts with the witness. There have been reports from those convinced they have seen the Virgin Mary and that she spoke to them, extended her arms as though to embrace them, to leave them messages given seemingly by telepathic means.

The most famous case of visitation comes from the experiences of a young girl named Bernadette Soubirous, who at the age of thirteen, in 1858, is said to have seen a vision of the Virgin Mary at Lourdes in France. According to Bernadette's own testimony, the Virgin Mary appeared to be a young girl no bigger than herself and dressed in white. She witnessed the vision a total of eighteen times, and gradually the apparition revealed more information about herself and the purpose of the visitation.

The vision informed Bernadette that a chapel should be built on the site. Later, she was told to wash her face in a stream, but there was no water nearby. Before witnesses, Bernadette then scrambled in the earth and water flowed forth. Apparently, soon after, a blind man was made to see after bathing his eyes with the water, and before long the site at Lourdes became the most important shrine in the world, visited by hundreds of thousands of pilgrims each year. (See Chapter 12)

A visitation almost as famous occurred in 1917, at Fatima in Portugal. Three young children from a peasant family saw the figure of a woman surrounded by a bright light floating above a tree in an isolated field near their native village. The woman told them that she was from heaven and that she would appear on the thirteenth day of each month. At the appointed time the following month, the children again saw the vision, and gradually, during the next few months, the story leaked out and adults began to accompany the three children to the site.

Eventually, a crowd of over seventy thousand people was attracted to the spot in the field near Fatima later that year to witness the spectacle. However, no one but the three children ever saw the image of the Virgin, although many in the crowd that October afternoon claimed they had witnessed the miracle of the sun "wobbling" in the sky after a downpour had soaked them. According to reports, immediately after the sun performed this trick, the clothes of the soaked crowd were found to be completely dry.

So, what is to be made of these claims?

The first effort at explaining these events is to suppose that the original witness or witnesses have become hysterical or self-deluded, and that they were able to subvert the thinking and the emotions of others.

The phenomena of mass hysteria and contagious hysteria are actually common occurrences and have been studied extensively by psychologists. It is an idea we have already encountered in the scientific investigation of ghosts, poltergeists, and alien abduction. The human mind is a very powerful instrument with great (and in many cases) still hidden and little understood potential. If an individual wants to believe strongly enough or has suffered an intense emotional disturbance, the mind can play very powerful tricks. It is then a question of the deluded creating further delusion in others, who may themselves be susceptible to suggestion. In the case of apparitions, these susceptible individuals have an intense religious faith.

It is certainly no coincidence that such experiences are frequently linked with religious events. Intense religious feelings are often the strongest that any human being ever feels, and may be even more powerful and deeply rooted than personal love for another human being. With this sort of force at work in the human psyche, other more mundane ulterior motives may be perverted by the subconscious mind.

In the case of Bernadette, there is a strong possibility that her deeply held religious convictions were exaggerated at a time when she would have been experiencing the natural chemical imbalances associated with puberty. If she believed intensely enough, perhaps the two forces fed off one another. Displaying profound faith is infectious in the right environment, and she could have convinced others the vision was genuine. From there on the myth became self-perpetuating.

The children at Fatima may have had other reasons for the strength of their conviction. Perhaps they formed a pact to try to become famous, and gradually began to believe in their own piece of fiction. The fact that no one else, not one of the seventy thousand witnesses who were eventually drawn into the net, could claim with conviction that they were seeing what the children saw, is strongly suggestive that there really was nothing to see. As for the strange events surrounding the behavior of the sun, anomalous meteorological incidences in which odd lighting conditions produce the illusion that the sun is moving unnaturally are more common than one might expect. The fact that the crowd had been soaked by a rain storm only moments before the sun was supposed to wobble adds weight to this claim because such conditions—bright sunlight through a wet atmosphere—are exactly those most likely to generate optical illusions and mirage-type effects. (See Chapter 9)

It would seem likely that visitations of the Virgin Mary are a combination of powerful self-delusion combined with a high level of mass suggestion, all within a highly charged atmosphere of religious fervor and exaggerated by a profound willingness to believe. A startling example of this occurred as recently as 1997 and precipitated a slightly different version of the apparition story; unusually, one not directly associated with religion.

When Diana, Princess of Wales, died and her coffin was positioned in a private room in St. James Palace, crowds queued around the block for days to write comments in a series of books placed on long tables in the public hall of the palace. The atmosphere among the patiently waiting crowds was charged with intense emotion. The population of Britain had been drained by the media coverage and the various twists and turns in the unfolding saga of the princess's tragic end. But then, for some people, it seems that the emotionally charged atmosphere simply became too much to take.

On the third day, as people lined up in the September warmth, someone claimed that they saw Diana's face in a painting in the great hall. The word spread, and within hours dozens of people swore they too could see the face of the princess. Then others announced that they had seen her in the glass windows along the corridor leading to the hall. People in the queue were interviewed by TV crews and newspaper journalists; all declared that they had seen a vision. But what was most striking about this incident was how the story became elaborated, almost like a game of "telephone." What began as an outline of Diana's head seen in a painting grew to a full-length glowing vision, an image of the "divine" Diana enshrouded in celestial light. The only surprise was that no one ended up claiming they had seen the princess standing beside them in line, alive and well, resurrected on the third day after her death!

An alternative to the religious experience of visitation are the many reported cases where statues of religious figures have been seen to bleed, to produce or absorb other liquids, and, even in some rare cases, to move.

Weeping Madonnas have been reported all over the world. Many have been explained easily as trickery. More often than not the fake is crude—a pump is placed inside the Madonna and the porous material of the statue absorbs what is often genuine human blood, which eventually finds its way to a part of the statue where the enamel has been scratched away—often the tear ducts or sometimes the palms of the hands.

More sophisticated fakes involve allowing the liquid to be absorbed by the porous material of the statue, then removing the pump and allowing capillary action to suck the liquid to the desired spot.

A few years ago a new idolatry craze hit the headlines. According to certain reports, statues in Hindu temples in India were observed to drink milk from a spoon placed at the mouth of the idol. News spread, and within days worshipers in temples as far afield as East London and New York were feeding milk to the images of their gods. As expected, the scientific community quickly came up with explanations. The high absorbency of the material used in making the statues soaked up the milk. Indeed, the liquid need not have been milk at all, any liquid would have done the same.

More difficult to explain is a famous case from Naples, Italy.

According to its guardians, for over six hundred years the blood of St. Gennaro has been preserved in a vessel used in an annual ceremony in the city. The ceremony involves a procession through part of Naples, during which the vessel of blood is swung on a ceremonial chain. At the beginning of the procession the vessel is seen to contain what can only be described as a congealed brown goop, but after the vessel has been swung around by the priest for an hour or two the brown gel turns into a red liquid that flows as easily as blood.

Many devout Catholics believe this to be a miracle, but the Church does not officially recognize this annual event, and it is certainly not sanctioned or deemed an article of faith by the Vatican.

In 1902 the blood was analyzed by scientists, who were only allowed to investigate the liquid through the glass (indeed even today, the local Church authorities will not allow even a tiny amount of the liquid to be removed from the vessel). Using a technique called *spectroscopic analysis*, in which light is passed through the glass vessel and the resultant spectrum studied, scientists were able to reach certain conclusions about the chemical makeup of the mixture. They found that the vessel indeed contained blood, but it was mixed with other unidentified contaminants.

Oddly, this conclusion seems to have comforted the believers, and even modern writers think that this discovery somehow adds weight to claims that this is a miraculous process. The investigator of the paranormal, Jenny Randles, has written: "This quashes some assertions that it is purely a chemical concoction."[4]

What is meant by this is unclear. Blood is, after all, a concoction of chemicals. A more recent investigation conducted in 1991 by an Italian chemist, Professor Garlaschelli, confirmed the results of 1902. The material in the vessel is indeed blood mixed with other materials. But Garlaschelli has also constructed a theory to explain what may be happening in the vessel during the procession. He has produced a blend of chemicals that exhibit what is called *thixotropy*, which is a property possessed by some substances allowing them to turn from solid to liquid when exposed to external forces.

Nondrip paint is a material that exhibits thixotropy. When the brush applies the paint to a surface, it is in the liquid state because of the movement of the brush and the force applied by the painter. When the force is removed, the paint quickly solidifies and does not run.

In the case of St. Gennaro's blood, the devout have erroneously concluded that the prayers of the faithful in some way change the chemical consistency of the blood of a long dead saint. In reality, a purely physical process changes the chemical composition of a complex blend of materials. The substance in the vessel may well comprise in part the blood of St. Gennaro, but combined with other substances in the vessel, it has taken on the rare property of being able to change consistency through thixotropy.

Closely linked to the experiences of those who claim to see visions of religious figures (usually the Virgin Mary) are the surprisingly large number of cases involving stigmata—spontaneous manifestations of bloody wounds more often than not related to what Jesus Christ may have suffered during crucifixion.

The earliest documented case of stigmata is attributed to St. Francis of Assisi. These began to appear shortly after he had a vision in the year 1224. Like modern-day stigmatics, St. Francis produced wounds in the palms of his hands and in the middle of his feet which bled profusely. According to legend, St. Francis's stigmata was so powerful, real nails could be seen in the wounds on his hands until after his death. These were apparently witnessed by the many hundreds of pilgrims who saw him lying in his coffin. Of course, such tales are difficult to verify seven hundred years later, and it is quite possible that the detail of the nails was faked by self-interested custodians of the St. Francis enigma.

Today, investigators of the phenomenon of the stigmata have pinpointed five common locations for them. Known as the "typical stigmata," these appear on the two hands, two feet, and in the side. The last of these is supposed to represent the point of entry of the spear stuck into Christ's body during the torture of the crucifixion. Sometimes, but less commonly, wounds appear above the eyebrows and around the head of stigmatics, which represent the placement of the crown of thorns on Christ's head.

There have been more than three hundred documented cases of stigmatism since St. Frances of Assisi. Until recently there were far more women displaying stigmata than men, but the gap is closing. This is thought to be because men are becoming less inhibited and more able to discuss what may be viewed as their feminine side, including (in admittedly rare cases) discussion of their own stigmata experiences.

The most famous modern stigmatic was Padre Pio, an Italian Roman Catholic priest who first received the stigmata while praying in September 1915. By 1918 he had all five of the typical stigmata and was observed bleeding profusely from his wounds on many occasions and by large numbers of witnesses.

Another stigmatic, a Bavarian woman named Therese Neumann, born in 1898, experienced dramatic stigmata in all five places and across her eyebrows every Friday for many years. Each Friday she would lose as much as a pint of blood, and according to reports, up to 3.5 kilos (eight pounds) in body weight. Even more miraculous was the fact that witnesses swore that her wounds were always completely healed by Sunday, thus mirroring the three-day passion of Christ.

Before considering some possible physical explanations for these experiences, it is important to consider the psychological aspects of such cases.

Again, we return to the theme of hysteria, and the enormous energy of religious devotion. We have seen elsewhere (see Chapter 7) that when properly trained, the mind can control the physical actions of the body. Adepts can overcome pain and accomplish the most amazing feats of endurance. And, in one sense, we can perceive religious obsession and the regimen some religious fanatics follow as a form of "training" or preparation of physical endurance. Constant denial is really a way in which certain individuals can achieve astonishing results which those who lead "normal," comfortable lives could hardly contemplate. It should not be surprising then to discover that these same individuals may be able to spontaneously generate odd effects within their own bodies.

This is supported by one glaring fact unearthed by research into this phenomenon—that stigmata first appeared (possibly with St. Francis) within only a few years of the earliest graphically painted representations of Christ's crucifixion. Until the beginning of the thirteenth century, artists did not show accurate anatomical details surrounding Christ's suffering. It was only around 1220, three or four years *before* the stigmata appeared on St. Francis's body, that the first vivid representations of nails driven through flesh were seen by the faithful. However, even more striking is the fact that in those early representations the wounds were always shown to be in the palms of the hands and the center of the feet. Yet it is now known that the nails that pinned Christ to the cross were driven through the wrists and ankles and not the places where stigmatics regularly display their wounds.

This does not imply that St. Francis or any of the other three hundred cases of stigmatism are fakes (although some may well be), but does it suggest that the wounds have been generated by some strange mechanism in the tortured and obsessed mind of the devout? That stigmatics produce these wounds in the places they believe they should appear?

Some researchers are convinced that many of the wounds displayed by stigmatics are merely self-mutilation. Dr. Eric Dingwall of the Society for Psychical Research conducted extensive studies of the phenomenon in the 1920s and reached the clear-cut conclusion that all stigmatics were victims of their own lifestyle. He cited in particular St. Mary Magdalen de Pazzi, who was apparently a stigmatic during the last years of her life in the 1580s. Dingwall found that St. Mary frequently starved herself and her nuns, went in for exhaustive self-flagellation, and that many of the nuns under her care suffered physical torture, both applied by others and self-inflicted during their devotions. He concluded that such a regimen could quite naturally lead to definite physical signs of abuse.

But Dingwall's conclusions do not satisfy modern researchers, and there have been cases involving people who had led an otherwise normal life. One example is the first case of a black stigmatic—a Californian girl, Cloretta Robertson, who first showed signs of stigmata when she was ten years old in 1972. Cloretta was not a devout Christian and came from a family that was not particularly religious. However, what was significant about her case was that a week before the bleeding wounds appeared in the palms of her hands, she had seen a television program about the crucifixion that moved her deeply.

The implication from this, and some other modern cases such as that of Therese Neumann, is that there is a strong psychological element in the precipitation of spontaneous wounds.

Professor John Cornwell of Cambridge University has put forward the theory that stigmata may be caused by a psychosomatic disorder called *psychogenic purpura*. This is a rare disease in which patients bleed spontaneously without displaying wounds. In some cases visible wounds appear later. However, the apparent link with stigmata goes further, because patients suffering from psychogenic purpura usually show symptoms after experiencing severe emotional trauma. Cases have also been known in which patients have been observed to bleed spontaneously following hypnotic suggestion.

There is little doubt that the majority of stigmata are self-inflicted injury and produced for effect—a physical manifestation of devotion, a fashion accessory for the psychotically pious. But it would also appear that some rare cases are genuine. However, they are not produced by "divine intervention." Instead, they appear to be the result of intense emotional involvement with the *image* and emotional wellspring of religion. In some rare individuals a biochemical abnormality blends with powerful emotional forces to produce a physical manifestation of their obsession.

Perhaps even more mysterious than stigmata is the final "miracle" to be considered in this category—the phenomenon of *incorruptibility*.

This is a rare phenomenon and involves the apparent preservation of dead bodies by purely natural means. Almost all documented cases of incorruptibility involve the remains of important figures in the Christian church, particularly saintly figures. The reason for the predominance of accounts involving saints or beloved religious figures is that these people are the most likely to be reinterred and their physical condition studied after burial.

One investigation conducted during the 1950s by the Jesuit researcher of paranormal phenomena, Father Herbert Thurston, involved analysis of the physical remains of forty-two long-dead saints. Father Thurston found that twenty-two of these were in a better condition than would be expected for their age.

Probably the best documented case of incorruptibility is that of St. Bernadette (Bernadette Soubirous), the young girl who claimed to have seen the Virgin Mary and established the shrine at Lourdes. After her vision, Bernadette became a nun and died young, at the age of thirty-four in the convent of St. Gildard in Nevers, France. In 1909, thirty years after she died, Bernadette's body was exhumed, and, according to an eyewitness: "Not the least trace of corruption nor any bad odor could be perceived."[5]

Often, incorruptibility is associated with other strange phenomena. Sometimes, instead of producing the rank smell of a decomposed body, the saintly corpse is said to give off a pleasant odor, usually described as "fruity." In some cases oil has been exuded from the skin of the dead body. Indeed, the body of Marie Marguerite des Anges is said to have produced so much oil that it was used to burn the lamps in the chapel of her convent. Ironically, before her death she had prayed that her body would be burned as a sacrifice to the blessed sacrament.

Incorruptibility has been observed outside the circle of religious figures. In Kiev there are seventy-three very well-preserved bodies naturally mummified and lying in open coffins. And in a set of catacombs in Palermo, Sicily, hundreds of bodies have been exposed to the air, in some cases for over two centuries, yet none have decomposed to anything like the degree they should have, and none are skeletal.

Although dramatic, the phenomenon of incorruptibility can be explained using basic biochemistry. A major clue to the mechanism at work here is the reports of fruity odors emanating from the dead bodies. Religious fanatics often submit themselves to extremely rigorous lifestyles, and the most usual way they do this is to starve themselves. As a result, the fat content of their bodies is exceptionally low. When any animal dies, the bacteria that live naturally inside it begin to break down the structures in the body. The first materials to be broken down are fats, and then the bacteria move on to the body's proteins. In the case of a very thin person, there is little fat to break down, so the bacteria quickly move on to the proteins. This is called the *deamination* of proteins, and the chemical product of this breakdown is chemicals called *ketones*, which have a fruity smell.

Another significant series of biochemical events explains how the bodies of some saints appear to be remarkably well-preserved even many years after their death. When the bacteria decompose the body, they begin in the intestines and move outward. But in the case of someone who has starved themselves, the bacteria have very little to feed on and many die or go into a dormant state. Meanwhile, the skin of the corpse starts to dry out and gradually hardens. Over a period of time, if the atmospheric conditions are appropriate, the skin dries to the point where it is almost leathery in texture and the surviving bacteria cannot break through. The eventual result of this is that the corpse appears to be perfectly preserved, but inside it has completely decomposed, the organs and connective tissue digested by bacteria.

So what is to be concluded from this analysis?

Certainly some cases of individuals suffering stigmata are genuine. Equally, incorruptibility—or at least outward, *apparent* incorruptibility—is also a well-documented and quite genuine phenomenon. Visions and apparitions are almost certainly due to hysterical responses and suggestion exaggerated by religious fervor. Descriptions of biblical "miracles" are unreliable because they have been handed down to us after generations of revision and reworking, delivered from an ancient time when the universe seemed to be beyond the very understanding of mere mortals. Almost all the extraordinary events described in both the Old and New Testaments can be explained by applying simple scientific ideas, a sprinkling of logic, and common sense, all viewed with a healthy degree of detachment. As soon as emotional commitment or "faith" come into the equation, any hope of a realistic solution is destroyed.

The universe harbors many secrets, and happily, humanity is on an endless journey to unravel these secrets; it is the excitement of unraveling these mysteries of life that make them worth the effort of investigation. But as we have seen with many of the phenomena described in this book, most of these mysteries require knowledge to unravel them truly. Any mystery can be easily explained if we are not discerning or if we are obsessed with a notion, but these are not true answers in the universal and most profound sense. They are merely self-satisfying solutions.

Such a false process can be illustrated by the old joke about the researcher who wanted to discover how spiders could hear sounds. He placed a spider in a box and made a loud noise at one end, causing the spider to run in the opposite direction. Next he took the spider out of the box, cut off all its legs, placed it back in the box and again produced the loud noise. When the spider did not move, the researcher concluded that the spider's ears were in its legs.

17 Searching for the Secrets of Life

A most wonderful majesty and archmajesty is the tincture of
sacred alchemy, the marvellous science of the secret
philosophy, the singular gift bestowed upon men through the
grace of Almighty God—which men have never discovered
through the labour of their own hands, but only by the
revelation and the teaching of others.

THOMAS NORTON, *The Ordinall of Alchimy*, 1477

Strands of long gray hair protrude from the base of the old man's
cap. His sweaty face looks incredibly thin as golden light from the
fire dances across his features. He stokes the fire and peers through
the haze into the receptacle glowing iridescent amidst the flames,
then steps back and sits on his stool, staring into the light. The rest
of the smoke-filled room is gloomy; the dim light of early morning
hardly illuminates rows of glassware, metal tools, and jars of mercury
lying in shadow upon wooden shelves. Beneath the jars are rows of
books, and in the half-light, handwritten hieroglyphics can just be
seen covering pages tinged in brown.

For many weeks the alchemist has worked alone, always at night.
So often he has fallen asleep only to awake suddenly to see demons
at his throat, beasts suspended in the air mocking him before they
disperse and fade.

Then suddenly, in the fire, he sees it: there, lying at the bottom
of the receptacle, a nascent glow, a glimmer of treasure. He leans
closer, his fingers narrowly avoiding the flames. Suppressing his excitement, he studies the globule of shining metal at the bottom of
the container. With a pair of metal tongs, he pulls the glass vessel
from the fire and holds it up to the glow of the flames and looks
through it. When he is satisfied he has made no mistake, he moves
his stool to a low bench at the back of the room, leans over the
empty page of his notebook and begins to describe the technique
that has enabled him to fulfill his dream. He scribbles fitfully about

his findings, staring as he does so at the ingot of material laying in a puddle of pure gold at the bottom of the crucible—the dreamed of *Philosophers' Stone*.

Alchemy has been practiced for thousands of years, and it has its followers today. Some say that the art has its roots in truly ancient times, before recorded history, and that such figures as Moses were adepts. However, this is almost certainly an example of the exaggerated claims that is part of the alchemist's stock-in-trade.

The first known alchemical work was probably a book called *Phusika kai mustika* (On natural and initiatory things), written by one Bolos of Mendes, which contains instructions for the making of dyes and the working of precious metals and gems. When exactly this was written is uncertain, but it probably dates from around 250 B.C. It does not describe transmutations or processes to produce what has become known as the "elixir of life" (these were later obsessions), and so in many respects it is atypical of the material alchemists studied in translated form in Europe some fifteen hundred years later.

Most alchemists from the Dark Ages to the beginning of the scientific era were inspired by the belief that alchemical wisdom extended back many centuries to ancient times, and that its roots could be found within a collection of ideas known as the "Hermetic tradition."

The literal meaning of this is: "body of occult knowledge," and it was believed to have originated in the mists of time and "bestowed" on humanity through supernatural agents. The very title, "Hermetic tradition," derives form the god Hermes, and a mythical figure known as Hermes Trismegistus (Hermes the Thrice Great) is credited with the composition of some of the most important early alchemical works. Venerated by alchemists throughout history, it was said of Hermes Trismegistus that he "saw the totality of things. Having seen, he understood. Having understood, he had the power to reveal and show. And indeed what he knew, he wrote down. What he wrote he mostly hid away, keeping silence rather than speaking out, so that every generation coming into the world had to seek out these things."[1]

Naturally, it was in the alchemist's own interests to further the misconception that his art had mysterious, ancient foundations, because it added even greater exclusivity and self-importance to his ideas. Furthermore, it became an essential feature that the alchemist's techniques were hidden or, as this writer describes it: "every generation coming into the world had to seek out these things."

In reality, almost all the techniques used by the alchemists of Europe, and indeed their most sacred texts, did not date back to the era of Old Testament times but originated instead in the city of Alexandria around 200 to 300 A.D.

The earliest theoretical basis for alchemy in the West stems from Aristotle's notion of the four elements and the concept that one element may, under the correct conditions, be *transmuted* into any of the others. Aristotle also believed that each element had special properties that were related to human emotions and characteristics. According to this philosophy: Fire is related to the blood, and therefore passion; Water is manifest in phlegm, too much of it produces laziness; Earth is found in black bile, which is associated with melancholy, while Air is present in yellow bile and is linked with the emotion of anger. According to this set of ideas, all matter can be transmuted into any other by changing the proportions of the four basic elements from which it is composed. For example, relatively valueless lead could be changed into precious gold simply by realigning the proportions of the four elements. Fire, Earth, Air, and Water, present in the lead.

In terms of technique, alchemy was greatly developed by the Arabs following the fall of Alexandria at the end of the fourth century. So much had been lost in the destruction of the city's great library that a mere outline of the subject survived, but from this a slightly altered form of alchemy developed. Early Syriac texts, which constituted the majority of the elementary works in alchemy, were soon translated into Arabic, and this teaching spread beyond the Near East. But the most important changes to alchemy came not so much through development of processes or special laboratory methods as from the "spiritual" and metaphysical foundations of the subject.

We may think of alchemy as being based upon two central

themes—the search for the Philosophers' Stone and the production of a supernatural elixir of life. Both of these concepts came not from the ancient texts supposedly handed down by figures such as Moses, but from ancient China. The Chinese, who may actually have been the first alchemists anywhere in the world, were certainly the earliest to try to produce a "magical" material with the power to change "base" metals into gold. They were also the first recorded seekers of a potion that could restore life or endow eternal youth. In fact they were so enthusiastic that they conducted experiments with various concoctions on convicted criminals—the original lab rats!

The search that probably began in Alexandria and was pursued by the Arab philosophers of later centuries eventually led to Europe. With the fall of Rome, Western Europe descended into the Dark Ages. It was not until the eleventh century that learning started to return, through the migration of Arabic philosophy and science and the beginnings of trade between East and West. As a result, many of the early alchemical works originating in Alexandria and later modified by the Arabs were translated into Latin and spread the art throughout the Continent.

In European history, many innovations are attributed to the great twentieth century philosopher Roger Bacon. In some ways Bacon was a man born far ahead of this time, and he is believed to have rediscovered gunpowder (probably first developed by Chinese alchemists over a thousand years earlier). He also drew designs for a telescope several hundred years before the Dutch astronomer Hans Lippershey reinvented it in 1608. Although he followed many traditional teachings, particularly conventional Christian doctrine, he was a practicing alchemist and wrote a treatise entitled *Speculum Alchimiae* (The Mirror of Alchemy), which was published in 1597. He is also known to have been one of the first to see the usefulness of experiment, and wrote three farsighted tracts, *Opus Majus*, *Opus Minor*, and *Opus Tertium*, which outline his philosophy and his experimental techniques in a range of disciplines, much of which we would now recognize as alchemical. Bacon's work established his reputation for posterity, but others viewed his ideas as too close to the occult and anti-Establishment (which in those days meant

they did not agree with Aristotle's accepted ideas about Nature).*
Seen by the then pope, Nicholas IV, as a subversive, Bacon as
imprisoned for life as a heretic.

Other famous adepts who mixed what would now be considered
"respectable" science with magic were contemporaries of Bacon,
Albertus Magnus and his pupil Thomas Aquinas. As well as developing notions of natural philosophy such as the aphorism "like seeks
like," and ideas about the nature of fire which were worthy for
their time, Magnus and Aquinas were also said to have produced
automatons that could speak and acted as domestic servants by use
of the Philosophers' Stone and a mysterious *exlicir vitae* (elixir of
life). But typically for commentators of the time, it was not enough
for a philosopher to understand how things worked, he had to perform tricks, feats of "magic," to prove himself. Albertus Magnus was
believed to be able to control the weather and to influence the
seasons. In keeping with the intellectual mood of the era, rather
than by formulating theories, it was with a spirit of wonder and a
belief in the occult properties at work in the world (what we see as
the forces of Nature) that dictated how life flowed—and it was within
this climate that the ideas of the alchemists flourished.

Instead of being content to help develop a coherent philosophy
or a "science," they pursued their own fantasies and their own desires, and helped to quell any attempt to formalize their art by
following these dreams in their distinctly individualistic ways. Only
rarely do alchemical tracts agree upon any method or technique.

Later, during the fifteenth century, the pursuits of the alchemist
were given a huge boost by the rediscovery of lost Hermetic texts
given the collective title *Corpus Hermeticum*. This treatise was written from the viewpoint of a seeker of truth who is led to the wonders
of the universe by an ominpotent being, and it begins: "Once upon
a time, when I had begun to think about the things that are, and
my thoughts had soared high aloft, while my bodily senses had been

---

*A note here to alleviate any misunderstanding—although I mentioned earlier that early
alchemical reasoning was based upon Aristotle's idea of the four elements of Nature,
Aristotle himself should not be thought of as an alchemist. Indeed, his almost entirely
wrong ideas about science were the bedrock of orthodox Natural Philosophy (science) for
about 1,400 years and the polar opposite of many concepts held sacred by the alchemists.

put under restraint by sleep—yet not such sleep as that of men weighed down by fullness of food or by bodily weariness—I thought there came to me a being of vast and boundless magnitude, who called me by name, and said to me, 'What do you wish to hear and see, and to learn and to come to know by thought?' 'Who are you?' I said. 'I,' said he, 'am Poimandres, the Mind of the Sovereignty.' 'I would feign learn,' said I, 'the things that are, and understand their nature, and get knowledge of God.' "[2]

In 1460, Cosimo di Medici, a Florentine duke had sent emissaries around the world in an effort to track down ancient manuscripts about the Hermetic arts. A monk came to him claiming that he had in his possession a work written by none other than Hermes Trismegistus himself and dating from the time of the ancient Egyptians. It was not until 1614 that the manuscript was found to be no older than the second or third century A.D., but during the intervening period it inspired several generations of alchemists throughout the Continent and beyond, and was probably the single most important factor in the massive growth of interest in the Hermetic tradition and the occult during the Renaissance.

By the sixteenth century there were literally hundreds of wandering magi who found sponsors from impressionable wealthy merchants and nobility. Others who became enamored of the art were already rich—the sons of wealthy men, who actually lost the family fortune chasing dreams of making unlimited amounts of gold. Often these seekers died in abject poverty having succeeded only in wasting their lives trying to hunt down their fantasies.

One example is the case of Bernard of Treves. Born in Padua to a wealthy noble family in 1406, he succeeded in squandering his entire inheritance during a lifetime dedicated to alchemical fantasies. Chasing one crazy method after another, tricked by a succession of thieves and fraudsters during a lifetime of travel through Europe, Bernard ended his days as a beggar on the island of Rhodes, at the age of eighty-five.

However, many of the great names of European Medieval and Renaissance philosophy have also been associated with alchemy, and during this prescientific era the distinction between what would later be refined into "science" and what was clearly "magic" was blurred.

It is also clear that many early technologies and elements of pre-Newtonian scientific knowledge were intermeshed with some of the more bizarre notions of the Alexandrian magi.

For some five hundred years, from the early twelfth century to the mid-sixteenth, Europe was the new center of the alchemical world, a place where sages and "wise men" traveled freely from state to state in search of elusive but greatly prized wonders. Many alchemists wrote of their adventures and their experiments, but almost always their recipes were coded so that others could not copy them without first gaining insights into the art and undergoing special initiation. Some spent their entire lives attempting to decode the works of others and adding their own interpretation to handed-down wisdom.

In some countries and at certain times, alchemists were tolerated, even encouraged and financed by the ruling monarch; in other states, alchemists and magicians were reviled and their practices deemed illegal.

In England, alchemists received mixed fortunes. In 1404, Henry IV made the practice of alchemy a capital offense because it was thought that if alchemists could succeed, they would disturb the status quo by producing vast amounts of gold that would destabilize the economy. However, Queen Elizabeth I employed alchemists in an attempt to boost the royal coffers. One of her favorites was John Dee, a gifted natural philosopher as well as a deluded alchemist and occultist.

One alchemist, Jon Aurelio Augurello, who lived in Italy during the fifteenth century, presented the then pope, Leo X, with his latest alchemical work, *Crysopeia*, which described the process of making gold. Having dedicated the book to Leo, Augurello was hopeful the pope would return the favor with a reward. He did—Leo recalled him to the Papal Court and with great pomp and ceremony drew from his pocket an empty purse and presented it to the penniless alchemist, saying that because he was such a great magician and could make gold, he would need a purse to keep it in.

Other rulers were more extreme, both as supporters and avengers. Frederick of Wurzburg maintained special gallows for hanging alchemists and used them frequently, while Pope John XXII, himself a practicing alchemist, actively encouraged others in the art.

Hundreds of alchemists wrote books about the techniques they used, but they deliberately obscured their meaning with codes or poetic language so that other alchemists could not copy their work. A good example is the writings of a female alchemist from the second century, Kleopatra, which begins: "Take from the four elements the arsenic which is highest and lowest, the white and the red, the male and the female in equal balance, so that they may be joined to one another. For just as the bird warms her eggs with her heat and brings them to their appointed term, so yourselves warm your composition and bring it to its appointed term."[3]

Of course, the other reason they hid their findings was to cover up the fact that they were totally unsuccessful in finding what they were after—unlimited amounts of gold. Quite simply, most alchemists were chasing sunbeams in their attempts to transmute matter.

Alchemists could never have hoped to succeed, because they were attempting to transform the basic fabric of matter by using nothing more powerful than a furnace and a mixture of simple chemicals. Transmutation is only possible today in the heart of nuclear reactors where large atoms are split into small elements in a process called "nuclear fission." It is now possible to produce gold from other metals, but the amount of energy needed (and therefore the cost involved) in doing this would be far greater than the value of the material produced by the end of the process.

The methods of the alchemists were very basic. During the fifteenth and sixteenth centuries, when alchemy was at its most popular in Europe, they usually began by mixing in a mortar three substances—a metal ore, usually impure iron; another metal, often lead or mercury; and an acid of organic origin, most typically citric acid from fruit or vegetables. They ground these together sometimes for up to six months, to ensure complete mixing, and the blend was heated carefully in a crucible, the temperature allowed to rise very slowly until it reached an optimum and kept there for ten days. This was a dangerous process that produced toxic fumes, and many an alchemist working in cramped, unventilated rooms succumbed to poisoning from mercury vapor; others went slowly mad.

After the allotted period of heating was completed, the material in the crucible was removed and dissolved in an acid. For many generations, alchemists experimented with different types of solvent, and in this way nitric, sulphuric, and ethanoic acid were all discovered (possibly by the Arabs of the fourth and fifth century). This dissolution process had to be conducted under polarized light (light that vibrates in only one plane), which they believed they could produce by using sunlight reflected by a mirror or working solely by moonlight.

After the material was successfully dissolved in the solvent, the next step was to evaporate and reconstitute the material—to distill it. This distillation process was the most delicate and time-consuming stage of the whole operation and often took the alchemist years to complete to his satisfaction. It was also another highly dangerous phase—the laboratory fire was never allowed to go out, and claimed many lives through the centuries.

If the experimenter was not consumed by flame and the material was not lost through poor control, then the alchemist could move on to the next stage, a step most clearly linked with mysticism. According to most alchemical texts, the moment when distillation should be stopped was determined by "a sign." No two alchemical manuals agreed upon when or how this should happen, and the poor alchemist simply had to wait until he deemed it the most propitious moment to stop the distillation and to move on to the next stage.

The material was then removed from the distillation equipment and an oxidizing agent added. This was usually potassium nitrate, a substance certainly known to the ancient Chinese and quite possibly to the Alexandrians. However, combined with sulphur from the metal ore and carbon from the organic acid, the alchemist then had, quite literally, an explosive mixture—gunpowder.

It was by reaching this stage that Roger Bacon probably made his discovery in the thirteenth century, and many an alchemist who survived poisoning and fire ended his days going up with his laboratory.

Those who managed to master all these stages of the complex and time-consuming process were then able to continue to the final stages, where the mixture was sealed in a special container and warmed carefully. After cooling the material, a white solid was sometimes observed, known as the White Stone and capable, it was claimed, of transmuting base metals into silver. The most ambitious stage—producing a red solid called the Red Rose by an elaborate process of warming, cooling, and purifying the distillate, could lead eventually, the alchemist believed, to the production of the ultimate substance, the Philosophers' Stone itself, the fabled material that could transmute any substance into pure gold.

All of these stages in the process were described in the literature allegorically and were enveloped in mystical language and secret, esoteric meaning. So the blending of the original ingredients and their fusion through the use of heat was described as "setting the two dragons at war with one another." In this way the male and the female elements of the substances symbolized by a king and a queen were released and then recombined or "married." This was the concept behind one of the most famous of all alchemical books, the allegorical romance, *The Chemical Wedding*, which on one level has been interpreted as describing the transmutation process.

As well as this physical aspect of alchemy, we should consider the psychological element, the purely "spiritual" dimension to the work of the alchemist.

The spiritual element of the experiment was also the key to the alchemist's philosophy. It is this that has led some writers on the subject to suggestion that, for many alchemists, the practical process was a side issue and their search was really for the elixir or the Philosophers' Stone *within them*. In other words, by conducting what seems a mundane set of tasks, they were actually following a path to enlightenment—allowing *themselves* to be transmuted into "gold." For this reason, the alchemist placed great store by the concept of "purity of spirit," and often spent many long years in preparation for the task of transmutation before even touching a crucible.

Carl Jung was fascinated with alchemy and wrote a great deal on the subject. He was particularly interested in the motivation of the alchemist, as well as what they were really seeking, and came to the conclusion that alchemical emblems bore a close relationship to dream imagery. This observation eventually led him to one of the most important breakthroughs in his thinking, the concept of the "collective unconscious." In this theory, Jung speculated that at a deep level of the subconscious mind, the psyche of a person merges with the collective psyche of humankind, so that all individuals share a common heritage of symbols or images. These he called *archetypes*, and they manifested, he believed, in our dreams and affect waking thought patterns subconsciously. Jung was very interested in his own dreams and studied them carefully. In some he saw alchemical imagery. "Before I discovered alchemy," he wrote, "I had a series of dreams which repeatedly dealt with the same theme. Beside my house stood another, that is to say, another wing or annex, which was strange to me. Each time I would wonder in my dream why I did not know this house, although it had apparently always been there. Finally came a dream in which I reached the other wing. I discovered there a wonderful library, dating largely from the sixteenth and seventeenth centuries. Large, fat folio volumes, bound in pigskin, stood along the walls. Among them were a number of books embellished with copper engravings of a strange character, and illustrations containing curious symbols such as I had never seen before. At the time I did not know to what they referred; only much later did I recognize them as alchemical symbols. In the dream I was conscious only of the fascination exerted by them and by the entire library."[4]

The alchemists, Jung believed, had been inadvertently tapping into the collective unconscious. This led them to assume they were following a spiritual path to enlightenment when they were actually liberating their subconscious minds through the use of ritual. This is not far removed from other ritualistic events—those exploited by faith-healers, the ecstasy experienced by ritualistic voodoo dancers, or charismatic Christian services. Jung said of alchemy: "The alchemical stone symbolizes something that can never be lost or dissolved, something eternal that some alchemists compared to the mystical experience of God within one's own soul. It usually takes prolonged suffering to burn away all the superfluous psychic elements concealing the stone. But some profound inner experience of the Self does occur to most people at least once in a lifetime. From the psychological standpoint, a genuinely religious attitude consists of an effort to discover this unique experience and gradually to keep in tune with it (it is relevant that the stone is itself something permanent), so that the Self becomes an inner partner toward whom one's attention is continually turned."[5]

To the alchemist, the most important factor in the practice was participation of the individual experimenter in the process of transmutation. The genuine alchemist was convinced that the emotional and spiritual characteristics of the individual experimenter was involved intimately with the success or failure of the experiment. And it is this concept, more than any other aspect of alchemy, that distinguishes it from orthodox chemistry—the scientific discipline that began to supersede it at the end of the seventeenth century. The alchemist placed inordinate importance upon this element of his work, and for many skeptics this pushed the subject into the realms of *magic*, and left it forever beyond the boundaries of "science."

Enthusiasts of alchemy claim there are many parallels between modern physics and the traditions of alchemy and refer especially to what they see as the anthropomorphic dimension of some of the latest ideas at the forefront of quantum mechanics. But these claims are quite unjustified.

The main point of confusion comes from the idea that, according to some versions of quantum theory, the experimenter plays a role in the experiment—that the experimenter can affect the outcome of an experiment (see Chapter 6). But although this may appear to be a link with the beliefs of the alchemist, there is no direct comparison between this result and the idea of the alchemist influencing the contents of his crucible.

Quantum theory is an exact, mathematical science based upon a collection of fundamental concepts that show rigorous consistency and relate closely and cohesively with other scientific disciplines. Most important, *quantum theory works*. Without it, we would have no lasers, television, satellite communications—the entire catalogue of modern technological trappings. The practicality of quantum theory is unquestionable, whereas the usefulness of alchemy in pushing back the barriers of cutting-edge science is nonexistent. Modern physics is demonstrable, and most significantly, *it is repeatable*. Although to the uninitiated the language of science is indecipherable (much like that of the ancient alchemists, some declare), it is never-theless a common and very strict language, consistent and communi-cable—unlike that of the alchemists of lore. And unlike these intrepid adventurers, the modern seeker of Truth, the physicist, does not hide behind a facade of mystical code; and works independently of religious feeling or emotional character.

Put simply, modern chemistry is very different from alchemy because it is not based upon a faith system, but is a unified subject with shared rules and mathematical integrity which sticks to empiri-cal knowledge and logical experiment.

Having said this, it is important to remember that all the efforts of the alchemists down the centuries, from ancient times until it began to wane in the seventeenth century, did actually pay divi-dends. The alchemists invented many techniques, including heating methods, decanting, recrystallization, and evaporation, and pio-neered the use of a vast range of chemical apparatus, including heating equipment and specialized glassware.

Successive generations of alchemists also refined the technique of distillation, based upon the earliest form of the method practiced by the magi of Alexandria almost two thousand years ago. Today, no chemical laboratory would be complete without distillation appa-ratus, alcohol could not be produced in large quantities without a still, and the same equipment on a much grander scale allows crude oil to be separated into its components in an oil refinery.

These practical applications of alchemy are very useful and have made a difference to the practice of modern science, but there is another aspect to alchemy that provided even greater real benefits. Alchemy had an enormous impact upon several important scientists during the seventeenth century, including Isaac Newton, Isaac Bar-row, and Robert Boyle. Of these, the most important was Isaac New-ton, who was no mere dabbler in the art of alchemy, but actually derived his famous laws through his obsession with the occult.[6] Quite simply, modern physics, which has its foundation in Newton's work, is not based purely upon scientific experiment and mathematics but has come about largely through Newton's alchemical experiments.

Newton was obsessively secretive about his interest in the occult because his enemies would have used it to illustrate a conflict with

his conventional scientific work, and because officially at least, alchemy was still illegal and carried the death penalty. But by the
time biographers came to consider his life, Newton was dead, the
need to hide his interest in the occult gone, and the subject was
discussed openly by academics soon after. His early biographers were
confused by the body of incriminating evidence found in Newton's
vast library and within his huge collection of papers and notebooks.
It was apparent that the most respected scientist in history, the model
for the scientific method, had spent a greater portion of his life
intensely involved with alchemy than he had researching pure science. They also confirmed what a few of Newton's close friends
knew during his lifetime—that he had expended a vast amount of
his time studying the chronology of the Bible, prophesy, investigating
natural magic, and, most of all, attempting to unravel the Hermetic
secrets—the *prisca sapientia*, all of which had greatly influenced his
scientific thinking and the evolution of his epoch-making
discoveries.

After studying the contents of Newton's secret papers—those documents, manuscripts, and notebooks ignored by early Newton biographers—the great economist and Newton scholar, Maynard Keynes
delivered a Royal Society lecture in 1936 in which he concluded
that Newton was "the last of the magicians, the last of the Babylonians and Sumarians, the last great mind which looked out on the
visible and intellectual world with the same eyes as those who began
to build our intellectual inheritance rather less than ten thousand
years ago. Isaac Newton, a posthumous child born with no father
on Christmas Day, 1642, was the last wonder-child to whom the
Magi could do sincere and appropriate homage."[7]

At the time of his death, his library contained 169 books on
alchemy and chemistry, including works by some of the most important names in the history of the subject, and it was said that he
possessed the finest and most extensive collection of alchemical texts
ever accumulated up to his day. Amongst these books was a copy
of the Rosicrucian Manifestos published in *The Fame and Confes-
sion of the Fraternity R. C.*, an English translation by the alchemist
Thomas Vaughan in 1652 which is heavily annnotated by Newton

himself.* He also read two important books about the Rosicrucian movement by a famous alchemist, Michael Maier, *Themis aurea* and *Symbola aureae mensae duodecim,* and made extensive notes on all three works.

In all, Newton possessed nine works by Maier, eight by the celebrated Spanish alchemist Raymund Lull, who was a contemporary of Roger Bacon, and four volumes by a Benedictine monk named Basilius Valentinus (who was a peer of the unfortunate Bernard of Treves). Along with these were works by Thomas Vaughan under his pen name of Eirenæus Philalethes; texts by the well-known sixteenth-century English alchemist George Ripley; those written by the great Polish adept, Michael Sendivogius; and, perhaps most important, one of Newton's first and most used purchases of alchemical literature—the six-volume collection *Theatrum Chemicum Britannicum* by another key English alchemist, Elias Ashmole.

An important figure in these books was the alchemist Paracelsus, a man whose name has become almost synonymous with early medical practice. Born near Zurich in 1493, he traveled around Europe searching for the secrets of the Ancients, squandering much of his talent and any money he earned along the way. Like most of his fellow seekers, he died in poverty, discredited by the intellectual establishment. What made him unusual was his interest in applying alchemy to medicine.

"[Alchemy's] special work is this," he wrote. "To make arcana [a celestial power Paracelsus believed to be contained in metals], and direct these to disease . . . The physician must judge the nature of Medicine according to the stars . . . Since medicine is worthless save in so far as it is from heaven, it is necessary that it shall be derived from heaven . . . Know, therefore, that it is arcana alone which are strength and virtues. They are, moreover, volatile substances, without bodies; they are a chaos, clear, pellucid, and in the power of a star."[8]

---

*The Rosicrucians were a secret society who believed they possessed supernatural powers. They caused quite a stir in seventeenth-century France, Germany, and England, but gradually faded from the scene. However, some believe that they continue to exist today and play a keen role in guiding world political affairs. One can only assume they are rather weary of the task!

Another key figure from alchemical lore prominent in Newton's library was Cornelius Agrippa. He was said to possess extraordinary powers of the mind and body, but he too was misled by impossible dreams and never managed to bring to reality these hypothetical abilities. A contemporary of Paracelsus, he traveled widely, working in turn for the Emperor Maximillian, King Francis I, and Margaret of Austria, and along the way turned down a generous invitation to join the court of Henry VIII. He wrote a number of books in his lifetime, some of which appeared in Newton's library in translated form or as part of collections. His most important book was the *Vanity and Nothingness of Human Knowledge*. Other works considered various means of transmutation, and it is these that probably interested Newton most when he began his own alchemical researches.

By comparison, the alchemists of the sixteenth and seventeenth centuries, men such as Michael Maier, Thomas Vaughan, Robert Fludd, Elias Ashmole, and others, were far more realistic in their approach and ideals.

Maier, who was born in Germany in 1566, was another academic who for many years held a respectable position as physician to Emperor Rudolph II. After the death of the Emperor in 1612, he traveled throughout Europe, establishing a network of contacts with other alchemists and philosophers. He came to England and for a time was a close associate of the English alchemist and Rosicrucian, Robert Fludd. Later, between 1614 and 1620, he embarked on writing a collection of books which proved to be highly influential in the alchemical world. These included a book of alchemical emblems or symbols called *Atalanta fugiens*, which contained esoteric text relating to a blending of alchemy with rationalism and orthodox religion. Some see this as one of the earliest models for the ethos behind the establishment of the Royal Society.

As well as owning the largest private collection of occult books in the world at the time, when Newton died, the million or so words on the subject of alchemy he left behind (more than he composed on pure science) serve as a clear indication that he not only read and researched, but was himself an active experimenter and chronicler. But what did he actually achieve in the art?

Newton's greatest contributions to science began early in his career. According to the history books, his great achievement—the elucidation of the theory of gravity—came in 1666, when he was living again at his mother's house in Woolsthorpe. It is true that Newton, along with the rest of the academic community, fled Cambridge during the plague years of 1665-66, and he did indeed return to live with his mother at their rural home. It is even possible that he did one day sit under an apple tree as he mused upon the meaning of gravity and could have seen an apple fall. This may have pushed his thinking along, but it is ridiculous to believe that the whole concept of gravity came to him then in one great rush. Today it is recognized that Newton probably made up this story to conceal the fact that he had used alchemy to help derive his famous theory.

The creation of the theory of gravity took Newton almost twenty years to elucidate, and did not really take shape until he wrote his great book, the *Principia Mathematica*, between 1684 and its publication in 1687. And during the twenty years taking us from the garden in Woolsthorpe to the appearance of this work, there were many influences that shaped the theory. Most important was mathematics. As a student, he formulated a relationship between the distance between two bodies such as planets and the force of gravity between them, called the *inverse square law*. Newton was not able to explain how this might operate.

In the seventeenth century the idea that an object could influence the movement of another without actually touching it was unimaginable. This behavior is now called "action at a distance," and we take it for granted. But people in Newton's day could not understand this and saw it as magic or an occult property. Through his experiments in alchemy, Newton was able to approach gravity with a more open mind than most of his peers.

Newton began investigating alchemy in about 1669, after he had returned to Cambridge and around the time he was appointed Professor of Mathematics there. He traveled to London to buy forbidden books from fellow alchemists, and carried out his private experiments, hidden away from the authorities and his rivals within the scientific community. His earliest experiments were very basic, but after reading everything he could about the practice, he soon pushed the art beyond the limits set by his predecessors. In true scientific fashion, he approached experiment logically and with great precision, meticulously writing up what he had discovered. Whereas the alchemists of lore fumbled around for years not really knowing what they were doing, Newton approached his work systematically.

Another great difference between Newton and his predecessors was that he was never interested in making gold. His sole purpose in studying alchemy was to find what he believed were hidden basic laws that governed the universe. He may not have realized that he would come to a theory of gravity through alchemy and other occult practices, but he did think there was some basic law or hidden ancient knowledge to be found from his researches.

The breakthrough came from alchemy when Newton observed materials in his crucible and realized they were acting under the influence of *forces*. He could see particles attracted to each other and other particles repelled by their neighbors without any physical contact or tangible link between them. In other words, he saw action at a distance within the alchemist's crucible. He wrote to his friend Robert Boyle about his discovery, describing what he saw as a "secret principle in nature by which liquors are sociable to some things & unsociable to others."[9] Soon, he began to realize that this might also be how gravity worked and that what happened in the microcosm of the crucible and the alchemists fire could perhaps also happen in the macrocosm—the world of planets and suns.

These influences came to fruition in the *Principia*, seen as probably the most important scientific treatise ever written, ironically, a book that came not just from Newton's genius for science, but his obsession with the occult and the arcane lore of the Ancients. And from the *Principia* came the Industrial Revolution, modern science, and much of the everyday technology we take for granted today.

And that is really where alchemy and its uses come to an end. Yet, some refuse to accept this. Surprisingly, perhaps, at the end of the twentieth century there remain believers in alchemy. Although discredited by the advent of empirical science, alchemy has survived and maintained a following throughout the Enlightenment, circumventing Victorian rationalism, the succession of technology, and the atomic theory.

As a final refuge for those who insist the world is flat and that NASA faked the Apollo moon-landings, the ancient art of alchemy is a warm haven, with a hall of fame filled with fellow eccentrics and men of great intellect. Ironically, these great men of previous ages produced something workable almost by default, and in so doing created some of the most important science in the history of human civilization.

18 We Are Made of Stars

Thank your lucky stars . . .

Of all paranormal concepts, astrology probably touches the lives of more people than any other. It is also the single most respected and documented supernatural idea among those inclined toward New Age or alternative philosophies. Yet, at the same time, to the scientist, astrology is the most vague and archaic pastime, rooted in primitive thought and misguided in the extreme.

To rationalists, astrology is an irritating puzzle, for when those we believe to be otherwise rational, intelligent people let slip at dinner parties that "there may be something in it," or that "maybe there are secrets about astrology that we do not understand," the rationalist begins to wonder if they have chosen the right friends. How many times have any of us been at a gathering when the subject of astrology comes up and someone has declared that they can tell the star signs of others in the room?

According to some statistics, ninety-nine percent of people know their star sign, and an estimated fifty percent of the population consult horoscopes regularly. One company that has set up a horoscope telephone line, called "Astroline," claim they have in excess of one million calls per year. The reason for the huge public interest in astrology and these frankly disturbing statistics is a complex issue, but one worth addressing.

First, astrology is now a mass market phenomenon. Ironically, for an age fascinated with the occult in a variety of manifestations, the Victorians were not terribly interested in astrology or horoscopes, and it was only revived with the advent of mass media. Originally, the newspaper horoscope, the favorite of all tabloid newspapers, was created in the 1930s to boost circulation, and at one point in the 1980s the horoscope page of the late mass media astrologer Patric Walker was said to have increased weekly sales of the British newspaper, the *Mail On Sunday,* by a staggering 200,000.

I will look at the differences between this form of astrology and what the serious practitioners deem the real "science" of astrology later in this chapter, but it is clear from the success of modern-day popular astrologers such as Russell Grant and Mystic Meg that there is a great public need for such people.

Part of the appeal is of course entertainment. Horoscopes are about everyone's favorite subject—"me"—and therefore they generate an undying appeal. Newspaper astrology is dynamic—it keeps offering new stories for the readers and it is always, always comforting. In an ever-changing world in which many people feel they have become less than a number, horoscopes offer a personal touch, something in which they can immerse themselves, something with which they can feel comfortable.

And, like many aspects of the paranormal, astrology is also undemanding. Real science, the sort that moves civilization forward, the sort that has made all our lives infinitely better than it was for our ancestors, is to many people hostile, frightening, and difficult. To understand the universe through science requires either training or the dedication required to read books. Astrology is an easy route to what some believe to be an enlightened view, it is thought to bestow "mystical secrets," what so many people believe to be the real "keys to the universe."

Even orthodox religion presents problems to those searching for something beyond science or beyond the "difficult" intellectual world. Religion makes just as many demands as science, but these are emotional and spiritual rather than intellectual or rational. Most orthodox religions require commitment to a complete bundle of ideas or a faith system, and many parts of this package may not appeal. To complete the problem for religion at the turn of the millennium, religious orthodoxy is perceived, especially by the young, as distinctly uncool. Western religion appears to ignore the flow of modern life, it does not seem to incorporate the needs of many people today and is a doctrine that has been marginalized by the broader sweep of history. To those who believe this, but still seek a spiritual aspect to their lives, astrology offers an easy path that is directed straight to the self.

Celebrities seem particularly attracted to astrology. The Queen-mother is said to have consulted astrologers, and the late Princess Diana is reported to have had long discussions with a clairvoyant and astrologer only days before her death in August 1997 and employed an astrologer regularly for years. Pop stars, Hollywood actors, and television personalities have also recently "come out" as fans and users of astrology, and belief in astrology has become something of a fashion statement. To cater to this need and to fuel the flames, bookstore shelves creak under the weight of books on the subject in all its possible permutations—love signs, astrology and the body, your horoscope, sex and the stars—all of it geared to massage the ego and calm the nerves of stressed-out modern people, an understandable need.

So, given the very real, human reasons for the enduring popularity of astrology, what is its basis? And is there anything in what the enthusiasts claim categorically to be a "science"?

Astrology is a truly ancient practice. No one is certain when or where it began. Practitioners assign overwrought significance to its supposed origins, placing it in the ancient realm of Atlantis and suggesting that the Ancients bestowed astrological knowledge on the Sumarians and the Babylonians. The truth is probably more prosaic: the original seed of astrology was cultivated about ten thousand years ago, but where exactly remains a mystery. Some theories suggest that monuments such as Stonehenge, which are around five thousand years old, could have been constructed for astrological purposes; others place the first dabblings earlier and in the Middle East.

The oldest relic linking humans with astrology comes from a document called the Venus tablet, or the *Enuma Anu Enlil*, dated to the age of the Babylonian civilization, sometime between 1800 and 800 B.C., and containing astrological references and talk of "omens." The tablet reads: "In month eleven, fifteenth day, Venus disappeared in the west. Three days it stayed away, then on the eighteenth day it became visible in the east. Springs will open and Adad will bring his rain and Ea his floods. Messages of reconciliation will be sent from king to king."

Early Greek civilization adopted astrology, and it was placed in high esteem by the philosophers of the time. There was no delineation between astronomy and astrology from the era of the Greeks almost until the time of Galileo, and the same mathematical tools were employed to study the movement of the stars as those used to help astrologers make prophesies and personal star charts.*

As we saw in the last chapter, modern European culture was spawned at the end of the Dark Ages by the arrival in Europe of Arabic philosophy and science in the eleventh and twelfth centuries. It was wrapped up with a fascination with alchemy and the ancient mysticism or Hermetic secrets, the *prisca sapientia*. By the Renaissance, around the beginning of the fifteenth century, interest in the ancient teachings and the writings of men such as Aristotle, Archimedes, Galen, Democritus, and others had reached a new peak as ancient texts were discovered in monasteries and private libraries and a new learning based upon the old ideas of two millennia earlier flourished. Our modern preoccupation with astrology had begun.

In one form or another an abiding interest in astrology survived the Age of Reason, but it fell into decline during the late eighteenth and early nineteenth centuries, partly, it is believed, because of the vogue for everything "modern"—the desperate and quite laudable clamor toward the age of technology. Isaac Newton and others had crystallized the concept of the universe as a mechanical thing—the planets held in orbit by gravity—indeed, a mysterious force, but one that could be understood using human ingenuity and mathematics. Suddenly the ancient gods and the flux of existence no longer seemed all-powerful and beyond us. As human civilization rushed headlong through the Industrial Revolution and into the era of modern medicine, modern astronomy, and a world ruled by Darwin, Marx, and Einstein, belief in astrology became terribly unfashionable, an embarrassment.

---

*Galileo was, incidentally, a practicing astrologer as well as one of the first modern astronomers. However, historians believe that he "played" with astrology simply to help pay the bills and thought little of it as a "science."

Many scientists and rationalists would wish things had stayed that way. Richard Dawkins has called astrology "meta-twaddle" and has declared that "there's a thing called being so open-minded your brain drops out." Immediately after the Industrial Revolution the upright citizens of the western world were content with their faith in God and machine, they had no urgent need to find spirituality elsewhere, and indeed, they would have considered the modern disrespect for orthodox religion as blasphemous and the search for solace in eastern religion, alternative philosophies, and astrology as fickle.

Today, the art of astrology is experiencing something of a renaissance, but there is astrology and astrology, and if we are to appraise the subject properly, we must have clear definitions of what it is all about.

To the serious astrologer, newspaper gurus of the subject—the Russell Grants and Mystic Megs—are nothing more than charlatans who are merely in it for the money. To them, the nonsense we see spouted out on television shows and in the tabloid press demeans what they consider to be a serious subject based upon very strict and highly complex rules. So, effectively, for the moment, we can ignore this aspect of astrology, consign it to the intellectual garbage can; which is actually nothing more than the "serious" professional astrologer would demand of us.

However, just because high-powered astrologers who claim to be masters of a complex craft say that their work is based upon ancient and elaborate mathematical concepts does not make it true. Indeed, constructing a complete birth chart for an individual requires an element of skill unnecessary for a tabloid columnist. But actually, even this so-called technical expertise requires little more than high-school-level mathematics. To draw up a birth chart, a half-dozen factors need to be considered—fewer than a simple engineering task such as building a wall.

The astrologer needs a basic grounding in geometry and a little trigonometry, as well as a smattering of simple astronomy. But this is nothing a reasonable intelligent twelve-year-old could not manage. To coin a popular but quite apt phrase: "It's hardly nuclear physics."

As well as this quite minimal mathematical ability, the professional astrologer employs a strong element of interpretation based upon ancient handed-down wisdom. Although the "serious" astrologer surrounds his or her trade with a veneer of expertise, and frankly makes too much of the mathematical skill involved and the training required to master the art of astrology, there is also the question of motive. Expert astrologers and critics are united in condemning tabloid astrology and horoscopes, but surely the motivations and the drives of the high-brow practitioners are little different. Do they not offer fantasies wrapped up in what they like to think is complex mathematics? The only real difference between them is the market to which they appeal. In England, self-proclaimed elitist astrologers such as Shelley Von Strunckel and Sally Brompton—who is a graduate of something called the Faculty of Astrological Studies in London—write books aimed at the "highbrow" end of the popular astrology market and act as personal consultants for the rich and the famous.

Such divisions in a single practice produce an amazing range of beliefs and subsets. It is possible to see more rival disciplines within the broader sweep of astrology than any other aspect of the paranormal. So, not only do we have the upscale practitioners and the "lowbrow" tabloid horoscope hacks, but there are those who believe astrology is the overriding power that shapes us, determines who we marry, what careers we take up, and our state of health. Yet, others see astrology as merely a means of interpretation, a little like tarot or the use of the I Ching.

Some astrologers believe that the stars influence our health. This was an idea first popularized by the alchemist, astrologer, and all-around mystic Paracelsus, who believed there was an intimate connection between the elements—the four substances of which, according to Aristotle, all matter was made—and the stars in the heavens. He linked the behavior of the elements, the movements of the stars, and the well-being of his patients in a triumvirate that led to an astrological alchemy that some modern astrologers still support.

Others totally disregard the advances of the past century and place far greater significance upon the influence of the stars than they do upon factors such as psychological development, genetics, or environment. An example is a book called *Life Cycles* by an

American astrologer, Rose Elliot in which it is claimed that child development has little to do with mundane factors such as nurture or nature but is controlled by astrological influence. According to Elliot, the child's behavior when they first experience "separateness" from their parents (often referred to as "the terrible twos") is not molded by anything so prosaic as the relationship between child and parent, the presence of a younger child, the personality of the child, the genetic characteristics of the child and parents, or the environment in which the child is raised. No, it is because the planet Mars returns to its birth position in the child's star chart at the age of two, and, because Mars is identified with war and aggression, it exerts a disturbing effect on the mood of the child. Obvious, really.

So, this is the background to astrology. It is a much divided subject, split into elitist and populist groups. Like many subjects and enthusiasms, it is quite naturally composed of people with very different opinions and views and those who place different emphasis upon certain aspects of the art. It is also a very ancient practice and has its roots in the mists of history, a practice that has been manipulated and molded for different ages, but has at its heart a set of ideologies that have remained unchanged since ancient times. The astrologer believes that the planets of our solar system and the stars themselves play a fundamental role in the way we develop as individuals and what happens to us in our daily lives; that celestial objects can control our destiny and that of the world at large. To the astrologer, world events are not governed by chaos theory, a succession of random events upon which we humans try to impress our individual wishes and desires with varying degrees of success, but are instead predetermined and set in motion and maintained in equilibrium by some intangible force exerted by the planets and stars.

Let us for a moment suspend disbelief if we are skeptics, and contain our desires and wishes if we are ardent enthusiasts. How could astrology work?

Unless the influence the stars have upon our lives is due to some as-yet-unknown force, we only have conventional forces we already know about to explain it. There could possibly be other forces at work in the universe, but as science builds a clearer picture of the way the universe works, the room for new forces and strange mechanisms diminishes (although it would be extremely arrogant of any scientist to assume that all forces and all mechanisms are known). However, for the purposes of trying to work out how stars and planets could have an effect upon our lives and personalities, postulating some unknown force gets us nowhere. If we say, then, that if astrology is a true mechanism controlling the way in which we as individuals interact with the universe, we have a limited number of options.

The most popular idea has long been that some form of gravitational force is at work and responsible for the claims of astrology, that in some way the gravitational force between distant planets and ourselves causes a mysterious link so that we are all subject to this mechanism. In other words, as the stars and planets move in their paths, there is a *flux*, or force, or a "system of energies," that makes us what we are and dictates what we do. This then leads to the actions of the individual and the future of nations.

The major flaw with this idea is that the force of gravity is extremely weak. In fact it is the weakest of the four types of natural forces—the weak nuclear, the strong nuclear, the electromagnetic, and gravitational force. It has been calculated that the gravitational force between a baby at the moment of birth and the midwife in the delivery room would be a million times greater than the gravitational influence of any planet in our solar system and an astronomically larger influence (literally) than the distant stars.

To see why this is so, we need only a brief consideration of Newton's discoveries about gravity made over three hundred years ago. Newton created a grand theory, the Law of Universal Gravitation, which demonstrates that every single material thing in the universe exerts a force upon any other material thing in the universe. So far, so good, for the astrologer. The problem is, not only is the force of gravity weak, it depends upon the distance between two pieces of matter and diminishes in power the further apart the objects are.

Newton showed that the way the force of gravity between pieces of matter changes over distance may be calculated using what is called an *inverse square law*. For example, imagine two planets A and B, orbiting a sun. Suppose A and B are of equal size, but the distance between A and the sun is half the distance between B and the sun.

This will mean that the force of gravity between A and the sun will be four times greater than between planet B and the sun. Similarly, a third planet C (of equal mass to A or B) which is three times farther from the sun than A, will experience a gravitational attraction (1/9) of that between A and the sun.

Everything in the universe adheres to this inverse square law. This is why the midwife (a relatively small object, but much, much closer to the newborn baby) actually exerts a far greater gravitational influence upon the child than a planet hundreds of millions of miles away.

Often, astrologers and enthusiasts who want to believe in astrology try to use Newton's law of gravitation to produce what they like to believe are explanations for the power of planetary influence, but they miss the point completely. The most familiar argument goes something like this: "The moon creates tides on Earth because of the gravitational interaction of the two objects—the Earth and the moon—why, then, shouldn't the moon and the planets also affect the human brain? After all, we are made up almost entirely of water."

The first part of this argument is quite correct. The gravitational forces that act between the Earth and the moon, and indeed the Earth and sun, create the effect of tides.* However, this is because we are dealing with very large masses. There is no comparison between the gravitational effect between the moon and such a small mass as a single human and that between the moon and the entire planet Earth.

So, what other forces could we consider as candidates to explain the claims of the enthusiasts? Well, there are actually very few. There is the *tidal force* between any two objects. This is linked to gravitational forces, but the effect of distance in calculating its influ-

---

It is also interesting to note that there is such a thing as *land tides*, again caused by gravity. But because we are dealing with solids these are much weaker effects. It is also worth noting that the moon presents the same face to the Earth at all times because of gravitational effects—the Earth being the larger body, long ago "despun" the moon. This has happened to all large moons in the solar system.

ence is even greater, so this force would have even less impact upon us.

Aside from these possibilities there are the other three fundamental forces, the weak and strong nuclear forces and electromagnetism. But each of these operate over short distances compared with the force of gravity. The weak nuclear force is responsible for the decay of unstable nuclei, electromagnetism requires bodies to be charged for any form of interaction to take place and accounts for electrical interactions and a large proportion of chemical processes. The strong nuclear force operates between subatomic particles in the order of 10^{-12} centimeters apart and has infinitesimally small influence at any greater distances.

However, these problems have not stifled the creation of many imaginative ideas to explain how astrology could work. The exciting idea that there may be an explanation for existence or the meaning of life in the stars, and the allure of finding how that could operate, has even drawn some otherwise respectable scientists into the net.

Professor Peter Roberts, author of the book *The Message of Astrology*, has suggested that the human body can somehow act as a conduit for a mysterious form of what he describes as "resonant planetary interaction." This is a pseudoscientific term conjured up to describe what Roberts believes is the influence of planets acting through some mysterious force that travels across space using a wave with a frequency sympathetic with some sort of "life force" or energy within all of us. Unfortunately he does not explain where this resonance originates, how it operates, or how it impacts on human beings, using anything other than the vaguest terms. So in effect it leads us nowhere. Roberts is joined by another scientist, an astronomer named Dr. Percy Seymour, author of a New Age tract called *Astrology: The Evidence of Science.*

The problem is, all this is again terribly vague and takes us little further than the astrologers who talk about some strange energy coming from the planets and stars. Using these muddled ideas about resonant planetary interaction, Seymour even has the temerity to try to dismiss the scientific objection to talk of external forces emanating from the cosmos influencing us at birth. In a recent newspaper article, the journalist (who was clearly on the side of astrology), said:

"Trotted out by detracting scientists time and again, the standard objection [to any form of force such as gravity influencing our lives at the moment of birth] is that the magnetic resonance of the planet is so slight that it would be swamped by the electrical equipment in the hospital, or at home by the likes of storage heaters. By way of rebuttal, Dr. Seymour proceeds as if ripping through a rather dim first-former's test paper. 'Firstly bear in mind the old trick of an opera singer shattering a wineglass. It only works when the voice resonates at the same frequency as the atoms of the glass. So, in the hospital, there's no question of planetary resonance being swamped because the electrical equipment will be operating at different—that is, the "wrong"—frequencies. If your radio is not tuned to a sole station, you won't hear it.' "[1]

These are neat analogies, but this "explanation" actually answers nothing and certainly is not an argument that rips through "a rather dim first-former's test paper." Dr. Seymour assumes that the frequencies generated by the machinery nearby would be the "wrong" frequencies, but how does he know? Where is the evidence? What single scrap of proof can he or any other enthusiast for astrology show to demonstrate irrefutably that there is a strange resonance that affects all our lives? Let alone at which frequency it operates. It is pure speculation—something I can only assume he would never allow in his official work as an astronomer.

Furthermore, what does "magnetic resonance" really mean in the sense these authors use it? Has anyone ever noticed the work of this mysterious force other than using known interactions between particles and electromagnetism, such as that at the heart of a very useful laboratory device called a *nuclear magnetic resonance spectrometer* (NMR)?

As Professor Seymour's description shows, magnetism is one other possible contender from the armory of the enthusiast often used in an effort to define the force that is supposed to enable planetary influence. But again, magnetism is an incredibly weak property that only operates over short distances—try it yourself with a compass and a small bar magnet.

In reply to this, enthusiasts point to the fact that migratory birds are known to be able to navigate using the lines of magnetic force around the earth. Why then, they argue, could magnetism not lie at the root of planetary influence in astrology? The answer is similar to that used to counter the astrologer's hijacking of Newton's law of gravitation. Migratory birds do not make interplanetary journeys. The lines of force they mysteriously and fascinatingly employ are powerful magnetic lines of influence that are produced by the huge iron core of *our planet*. Besides, if astrologers really want to insist that there is some sort of magnetic effect at work, then that influence would be dominated overpoweringly by the magnetic field of the Earth itself. In a sense, this is analogous to the comparison between the influence of the midwife and the planets—the impact of the magnetic field of the Earth would be vastly greater than any intangible magnetic force that has somehow reached us from the other planets of the solar system.

This, then, is a summary of the ways enthusiasts of astrology suggest the planets could influence us, and the counterarguments of the scientists, but there is much more to the debate between science and astrology. There are matters of a far more mundane nature—the use of experiment and the advances in astronomy during the past three hundred years.

Let us look at the first of these. What has been gleaned from scientific experiment?

The most famous attempt to quantify what the enthusiasts claim and to use statistical analysis in an attempt to reach conclusions about astrology comes from the work of the psychologist and statistician Michel Gauquelin, who summarized his endeavors in a book called *Dreams and Illusions of Astrology*.[2]

Gauquelin was interested in the idea that celestial influence may, according to the believers, play a major role in the personality of individuals, and decided to see if this was apparent by matching star signs to profession. To do this, he took the birth data of 576 members of the French Academy of Medicine. To his surprise, he found that, in what he considered to be a statistically significant number of cases, the individual's birth charts had either Mars or Saturn just risen (this is called the "ascendant") or those planets were at that moment in the midpoint of the sky (the mid-heaven). To see if there really was anything in this, Gauquelin next repeated the process with a similar sample of mixed professionals taken at random. He found no correlation between the birth charts and the careers these people had followed.

Now intrigued, Gauquelin followed the experiment through with other groups to see if there was an emerging pattern. He found what he claimed was a disproportionate number of sports champions who were "Mars people"; in other words, those who had that planet prominent in their birth charts. Meanwhile, he found that artists and writers tended to be dominated by the moon, and that Jupiter was prominent in the natal data for military leaders, journalists, and politicians.

Michel Gauquelin was never an advocate for astrology (he died in 1991) but came to the conclusion that there was some unknown scientific principle at work here, some odd link between humanity and the mathematics of nature. He certainly never subscribed to the view that the apparent connection he had stumbled upon had anything to do with the direct influence of these planets via some mysterious force operating over interplanetary space. Of course, the astrology fraternity has lionized Gauquelin and adopted him as one of their own. Time and again the man's findings are paraded as "proof" that astrological principles are correct, that in fact we are mere creatures without any self-will or independent thought.

Equally obvious has been the reaction of the scientific community, who marginalized Gauquelin when he was alive and today treat his results with contempt. Some even refer to it using the jargonese term "non robust," which means it does not really stand up to intense scrutiny. And indeed, it would seem the reliability of his findings may be brought into question.

The subject of statistics is a notoriously slippery customer, and many believe it can never be relied upon to show clear evidence for anything unless it is backed up with other experimental proof or independent analysis. One of the crucial aspects of the discipline of statistical analysis is the need for a correct *sample size*. What this means is that statistical analysis of anything is utterly useless unless enough material is involved in that analysis. For example, imagine we had no idea of the chance of obtaining heads or tails when we toss a coin. We decide to find out, but we're in a hurry, so we can only toss the coin ten times. As chance would have it, that day we toss the coin and find that it comes up heads seven times and tails three times. We might then conclude, quite wrongly, that the chances of heads is always seventy percent and tails thirty percent.

A more rigorous experimenter, or statistical analyst, would spend a few days doing the same experiment and toss the coin, say, 5,000 times. They would certainly obtain a very different result from their colleague and get, within perhaps a couple of percent, 2,500 heads and 2,500 tails, showing the probability of each to be fifty percent.

The only difference between these two experiments is the sample size. The same principle applies in any statistical analysis. If pollsters sampled ten people before an election, they would almost certainly get a very different result than if they interviewed 100,000 voters the same day. The same applies with Gauquelin's work. In each of his studies he analyzed the natal charts of about five hundred people. This is not enough, and would be considered an inadequate sample to give any truly meaningful result.

Furthermore, in the case of the first trial, he also allowed for four separate criteria—he matched medics with four planetary events: Mars in the ascendant or in "mid-heaven," and the planet Saturn in the ascendant or mid-heaven. This further reduces the significance of the results.

This is not a trivial objection on the part of skeptical scientists. If paranormal investigators want to be taken seriously and want their claims to be accepted as supportable and verifiable, they have to stick to the same rigorous rules science applies to itself, or else their ideas remain mere speculation and guesswork. But even ignoring this, further doubt now falls upon Gauquelin's work because of more recent statistical analyses that contradict it. An analysis of one in ten returns from the 1971 British census (some three million people) showed absolutely no correlation between professionals and star signs. Even more embarrassingly for the believers of astrology, the then president of the Astrological Association, Charles Harvey, made some predictions before the results were published. He claimed that most nurses would tend to be born under feminine signs, while union leaders would have been born under masculine ones. Professor Alan Smithers of Manchester University, who studied the results, found that this was actually borne out, but also announced that there was a disproportionate number of miners born under Scorpio and Capricorn—both feminine signs.

As a final piece of evidence in this debate, it is conveniently forgotten that Gauquelin himself also produced results that completely contradicted his initial findings. In a test involving two thousand army generals, his previous discoveries led him to believe that their natal charts should show a predisposition toward the sign Aries. In fact, the generals were found to have been born within a random range of signs. Furthermore, it is significant that this time the sample was larger, and therefore statistically more sound, and the test was based upon linking a single rank with a single astrological sign.

Coupled with this was a very simple study that looked at the personalities of individuals born at the same time in the same hospital. It was found that each of them developed into quite different individuals and pursued very different careers, married very different people (themselves born under different signs from those predicted by astrology), and suffered different physical illnesses.

More prosaic experiments have also produced revealing results. Analysts long ago noticed that many horoscopes contained a large number of very vague statements. Examples include such waffling as: "You have considerable hidden talent that you have not yet used to your advantage," and, "While you have some personality weaknesses, you are generally able to compensate for them." Critics of astrology have dubbed such declarations "Barnum statements" after the American showman who coined the phrase, "There's a sucker born every minute."

Most strikingly, surveys have shown that when a sample of people are shown Barnum statements from horoscopes, ninety percent of them believe the statement applies to them, and can link what is said in even the most crass tabloid horoscope to events in their lives or their hopes and aspirations. The fact that on occasion these horoscopes have been either deliberately fabricated or written by hard-pressed journalists makes no difference at all. One research scientist, Geoffrey Dean, has also noted that when horoscopes contain succinct but slightly more specific personality references, such as "you have a good imagination," they are seen as less relevant to individuals reading them than the horoscopes full of Barnum statements.[3] The reason for this is clear: with a Barnum statement, anyone can make of it what they will.

Other experiments have been, if anything, even more revealing. In a set of tests relating to his work on linking personality with properties of natal charts, Michel Gauquelin placed an advertisement in the magazine *Ici Paris* offering free horoscopes to anyone who responded to the ad. He received 150 requests and duly posted the horoscopes. He then followed this up by asking each applicant what they thought of the horoscope they had been sent. Ninety-four percent said they believed the horoscope accurately fit their personality. What Gauquelin did not tell them was that they had all received the same horoscope—that of Dr. Petroit, an infamous French mass murderer.

So much for experimental and statistical analysis of what astrology claims; but what of the so-called "science" of astrology? Here, I'm afraid, this arcane study again fails to deliver.

First, we have to consider the fact that the constellations as we see them from Earth are pictures primitive humans contrived to help them understand the universe a little better. The stars that make up these constellations are not really grouped together; indeed, most are hundreds or thousands of light-years apart, and it is only from the perspective of someone here on earth that they seem to take on patterns such as the Plough or the Great Bear.

The second anomaly in the arguments put forward by astrologers is the matter of when exactly the mysterious "astrological force" is supposed to take effect. Is it not obvious that in terms of impacting on the character of the embryonic human, the moment of conception would be far more significant than the moment of birth—the point at which this new human being simply leaves one environment to join another?

But even if we ignore these problems, what is to be made of the fact that ancient astrology, which has remained ostensibly unchanged for many thousands of years, is based upon the premise that there are only six planets in our solar system? The ancients observed only Mercury, Venus, Mars, Jupiter, and Saturn. The other three planets of the solar system have all been discovered during the past 250 years—Uranus was discovered by Sir William Herschel in 1781, Neptune in 1845, and the most distant planet, Pluto, was first observed as recently as 1930.

Astronomers point out that if these three planets were unknown, then surely all natal charts drawn up before 1930 were incorrect, even if the celestial influence claimed by astrologers is real. When questioned on this, astrologers become strangely tight-lipped. If anything, their most common response is to say that the discovery of these planets makes no difference. When pushed on the matter, the popular astrologer and author of many books on the subject, Linda Goodman, has claimed intriguingly that "a planet does not have any astrological influence until it is discovered."[4] A statement one would have thought undermines the entire premise upon which astrology is built.

Clearly, astrology should not call itself a genuine science. Typically, it played no role in the detection of Uranus, Neptune, and Pluto, and it has offered not a scrap of useful material toward the discovery of *anything* tangible. In fact, many practitioners of astrology are proud of the fact that the central tenets of the subject are rooted in ancient understanding. Linda Goodman has stated: "Alone among the sciences, astrology has spanned the centuries and made the journey intact. We shouldn't be surprised that it remains with us, unchanged by time—because astrology is truth—and truth is eternal."[5]

Sadly for the enthusiast, astrology cannot be both "a science" and "unchanging"; the two are mutually exclusive. The essence of science is experiment and a willingness to question even long-established tenets of the subject. Without this, science would be a dead subject, as dead as astrology.

Two further points reinforce this view. First, there is the matter of the planet "Vulcan." During the 1850s the codiscoverer of Neptune, Jean Joseph Leverrier, calculated that there should be another planet within the orbit of Mercury—the closest body to the sun. He accounted for this by noting that the orbit of Mercury was not what would be expected if it was the only planet in the vicinity. We now know that there is no such planet as Vulcan and that the effect Leverrier had observed is actually a consequence of general relativity and not the presence of another celestial body within the orbit of Mercury. But this has not kept astrologers from getting it wrong again.

Perhaps because they realized they had been wrong-footed over the discovery Uranus, Neptune, and Pluto, they overcompensated. Hearing about the imagined planet Vulcan, it was quickly incorporated into their false science. Take, for example, Linda Goodman again, this time in her *Star-Signs* for 1968 (long after scientists had dispatched Vulcan to the land of Nod): "It's important to mention here the still unseen planet Vulcan, the true ruler of Virgo, since its discovery is said to be imminent . . . Many astrologers feel that Vulcan, the planet of thunder, will become visible through telescopes in a few years."[6]

The final thing to consider in this discussion of the foundations of astrology is the question of the relevance of the twelve signs. Astronomers have long known of a phenomenon called *procession*. This is another name for "wobble," and is exhibited by any rotating body. As the Earth rotates, it *processes*, which means that for observers on the Earth, the relative position of the sun and the constellation changes over a period of a few centuries. Now, of course, this relative position of the sun to the constellations is the essence of astrology—a star sign is literally the constellation in which the sun was positioned at the moment of birth. The constellations and the dates of the calendar were fixed two thousand years ago, giving us certain dates for certain star signs. For example, Sagittarians are born between November 23 and December 21.

But during the past two thousand years the relationship between the dates and the star signs has shifted by at least one sign, so that Sagittarians are actually Scorpians, Aquarians are really born under Capricorn, and so on. What this means for the personality profiles that many astrologers gleefully link to star signs, I'll leave the reader to ponder.

Astrologers ignore this latest blow to the art and claim it is irrelevant; which, if we needed any further proof, demonstrates that astrologers are certainly not scientists. Scientists do not ignore verifiable, repeatable experimental evidence.

Those astrologers willing to defend the issue do so by suggesting that the signs "remember the influence of the constellations that corresponded to them two thousand years ago."[7] But how do these "scientists" explain why the old correspondence of constellations did not remember the relationship between stars and dates that would have existed longer ago than two thousand years? After all, the Earth did not start processing only two millennia in the past.

So, what are we to make of all this? Clearly, to the scientist and, to be honest, anyone who places their worldview in the realm of reason and logic, observation and experience, cannot take astrology seriously. Putting aside the matter of how astrologers view themselves, how they delineate between what they like to believe are serious astrologers and commercially minded pundits and hacks, there is no rational basis to astrology and it simply does not work. Despite what one's friend may claim after a few glasses of wine, there is absolutely nothing to astrology. The world does not operate in the way astrologers fantasize it does. Our individual characters are molded by two things—nature and nurture (genetics and environment).

And that is how the world should be. We have no need for anything more esoteric. Genetics and environment (experience) are amazing enough in their own right and give the world all the variety, excitement, and wonderful differentiation we could ask for. Together, they give us a very colorful world full of individuals, good, bad, ugly, and beautiful. With this, who needs astrology?

But beyond even this is a set of ideas and facts far more elevating and inspiring than anything irrational astrologers can offer—"real astrology." For, yes, in one sense, at least, we are linked to the heavens, but in a way no astrologer would have ever discovered or be able to explain.

The stars that fill the cosmos, the tiny furnaces we can see each clear night, were not all made at the same time. Some, like our sun, are stars in their prime. Others are forming at this moment, and there are many, many stars older than our sun.

Some stars have long since expired. When a star dies it can do so in different ways, depending upon what type of star it was in the first place. Some swell up, then explode, spewing out vast quantities of matter and energy into the universe. These events are called *supernova* and they, in part, account for why we have a wide variety of different elements in nature. Some material from an ancient supernova ended up in the gas ball that made our sun, and the Earth was formed as a globule of plasma (superhot gas) that broke away from the cooling sun some 4.5 billion years ago.

So, some material that had once been in another star found its way into the material that made our Earth. And, for all our grand ideas about ourselves, we are, in essence, the physical matter from which we are made, and we are made from elements and compounds derived from the Earth—we are all part of an ecosystem. Except that this ecosystem is not self-contained and restricted to the material within our immediate environment; there is a constant flux between the Earth and the sun, the Earth and the creatures living here, and between creatures on the Earth.

Inside you and me are atoms that were once in the heart of a star perhaps thousands of light-years from earth. And one day, a tiny part of you or me, or the book you are reading, will find its way into the heart of another sun and from there to the body of an alien being, an alien book, perhaps even the equivalent of this full stop.

19 Fire From the Sky

> I would rather believe that Yankee professors lie than that
> stones fall from heaven.
>
> THOMAS JEFFERSON, 1807

The wasteland of the Tunguska region of Siberia is very sparsely populated today, but nearly a century ago its thousands of square miles was home to no more than a handful of people. That is why, on a fresh June evening in 1908, there were very few witnesses to the most powerful explosion in modern times. That evening, an object about a hundred meters in diameter and weighing around 100,000 tons exploded some six kilometers above the ground.

Of the few eyewitness accounts, most came from a trading post seventy kilometers from the epicenter of the explosion. One report reads: "I was sitting on the porch of the house at breakfast time and looking toward the north . . . suddenly the sky was split in two, and high above the forest the whole northern part of the sky appeared to be covered with fire." Another eyewitness recounted: "Suddenly before me I saw the sky in the north open to the ground and fire pour out. We were terrified, but the sky closed again and immediately afterward bangs like gunshots were heard. We thought that stones were falling from the sky, and rushed off in terror, leaving our pail by the spring."[1]

No one really knows what the object was. The most likely explanation is that it was the nucleus of a stray comet that had crossed the path of the Earth as we make our journey around the sun. Another explanation is that it could have been a meteorite—a large chunk of rock, or possibly a small asteroid—that had originated in our solar system. There are even those who find it necessary to add this event to the UFO conspiracy theory and suggest it was a large alien craft that exploded that night.

Whatever it was, it caused enormous devastation. Estimates of the power of the explosion vary, but when the region was explored a few years later it was found that in a few seconds over two thousand square kilometers of forest—about the area of greater London—had been totally obliterated. Seismographs around the world recorded the event, and tremors were felt thousands of kilometers away on the other side of the continent. Estimates vary, but it is safe to say that the explosion was the equivalent of at least a twenty-megaton bomb, or about a thousand times more powerful than the atomic bomb dropped on Hiroshima in 1945.

As it stands, the Tunguska incident, as it is commonly known, serves as a warning for what could happen; and illustrates how, when it comes to collisions with extraterrestrial objects, our fate as a race is in the lap of the gods. But as John and Mary Gribbin have pointed out in their study of near-earth collisions, *Fire on Earth*: we should not simply consider only the catastrophic outcome of such events; if the explosion at Tunguska had occurred four thousand kilometers to the west, in St. Petersburg, human history may have been very different, for the population of the city would have been killed in an instant, including in their number a young political activist, one Vladimir Ilich Ulyanov-Lenin.

Thankfully, such events as the Tunguska incident are very rare, but not so rare that we can all feel completely safe—we have experienced too many close encounters in recent decades to be complacent, and, according to some scientists, we should be doing more to protect ourselves. Just this century there have been at least five close calls—in 1937, 1968, 1989, 1993, and most recently in 1996. We have no defense system against the possibility of another such disaster and we remain wide open to impacts.

The energy released by such explosion is truly awesome. It is an easy matter to calculate the energy contained in an approaching asteroid or comet. The material from which the object is made is irrelevant, its energy would be the same if it was made of ice, iron, or french fries. The kinetic energy can be calculated using the equation:

$$K.E.=\frac{1}{2}\ MV^2$$

This is:

Kinetic energy of the object=½ the mass of the object (m) times
its velocity (v) squared

So, a relatively small object weighing about a mere million tons
(1,000,000,000 kilograms) and traveling at a leisurely 50,000 kph
(14,000 meters per second) would produce:

$$K.E. = \frac{1}{2} \times 1,000,000,000 \times 14,000$$
$$K.E. = 7,000,000,000,000 \text{ joules of energy.}$$

That's seven thousand *billion* joules of energy, or the heat produced by a two-bar electric heater for every man, woman, and child
on Earth radiated from the point of impact in a fraction of a
second . . . And this object would be considered a small asteroid.

So far in recent years no meteorite, comet, or asteroid has landed
in a densely populated area, but as humanity spreads into almost all
corners of the Earth, it is thought to be only a matter of time before
a catastrophe greater than any earthquake, eruption, or tidal wave
could strike. And, if the object were large enough, it would not be
a mere local disaster (a few million people killed), it could mean
the end of civilization itself.

The threat comes from three main sources—comets, meteorites,
and asteroids. Comets are celestial wanderers. But that does not
mean they travel randomly around the universe. In fact, they follow
very definite paths. Some, such as Halley's comet, travel through
orbits around our sun that bring them back into the inner solar
system with relative frequency—in the case of Halley's comet, every
seventy-six years. Others have much longer orbits, so they come
around every few thousand or every few tens of thousand years. It
is these that could account for the many known close encounters
the planet has experienced in the past. There is even a suggestion
that there might be a gigantic "doomsday" comet that has an orbit
bringing it into the solar system at long intervals and may account

for some of the great disasters of legend—the biblical flood and the sinking of Atlantis, to name but two.

Using current technology, such a comet could not be detected from Earth until it had approached within a few weeks journey time. The Hubble telescope might be able to detect such an object if we knew where to look, but at present it would be like trying to find a specific individual grain of sand on a beach—an impossibility.

Although comets have been observed since ancient times, and astronomers since the eighteenth century have been able to plot the orbits of a few, it was not until the 1950s that a theory began to emerge describing their origin. There are two "belts" or "regions" where there are high concentrations of comets. The *Oort Cloud*, named after its discoverer, lies about 100,000 astronomical units (or AUs) from the sun. An astronomical unit is the distance between the sun and the Earth—93 million miles. The Oort Cloud is thought to contain several billion comets, most of which stay within the vague limits of the belt and travel at only a few hundred kilometers per hour. During the long history of our planet, one of these comets has occasionally strayed much closer and entered the inner solar system.

Another belt begins close to the orbit of one of the outer planets of the solar system, Neptune, and stretches outward to about 100 AUs. This belt contains an estimated billion comets and may account for many more of those that stray close to the Earth from time to time.

Comets are mainly comprised of a chunk of ice and frozen carbon dioxide, ammonia and methane, along with grit and dust. The mass is concentrated in the nucleus, and as the comet approaches the sun and warms up, some of the nucleus sublimes and forms a gaseous tail.

The second type of extraterrestrial object to cause trouble are meteorites. These are just chunks of rock that sometimes contain organic matter. As they hurtle through space, they are known as "meteors," and only acquire the name "meteorite" when they enter the atmosphere. Like all celestial objects, as they approach the sun and come under its gravitational influence, they take up a regular orbit. If this orbit crosses that of the Earth, then the meteor may enter the atmosphere, and if they're large enough, they survive the huge heat generated upon entry and may reach the ground.

Scientists estimate that something in the region of 75 million meteors of different sizes enter the Earth's atmosphere each year. Most of these are tiny, grain-sized pieces of rock that burn up very quickly. But between 1975 and 1992 there were 136 air bursts caused by meteors several meters across. All of these exploded before they reached the ground, but if they had survived the heat and reached the ground intact, they would have been powerful enough to devastate a town or part of a city. And, once in a while a far larger object will cross our orbit around the sun, and the resulting meteorite has the potential to be a "city crusher" or even a "civilization-killer."

But more worrying than meteorites are asteroids. Like comets, asteroids lie in a belt, but their orbits are generally far closer to us. The asteroid belt is concentrated between Mars, which occupies an orbit with an average distance of about 85 million kilometers from Earth (200 times the distance to the moon) and Jupiter, which orbits the sun at a distance of about 550 million kilometers from earth and contains several million asteroids. But of more relevance to those worried about near-Earth encounters, there are almost a hundred *Apollo asteroids*, known to occupy a region far from the main belt, and many of these intersect the orbit of the Earth.

New asteroids are being discovered all the time, but most are relatively small. Although there are only a few asteroids larger than a hundred kilometers in diameter, around half a million are more than one kilometer across—certainly large enough to cause widespread devastation on earth.

So, having reviewed the sort of objects that can come into close contact with the Earth, what has happened in the past to suggest that we may be in danger from these things in the future?

The most famous example of a collision site is the Barringer crater in Arizona. This is a massive hole 1,200 meters wide and 180 meters deep that was produced about 25,000 years ago by a meteorite or a small asteroid sixty meters wide and weighing around a million tons. The explosion that produced the hole would have been about the same power as the explosion at Tunguska. But rather than it being a twenty-megaton air burst, this was a twenty-megaton ground level explosion.

The Barringer crater is one scar among many on earth. The evidence of meteorite and asteroidal collisions is to be seen everywhere in the solar system—the craters of the moon and similar marks on the surface of all the inner planets have been produced over eons of time when meteors, undiminished by their surfaces burning away in an atmosphere, created vast impact marks. On Earth, many meteorites and asteroids have landed in the sea and left marks thousands of feet under the surface of the ocean. It has been suggested that many "rim" features have been caused by collisions. Examples include huge "pockmarks" like the Wash in Eastern England, the Gulf of Tarentum in the heel of Italy, and the vast Hudson Bay in Canada.

However, the most important collision in history may have been one that occurred 65 million years ago and precipitated the extinction of the dinosaurs, allowing mammals to become the dominant species and opening the evolutionary road for Homo sapiens.

The notion that the dinosaurs might have become extinct because of a collision between the Earth and a large celestial object had been considered for some time, but there was no hard evidence for it, it was merely a neat idea. Then, in 1978, oil prospectors discovered a massive geological structure four thousand feet beneath the Gulf of Mexico called the Chicxulub structure. This crater dwarfs Barringer, and the thought of what may have caused it and the effects it had offers a chilling prospect. The 65-million-year-old Chicxulub structure is 110 miles across. Reaching out from the center are ripples of solid rock deposited there by the impact, and around the Gulf are to be found large deposits of minerals commonly associated with meteors; in particular, there are high concentrations of the element iridium. These mineral deposits would have been dumped on the Mexican shoreline by a tidal wave the height of the World Trade Center.

The fact that the meteorite that caused the impact crater at Chicxulub landed in the sea would actually have made little difference to its destructive power. If you think about the dimensions of the oceans, you begin to realize that from the perspective of an incoming meteor, the oceans and the seas are a wafer-thin covering that would hardly cushion the power of the collision. The tidal effects would also be a secondary destructive force after the initial impact.

The probable scenario that destroyed the dinosaurs began with such a monumental impact. The shock would have jolted the planet on its path around the sun, but even such a tremendous explosion would have caused only a relatively minor orbital disturbance. The greatest damage would have been caused by the dust and water vapor thrown into the atmosphere by the force of the collision, altering the climate of the planet. The actual effect would have been a shift of perhaps a few degrees in temperature, enough to wipe out many cold-blooded species, most especially the lumbering dinosaurs.

Geologists describe five major extinctions in the history of the Earth—what they dub "the Big Five." The extinction of the dinosaurs around 65 million years ago was the most recent, and it is suspected that some of the earlier extinctions were also precipitated by similar catastrophes.

In more recent times there have been some serious catastrophes that seem to have been initiated by collisions between the Earth and extraterrestrial objects. One astronomer who specializes in asteroid and meteorite collisions, Duncan Steel of the Anglo-Australian telescope in New South Wales, has pointed out: "On average there is one impact capable of causing global catastrophe with a very large fraction of humankind perishing about once every 100,000 years."[2]

It is no mere idle speculation to suggest that asteroid or comet impacts could have caused the destruction of at least one ancient civilization, even if evidence for such events are circumstantial. One example is the destruction of Mycenean civilization, the precursors of the Greek city states. An account of the destruction comes down to us from Solon, a law-giver of sixth century B.C. Athens, known to us through the historian Plutarch. Solon described the destruction of the Myceneans as having "the air of a fable, but the truth behind it is a deviation of the bodies that revolve in heaven round the Earth and a destruction of things on earth by a great conflagration. Once more, after the usual period of years, the torrents from heaven swept down like a pestilence . . ."[3]

Although this could be dismissed as romantic fiction of the time, there are actually several pertinent points in this brief passage. The notion of a "deviation of the bodies that revolve in heaven" shows a remarkable grasp of celestial mechanics, because although it is not necessarily a "deviation" that causes a collision, more a crossing of paths, the essence of what happens is there in the description. Perhaps more significant is the last sentence, beginning: "Once more, after the usual period of years . . ." Not only does this show that the ancient Greeks had some understanding that meteorite showers and the passage of comets across our skies follow a regular pattern and reappear at intervals, but that other catastrophes dating back further than the destruction of Mycenean culture could have been attributed to similar conflagrations. The entire notion of cyclical disaster is a strong thread in many ancient texts, including the Bible. This could be accounted for by superimposing quite natural climatic and even seasonal rhythms upon the development of civilizations, and their ability to ride out lean times as well as good, but this pattern could also be attributed to a long record of disasters precipitated by "fire from the sky."

Moving closer to our own time are the events of the mid-fifth century. In their book The Cosmic Winter, authors Victor Clube and Bill Napier suggest that there was a series of disasters between the fifth and sixth centuries that wiped out cities and killed tens of thousands. They suggest that the period, prominent in legend as the time of Arthur and the Knights of the Roundtable, is remarkably lacking in lasting chronicles. This is particularly odd because the period immediately preceding it—the departure of the Romans from Britain—is extremely well-documented, and the era after Arthur is equally well-covered by historians, including the Venerable Bede, who wrote the Ecclesiastical History.

Tales from the Arthurian period were first brought to general attention by Thomas Malory writing in the fifteenth century with his Le Morte d'Arthur, whereas a chronicler who would have been a contemporary of Arthur, one Gildas, who wrote a treatise called The Ruin of Britain, did not even mention the man who was supposed to be the king of a significant portion of what soon after became England.

The supposition is that Arthur did not actually exist but was created by the writers of the time to give the public some good news to counteract the catastrophes befalling them. And the times were indeed hard. There are several accounts of major catastrophes involving what we would attribute to collisions with celestial objects. From Gildas we have: "The fire of righteous vengeance, kindled by the sins of the past, blazed from sea to sea. Once lit it did not lie down. When it had wasted town and country, it burned up the whole surface of the island until its red and savage tongue licked the western shore. All the greater towns fell. Horrible it was to see the foundation stones of towers and high walls thrown down bottom upward in the squares, mixing with the high alters and fragments of human bodies."[4] And from contemporaneous Chinese chroniclers we have an account of a destructive comet that caused "the sun to become dim" and created a darkness that "lasted for eighteen months."[5] At the very time Arthur was meant to be fighting his battles for supremacy over England, in some regions of China eighty percent of the population were dying, and the capital of the northern empire, the Wei civilization, was deserted in the year 534 A.D.

Moving forward in time to our century, there have been several frighteningly close encounters. In 1937 an asteroid called Hermes, a chunk of rock weighing some 400 million tons and traveling at an estimated 80,000 kph, missed us by a little under 750,000 kilometers. That may sound a safe enough distance, but it is only about twice as far as the moon. If the asteroid had struck, you may not have ended up reading this because the impact would have had the force of 20,000 megatons of TNT, or one million Hiroshimas, producing a crater thirteen kilometers in diameter.

In 1968 another asteroid, Icraus, passed within 6.4 million kilometers—still in our own backyard. In February 1989 a similar object to Hermes passed within a million kilometers of our planet. When this asteroid was discovered by Dr. Henry Holt, an astronomer working with the gigantic Mount Palomar optical telescope in California, he was quoted as saying: "If this one had appeared only a few hours earlier, it would have nailed us."[6] This is absolutely accurate: The impact would have occurred because the orbit of the Earth would have been crossed by the trajectory of the asteroid. If the rock had traveled a little slower or if its trajectory were only slightly different, it would have arrived at the same point in space as the Earth and at the exact same moment—with catastrophic global consequences.

Four years after this close encounter, in 1993, there was an even closer and potentially even more deadly one. An asteroid estimated to be between ten and fifteen kilometers wide passed within 140,000 kilometers, half the distance to the moon. A collision with such a monster would have almost certainly meant the end of civilization. An object this large would have made a crater fifty kilometers wide and thrown trillions of tons of dust into the atmosphere. The few who would have survived the impact, the aftershocks, and earthquakes, would have been subjected to a "cosmic winter" lasting many years.

And there has been another since then. In May 1996 a meteorite some 1,200 meters in diameter passed us at a distance of 400,000 kilometers, a little beyond the orbit of the moon. Traveling at an estimated 85,000 kph, its impact would have been equivalent to the detonation of several hundred thousand Hiroshimas.

These close encounters are not only terrifying because the objects involved came so close, but perhaps even more disturbing is the fact that none of them were observed before they arrived in our region of space. In relative terms, these asteroids and meteors are tiny pinpricks of rock hurtling through space at tens of thousands of kilometers per hour, so they are virtually impossible to track, and appeared from nowhere and completely without warning. It is a chilling thought that if one day such an object does arrive at a point in space at precisely the same moment as the Earth, we will not know about it until it happens.

So, what are the real chances of such an event? Are we worrying unnecessarily or is it something that we should take seriously? Statistics are often suspect, and applying them to this problem is every bit as flawed as trying to judge the outcome of elections or guessing the flow of lottery numbers over a year of draws, but at present there is little else to go by. Staggeringly, it is said that any one of us is more likely to die from the consequences of a collision between the

Earth and a celestial body than in an air crash or a more common natural disaster such as an earthquake or a flood. But how sound is the reasoning that gives us this result?

I mentioned earlier that statistically, on average, there is an impact capable of causing a global catastrophe every 100,000 years. No one can say for certain when the last big one occurred. The Barringer crater was caused by a relatively small impact 25,000 years ago. If it occurred today, it would not mean the end of civilization but would probably kill millions of people. The same can be said for the comet collision in Siberia in June 1908. An impact that has caused a mass extinction has not occurred since the monumental event that destroyed the dinosaurs 65 million years ago. So, taking one line of reasoning, we are very much living on borrowed time. But as the astronomer Duncan Steel has pointed out: "These objects do not follow timetables like buses, so the question 'When was the last one?' is of no significance. All we can say is that the probability of a catastrophic impact is one in 100,000 per annum."[7] What bumps up the statistics to make death by asteroid more likely than other more prosaic means is the fact that when these impacts do occur, they are totally devastating. So, although they are mercifully very rare, collisions between the Earth and large celestial objects are all-embracing.

But how real does this make the danger? The answer to that depends upon how you view statistics. The fact that there is more chance of an asteroid destroying civilization on Saturday night than you or I winning the national lottery that evening is one way of looking at it. On the other hand, the fact that we have not had a civilization killer collision for 65 million years perhaps means our luck will hold out for many more millions of years.

Recently, a growing number of scientists have begun to think there may be a very real danger from asteroids and comets. Questions have been raised in the U.S. Congress and in Parliament in Great Britain, and committees and groups have been organized to investigate the phenomenon.

In Britain, an army officer, Major Jay Tate, has compiled a report on the question of near-earth objects and comet encounters, and there are plans to establish a small investigative team to look into the phenomenon and to suggest possible practical solutions.

An international group called Spacewatch, created to monitor observable objects, concluded that there were hundreds more asteroids crossing Earth's orbit than was suspected prior to a survey they conducted in 1991. These asteroids, they claim, are part of Earth's own mini-asteroid belt—a group of asteroids that orbit the sun following relatively erratic paths, but within a few million miles of the planet's course.

David Morrison, one of the founders of the Spacewatch program, believes that governments must investigate the problem of Earth collisions and set in motion some form of protective system. After an exhaustive study of the phenomenon and a workshop held at NASA in 1992, Morrison and his colleagues proposed a scheme called the "Spaceguard Survey." Their plan is to utilize a network of six new ground-based telescopes, each with an aperture of 2.5 meters—about half the size of the huge telescope at Mount Palomar (which was until recently the largest in the world). These telescopes would be able to plot the course of all celestial objects within the inner solar system and afford a warning time of a few months. The scheme would cost no more than $50 million to establish and have running costs of up to $10 million per year.

Perhaps it should come as little surprise that when the Spaceguard scientists took the idea to Congress they received short shrift. The warning time—a mere few months—was the sticking point. For congressmen without technical know-how, and living with a heavy burden of complacency, such protection was as good as useless.

This decision was, many believe, misguided. Any warning, they say, is better than nothing, and the running costs would be equivalent to 4 cents per U.S citizen per year—a tiny fraction of what most people spend on life insurance. But at present the U.S. government vouchsafing only a paltry $500,000 per year on investigating the phenomenon. This, Morrison points out, leaves the investigation of near-earth objects (NEOs) conducted by "fewer people than it takes to run a single McDonald's."[8]

It is possible to see where the skepticism derives and why those holding the purse strings are less than impressed by the potential dangers. To a politician, the short term is everything. He or she may have a niggling fear for themselves and their descendants over whether there is a real danger from extraterrestrial objects, but the demands of balancing the books and getting reelected override these.

There is also a degree of "technocomplacency" in all of us. We are the dominant species on Planet Earth. We have constructed a great civilization, we can put humans on the moon, live in space for months, conquer most diseases, generate energy from the fundamental particles of which the universe is made—how could we possibly be obliterated by a lump of rock? Deep in the hearts of most people, especially those without the benefit of a scientific or technical training, is the conviction that we are indestructible, that somehow our technology could cope with such an event. The brutal fact is: The last close encounter came like a bolt out of the blue, unnoticed until it had skimmed past us.

Hard as it may be to accept, we could do absolutely nothing about an object that was on collision course with us unless we had at least a year to prepare. There have been suggestions that the human race could turn the nuclear weapons we once pointed at each other toward space and create a genuine "space guard" against incoming objects. That may be feasible one day, but there are no plans to use the stockpile of weapons left over from the cold war in this way just yet. Others have talked vaguely about using a form of "Star Wars" weaponry to defend ourselves. That again would be feasible within a few years, given endless supplies of cash and manpower. The "Star Wars Project" was halted during a Republican administration with the cold war still very chilly. It is hardly likely that it would be resurrected to fight an invisible enemy that may not present any tangible danger for millions of years. If the U.S. government cannot be persuaded to part with a relatively trivial sum to create Spaceguard, as already proposed by Morrison and his colleagues, then neither they nor any other power would be inclined to spend a substantial percentage of their GNP on a defense system against asteroids and comets.

So, this is where we find ourselves today. Many believe there is a real threat overhead, others dismiss the danger and point to far more pertinent problems—world famine, disease, illiteracy, AIDS, wars, and relatively frequent natural disasters. Perhaps one day the human race will be lucky enough to experience a near-miss so tangible that governments are shaken out of their complacency and made to realize the potential horrors. Until then, we just have to sit with our fingers crossed, and hope our luck holds out.

20 Our Brethren Among the Stars?

> Evolution is the law of policies: Darwin said it, Socrates endorsed it, Cuvier proved it and established it for all time in his paper on "The Survival of the Fittest." These are illustrious names, this is a mighty doctrine: nothing can ever remove it from its firm base, nothing dissolves it, but evolution.
>
> MARK TWAIN, *Three Thousand Years Among the Microbes*

To many people, scientists and nonscientists, the question of whether or not there is life on other planets is one worth little consideration. To them the answer is clear: "Yes, of course there is life in great abundance—we live in a universe overbrimming with all manner of alien beings."

In this chapter, I want to start with the assumption (false or otherwise) that this is true and to move onward, to an investigation of what forms alien life might take and what we now understand could govern the evolution of extraterrestrial life.

This is a legitimate area of scientific study and one pursued daily by genuine, salaried scientists, men and women who call themselves *exobiologists*.

What we know of alien life is of course based upon pure conjecture, but guided by the laws of science we already understand and apply regularly to more prosaic matters. The only model we have is life on earth, for this is the only world we are sure harbors life. This is of course a limitation, and we will not know whether the theories of the exobiologists are even close until we encounter other life in the universe in one form or another. But I think anyone who has reached this far in the book will agree that the exploration of this subject is at the very least fun and may lead to some interesting ideas.

To cover this subject we need to ask the questions exobiologists ask and try to formulate sensible answers based upon the very latest information available. To do this we will need to bring together a collection of disparate disciplines, just as exobiologists do, and try to sift ideas to see whether they are "right" or "wrong." Sadly, we will not be able to reach any definite conclusions, just probabilities based upon what is presently known.

The first question we need to ask is the most fundamental of all: What is life?

At a glance, the answer might seem obvious. But actually, "life" is a tremendously difficult concept to define.

We can start by suggesting that all living things grow and move, but this does not help much. Crystals grow, producing regular patterns and repeated simple units which might be compared to cells; any liquid can flow. So, in themselves, these abilities are not enough to distinguish between animate and inanimate, or living and nonliving.

Perhaps a more sophisticated answer is to say that all life uses energy. But this too is inadequate. All machines use energy. A slightly more useful definition might be to say that all living beings can *control* energy; but then so do some advanced machines such as those developed in recent years that use "intelligent" software.

Is the argument that only living things process and store information an alternative? The answer is no, because even the simplest word processor does this.

So, how can we pin down the quintessential factors that separate the living from the inanimate?

Using the old-fashioned schoolbook definition, that all living things exhibit the three F's—fight, flight, and frolic—leads us into further difficulties. Lightning might be seen as capable of "flight"— it is repelled by certain matter and attracted by others—and the word "frolic" really means "to reproduce," which is certainly not limited to living things; after all, flames "reproduce," as do certain types of crystal.

A much better definition is to say that all living things reproduce *and pass on genetic material* or inherited characteristics to their offspring, and that this material has undergone some degree of mutation. In other words, they have taken part in the evolutionary process via natural selection, they haven't simply produced exact copies of themselves. But perhaps the final word on the matter should go instead to the late Carl Sagan who, shortly before his death, defined life as: ". . . any system capable of reproduction, mutation, and the reproduction of its mutations."[1] What he meant by this was that life was represented by any entity that allows for variation from generation to generation using the mechanism of evolution through natural selection, an entity that can pass on its characteristics, reshuffled by the processes of reproduction so that those characteristics will not appear to be exactly the same in the next generation.

But even this is actually not a totally satisfactory definition for several reasons, not least of which is the question of whether a cloned creature (which was produced without sexual reproduction and is a perfect copy of its parent) is actually alive. Dolly the sheep looks very much alive, but would not fit the above definition.

So, if we want to move on in this exploration, we have to accept a definition that links life with the ability to reproduce through a mechanism that allows for mutation. This is because the only way we know that life can evolve is via this route. Dolly the sheep may be a fully functioning sheep that can do anything any other sheep can do, but she and any descendants produced via cloning will play no role in the evolutionary development of the species of which she is a member, although her naturally produced descendants could, of course.

So, it is clear that evolution and life are linked, Indeed, any biologist would support the idea that without evolution there can be no life. Whether that life is on Earth or a planet orbiting Sirius, evolution will be a fundamental process.

How would evolution work on an alien plant?

Well, we cannot be sure, but it would seem likely it would operate in the same way it does on Earth. And to understand this, we have to explore the link between evolution and genetics.

Evolution relies upon a complex series of operations involving gigantic organic molecules such as DNA (deoxyribonucleic acid) and RNA (ribonucleic acid)—commonly known as the "molecules of life"—and a set of smaller building blocks, molecules called nucleotides, that form these massive structures, as well as the proteins, enzymes, and other biochemicals needed to run cells and sustain our existence.

Of course, we can ask how we know that life on other planets is based upon DNA, or even reliant upon the element carbon. But we have to make some assumptions. If we keep questioning at every level, we hit a wall of incomprehension and can go no further. So we have to apply some basic principles.

Scientists accept that there are certain fundamental axioms—concepts and theories that seem to lie at the heart of the universe. The theory of relativity appears to be one of these, Darwinian evolution is another. A still more fundamental concept is the "principle of universality," or the idea that the universe is homogeneous. Put into everyday terms this could be translated as: "what happens here, happens there."

How do we know this is true? Well, we can observe distant stars from Earth and determine that the chemicals present in those stars are the same as those we find on Earth and in our own star, the sun (but in different proportions).

So where does this lead us in our exploration of the laws governing alien biology?

First, we have to clear up what we mean by "alien life." The chances are there is an infinite variety of alien life-forms in the universe. There may well be many ways in which life could develop and change. It would seem likely that some form of mechanism always has to allow for evolution, or else the life-form could not develop, but it is conceivable that this life-form may not be based upon DNA. If this was the case, it is equally probable that we would not recognize such life-forms and would almost certainly not be able to communicate with them. Therefore, let us restrict this chapter to an exploration of DNA-based life.

For life to have evolved in a recognizable form, it would almost certainly be based upon the element carbon, which would then allow for a biological framework involving DNA and therefore a system involving evolution via natural selection (Darwinian evolution). This is a limitation, but as we know from living on Earth, where all life-forms are carbon-based, it still offers the potential for a massive diversity of life.

So, why carbon? Thanks to the principle of universality, we can safely say that carbon is the only element that will lie at the center of a biological system yielding life as we know it. The reason for this is that carbon possesses unique properties. In many ways it is similar to other elements, but in one vital respect it is different: it is the only known atom that can form the backbone of really large molecules, called *organic* molecules, and even larger conglomerates called *biochemicals*. It also has an almost unique ability to form long chains and rings of atoms around which other atoms can be attached. I say "almost" because other atoms can form chains and rings, but they do not show anything like the versatility of carbon. The best example, silicon, shares some of the characteristics of carbon, but because the bonds formed between silicon atoms are not as strong as those between carbon atoms, it can only form stable chains up to five or six atoms in length and is unable to form multiple bonds or cyclic structures, which carbon does easily.

Because of these facts, carbon is unique. It is the only atom able to form huge molecules—the building blocks of life.

And, because of the principle of universality, we know that this is true not only here in our local environment but is a fact of life everywhere in the observable universe. We know there cannot be another atom like carbon, one we have inadvertently overlooked, because such an atom would not fit into what is called the periodic table—a scheme in which all the different types of atom in the universe have a strict position and interconnect in a precise pattern. This periodic table, devised over a century ago by a Russian chemist, Dmitry Ivanovich Mendeleyev, allocates a position for all the elements, and it is inconceivable that some odd element perhaps found only in the Horsehead Nebula could be squeezed in. Over the decades since it was first established, all the gaps in the periodic table have been filled, and scientists have added elements at the end of the table (unstable and very short-lived atoms found only in extreme situations such as the heart of a nuclear process), but they could never discover any other previously unknown element that somehow fits into the middle of the scehme. (See Figure 20.1)

Thus, because of the homogeneous nature of the universe, we

1																	2
H																	He
1																	4
3	4											5	6	7	8	9	10
Li	Be											B	C	N	O	F	Ne
7	9											11	12	14	16	19	20
11	12											13	14	15	16	17	18
Na	Mg											Al	Si	P	S	Cl	Ar
23	24											27	28	31	32	35.5	40
19	20	21	22	23	24	25	26	27	28	29	30	31	32	33	34	35	36
K	Ca	Sc	Ti	V	Cr	Mn	Fe	Co	Ni	Cu	Zn	Ga	Ge	As	Se	Br	Kr
39	40	45	48	51	52	55	56	59	59	63.5	65.4	70	72.6	75	79	80	84
37	38	39	40	41	42	43	44	45	46	47	48	49	50	51	52	53	54
Rb	Sr	Y	Zr	Nb	Mo	Tc	Ru	Rh	Pd	Ag	Cd	In	Sn	Sb	Te	I	Xe
85.5	87.6	89	91	93	96	99	101	103	106.4	108	112	115	119	122	127.6	127	131
55	56	57 58 72			73	74	75	76	77	78	79	80	81	82	83	84	85	86
Cs	Ba	La — Hf			Ta	W	Re	Os	Ir	Pt	Au	Hg	Tl	Pb	Bi	Po	At	Rn
133	137	71			181	184	186	190	192	195	197	201	204	207	209	210	210	222
87	88	89															
Fr	Ra	Ac															
223	226	288															

Figure 20.1
The Periodic table of the elements—every place is filled

can assume that carbon is the only atom that can form molecules large enough to act as the "molecules of life"—the enormous structures such as DNA (deoxyribonucleic acid) and RNA (ribonucleic acid), or even the smaller building blocks, the nucleotides that form these massive structures, and the proteins, enzymes, and other biochemicals needed to run cells and sustain our existence.

These biochemicals lay at the heart of the field of genetics (see next chapter) and are linked with the meat of the discussion in this chapter—the question of evolution. But how does genetics link with evolution?

At the start of this chapter we reached a definition for "life" as a system that can reproduce and pass on mutated information from generation to generation or a commodity involved with evolutionary change via natural selection. Evolution works through reproduction.

However, this creates what seems on the surface to be a chicken and egg scenario. The genetic code is carried by DNA, but if the ability to undergo evolution is a requirement of "life" and this process itself requires an elaborate set of processes involving DNA, how did "life" originate in the first place? To put it another way: Any entity that can evolve—or by our definition be "alive"—has to be complex enough to possess the genetic material with which it can evolve. But how could an organism reach this level of complexity without evolving?

The fundamental problem comes down to the question: How could simple aminoacids that were probably around in the primeval soup on earth some four billion years ago have changed into what we see as *biological material* or the simple living matter that then evolved into more advanced forms, and eventually, us?

We are not so much concerned with the second part of this question—the grand sweep from single-celled organisms to twenty-first-century humans. For the purposes of this discussion it is the change from nonliving or what is called *prebiotic* material to living cells that lies at the core of the problem, the jump from unorganized RNA to a bacterium.

There are two theories that attempt to explain how this occurred. The first of these is the RNA-*world hypothesis*. This suggests that somehow a small quantity of a type of RNA was produced on the early Earth which had the ability to perform a number of roles on top of the functions it demonstrates today. The RNA that is postulated would have been able to replicate (make copies of itself) without the presence of protein (presumably it would use some of the protein within its own structure), and would also be able to catalyze every step of the protein production process.

This may seem unlikely, but recently scientists have found molecules called *ribozymes* which are RNA catalysts, or enzymes made from RNA. However, these molecules are still some way from RNA capable of self-replication.

The other contesting theory to explain the jump from prebiotic to biological material is that of biologist A. Graham Cairns-Smith of the University of Glasgow, who suggests that the organic agents that lead to the formation of living things actually evolved from inorganic materials.

Now, this may seem startling. After all, to most of us there is a vast difference between organic and inorganic materials. All living things are organic—the food we consume, the animals and plants that fill the planet. Inorganic materials include the rocks and stones, the gases that constitute our atmosphere, things generally deemed "inanimate."

Cairns-Smith points out that, like the biochemical system using DNA and RNA, complex *inorganic* systems are capable of replicating and passing on information, albeit in a much simpler way. In the system that operates in the modern biosphere, DNA carries a code in the form of an almost unimaginably complex set of instructions that is the blueprint for reproduction, while RNA and the proteins play their respective roles in bringing this about. What Cairns-Smith proposes is that around four billion years ago a simpler system operated which initially did not need RNA, DNA, or even proteins.

In his system, the first step was to produce what he calls a *low-tech* set of machinery using crystal structures present in clays. These clays, although far less complex than a DNA molecule, can create a self-replicating system in which information is passed on from one "layer" to another, mirroring the way DNA replicates.

From this low-tech start, Cairns-Smith supposes that a gradually more complex system evolved which incorporated organic molecules. These unsophisticated systems developed over time into the *high-tech* machinery we have today in which DNA, RNA, and proteins facilitate genetics and allow the evolution of living things via natural selection.

This leads us toward a model for how life may have begun on Earth, and would probably do equally well for any DNA-based life-form on other worlds throughout the universe. Given these possible mechanisms for how life began, how would life on other worlds have developed? Could we expect biological processes to have followed a similar route to the way they operated here? And if so, would it lead to human-type beings or very different creatures?

To find answers we must turn to the discipline of *developmental biology.* This is a blend of several linked fields—evolutionary biology, paleontology, and genetics—which considers the way in which creatures on earth have evolved from simple forms dating back billions of years to the fauna and flora of today.

Developmental biologists start with two other disciplines from biology. First, they need to consider *genetic retracing,* which involves analyzing how genetic material has changed over long periods. Genetic characteristics are of course one of the key factors in determining the nature and diversity of life, and it is possible to trace back the way genetic material has altered both within species and across different species. In this way, biologists can reach conclusions about common ancestors of modern species, which can lead to a "family tree" dating back many hundreds of millions of years. The other approach is *evolutionary biology,* which looks at the range of modern animal structures, or the "body plans" of animals—the fundamental groupings of different animal types we see all around us. From these it is possible to work backward using computer models to determine the original plans.

Neither of these techniques is straightforward. Evolutionary biologists use powerful computer models that utilize a vast collection of parameters—information gained from paleontological finds—to try to determine how "function" ties in with "design," and how organisms have adapted to their environment. Genetic retracing is a complex science because genes (like species) evolve at different rates and the evolutionary lines can branch in elaborate ways.

Life probably first appeared on earth around 3.85 billion years ago, and though we are not sure how the change from prebiotic material to "living" organisms occurred, it is clear that once it did, life flourished and evolved on this planet. But it was certainly no simple development.

For the first 2.85 billion years, life on this planet consisted of single-celled organisms such as bacteria. About one billion years ago the first very simple multicellular creatures, the algae, appeared for the first time. Then, about 550 million years ago, during what is called the *Neoproterozoic era,* more advanced organisms began to appear. These simple creatures probably resembled modern-day sea pens, jellyfish, primitive worms, and sluglike animals, all of which have left faint fossil remains and markings.

But then suddenly, around 530 million years ago, there was a complete, and in a geological sense, "sudden" change in the evolutionary development of life on Earth. Throughout the Neoproterozoic period, the Earth had been populated by animals that displayed a relatively small collection of different body plans, but then everything abruptly changed. Within a short space of time the Earth was populated by a vast range of different creatures.

This transformation is called the *Cambrian explosion*, a burst of activity in the life of the fauna of this planet. Before it, the Earth was home to only relatively few simple organisms; after it, Earth was populated with organisms that, although still simple, evolved into almost every known type of shelled invertebrate (clams, snails, and arthropods). In time these gave rise to modern vertebrates, then mammals and humankind.

Most important for this discussion, after the Cambrian explosion, all the basic body plans of all animals on Earth had been established. From that point on all evolutionary steps (including what some consider one of the most dramatic—when some animals left the sea to live on land) merely required subtle refinements of the basic animal types established during the Cambrian explosion.

Staggeringly, this single dramatic change in the history of life on Earth produced just thirty-seven distinct body plans, which account for absolutely every animal on the planet.

Furthermore, it is now known that almost all living things share a collection of "regulatory genes" which determine the essential body plan of a creature. Most animals start from a single cell, the fertilized egg or zygote. This cell then divides and multiplies, and specialized parts form—the organs, glands, skin, bone, muscle tissues. But every organism has a set of common genes controlling the process of protein formation that leads to the formation of these parts. As this process continues, the genes become more and more specialized.

The simplest genetic command determines the body axis of an embryo—which end becomes the head and which the tail, which is the back and which the front. This instruction is the most basic and is an example of a shared characteristic between almost all species. Further along the process, a collection of genes determine whether a head is developed outside the trunk of the body (as in most animals), while others instruct the growth of limbs. In species as different as, say, bats and eels, these instructions will be very different, but for a sheep and a dog, or even a sheep and a human, they will be much more similar. It is only when we consider comparatively specialized functions and characteristics of a creature that we see very clear differentiation between creatures. Even the bat's wings and the forelimbs of a sheep can be thought of as controlled by a similar set of regulatory genes.

The origin of this process is ancient. The DNA sequence for this process—a vast and highly complex set of instructions—was to be found during the *Precambrian period*, before the Cambrian explosion 530 million years ago. Indeed, it had to be in place to allow the explosion to happen.

The DNA sequence that does this is found in a collection of regulatory genes called *Hox genes*, which are usually found in a cluster in animal chromosomes and therefore often referred to as "Hox clusters." Amazingly, these clusters may be thought of as "templates" for the animal of which they are a part—the genes are literally arranged in the gene cluster in the precise way the animal part they control is positioned in the growing embryo. Genes that control the development of the head are at one end, the wings or legs partway down, and the rear end holds the genes that control development of the lower parts or rear of the animal.

What does all this mean for the exobiologist?

Well, it shows that only a few dozen patterns or layouts (body plans) are needed for a vast array of different species, and that if we restrict our vision of alien life to those based upon DNA, we still end up with a universe offering an incredible variety of shapes and forms. And what does this lead us to conclude about the shapes and forms of aliens on other worlds? Are we to expect little green men (the LGMs beloved of pulp science fiction circa 1920), or bug-eyed monsters straight out of a 1950s B-movie?

A way of resolving this may lie with the currently controversial idea of "convergence." This boils down to the idea that many very different starting points produce a limited number of solutions to a task.

An everyday example is an aircraft. The "task" is to build an efficient flying vehicle for a reasonable price that will move safely through the air a small group of human beings from A to B at high speed and in relative comfort. Now, before aircraft were invented, people may have thought there was a vast range of ways of doing this. Indeed, the experiments to make the first planes were extremely varied, and today, a century after the first heavier-than-air machines were flown, there is in fact an array of different flying vehicles. But these fall into a limited number of types (analogous to body plans in the animal kingdom), which, on the surface, look very different from one another.

In fact, differences between aircraft are relatively superficial and come down to appearance, size, and subtle refinements in style, layout of the interior, and alterations to fit specialized tasks. At their root, aircraft are all metal cylinders with doors, wheels, wings, engines, and tails; they all use fossil fuels; they have a front and a back, and seats on which humans sit; and they all travel through the skies. (The exception is a helicopter, but even this has many of the same characteristics.)

In the same way in which humans have solved this task, nature deals in a limited number of different ways with the design problems it faces. But most crucially, it always does so in the most efficient way it possibly can.

Scientists refer to this process as "perfectibility." And it works excellently here, so it could work equally well on perhaps the majority of inhabited worlds.

So, what limitations are there to the types of DNA-based lifeforms we could hope to find one day on another world?

To establish any kind of informed answer we must look at two distinct factors. First, we have to consider the evolutionary factors on an alien world, and second, the environmental conditions.

The key to the evolution of life on Earth were the great events of the Cambrian explosion. Only by looking at this singular example can we formulate an idea of what could have been happening, and might still be, on other worlds.

There are competing theories to explain what precipitated the events on earth 530 million years ago. The first is that the time was right—that nature had experimented with various evolutionary mechanisms and eventually hit on the right way forward. In saying this, I do not want to imply that nature in any way "planned" this or "knew" what to do, that it was in some way "guided" by an external agent, be it God or an alien in a shiny spacecraft deliberately setting in motion a process that would lead to the production of a dominant species like Homo sapiens. Natural selection does not operate by any sort of plan. It is driven by success and buffeted by random events, and needs neither a God nor alien intelligence. Rather, in one sense this explanation for the Cambrian explosion is linked to convergence—nature had simply found a path from point A to point B by trial and error.

If this is the case, then we can feel confident that a very similar process will have occurred on any number of alien worlds. It would be a fundamental and very simple process, therefore universal, a consequence of life reaching a certain level of complexity, a point from where it is spurred on to the next stage.

A rival theory offers a less rosy picture for the chances of there being advanced life in abundance, because it suggests that the Cambrian explosion was precipitated by some unknown freak ecological event such as a giant meteorite or asteroid collision, certainly an event at least as significant as the one now believed to have eliminated the dinosaurs.

Another alternative is the possibility that something caused a massive change in atmospheric conditions on the planet. An increase in the percentage of oxygen in the atmosphere would have almost certainly triggered a dramatic surge in biological activity on the surface and could explain why a vast array of fresh life-forms appeared during such a brief period.

At present nobody can say which of these theories is correct, but the answer will have a great bearing upon the type of universe in which we live. One theory claims to offer the key to a highly populated universe, the other a bleaker (but not totally lifeless) prospect. It might be argued that it happened here, so, with what are certainly a large number of planets to choose from, it could happen elsewhere.

Beyond the initial spark of evolution from the Cambrian explosion, there is a long hard road to what we call an advanced life-form, one capable of developing a civilization. But if, for the purposes of our discussion, we assume the Cambrian explosion was inevitable, can we say anything about evolutionary routes on other planets?

Once again we can only make comparisons with the one example we have—the evolution of life on Earth. To arrive at an answer to the question of whether intelligent life could have evolved on other worlds we have to look at what "intelligent" means as well as its relevance to the ability to form a civilization.

On Earth, the only animal with any form of conscious social interaction or intellect (as distinct from pure intelligence) is Homo sapiens. We are the only animals on the planet to keep records, to have developed a recordable language (a form of writing), to have built a civilization based upon trade, and, crucially, to plan, to have a concept of our place in the world and the flow of generations of our species.

So perhaps the first question to ask is: What is it about us that makes us different to the other species on the planet? If we can arrive at an answer to that, we might be able to extend the principle to extraterrestrials.

The difference between us and other species, at least in part, comes down to brain size. We have very large brains for our bodies. If the human brain was unfolded and spread out, it would cover four sheets of typing paper. By comparison, a rat's brain would barely cover a postage stamp. However, it is not just a matter of size, but the way the brain is used. Dolphins have very large brains, but it is believed that most of their brain capacity is involved with managing their complex sonar system.

The reason brain size is so important for humans is its use in developing the incredibly complex skill of language. And language is a basic (but not the only) requirement for civilization and social development.

In the case of human history, there was a "sudden" fourfold increase in brain size between 1.5 and 2.5 million years ago. It is thought that before this point, early human ancestors had a brain capacity comparable to that of a chimp (about one-quarter ours).

What caused this rapid development remains a mystery, but it marks another key turning point in the development of the human race, another crucial jump along the evolutionary road. The most likely explanation is that our ancestors were faced with a severe "challenge" to their continued existence.

The best candidate for this is the advent of the most recent Ice Age, the *Quaternary Ice Age*, which is believed to have acted as a "filter" for many species, including Homo sapiens. In other words, early Homo erectus, the immediate progenitors of the first Homo sapiens, learned a great deal from their experiences during this time. Biologists have reached the conclusion that the more intelligent land-based animals are omnivorous. This, they believe, is because omnivores can adapt their tastes to find diverse sources of food, and the effort to search out new resources is also a learning process which helps the animal develop skills not common in carnivorous or vegetarian animals. Similarly, the increase in brain capacity precipitated by the Ice Age came about because of the demands this placed on early humans. Those who survived this change in the environment did so by learning to find and use new resources, which led them to gradually develop social skills, to create communities, to develop language, and to eventually take the first steps toward civilization.

And with language comes what we call "intelligence." If we define intelligence as the ability to communicate and process ideas, then a huge leap in human evolution came about with the development of syntax, and from that the ability to string together meaningless sounds (phonemes) to make "meaningful" words. This ability enables us to create sentences and to communicate abstract ideas, to plan, to create social rules, taboos, and hierarchies. Language really is the cornerstone of civilization.

But would there necessarily have been events comparable to the Ice Ages on other worlds?

It would seem very likely. Although there are a number of plausible theories to choose from, nobody knows for sure why the series of Ice Ages occurred on earth and whether these are linked to very common natural processes in the life of a planet. But it would seem reasonable to suppose they are common to a good percentage of worlds. And of course, Ice Ages may not be the only form of challenge an embryonic dominate species might be offered. Other worlds may suffer environmental changes precipitated by comet or asteroid collisions, volcanic activity or short-lived irregularities in the behavior of the planet's sun.

Brain capacity and brain application are only part of the developmental formula, albeit crucial ones, to be taken into account. Another extremely important evolutionary factor is physical versatility. And this leads to the question: Could only humans have created a civilization on Earth? Why, for example, have dolphins not achieved the same status?

Many people tend to think of dolphins as highly intelligent creatures, and by some definitions they almost certainly are. But they exhibit a form of intelligence that appears to be very different to ours, one that has not led them to the creation of what we understand as a "civilization" or a "society."

The simple truth seems to be that dolphins did not have a chance of competing with humans because they live in an environment that makes it very difficult for an intelligent animal to create any form of infrastructure. And this is due to several complex factors.

Dolphins do not have digits with which they can manipulate materials, and they certainly have not developed opposable thumbs, one of the most important distinctions between human and nonhuman primate development on earth. Dolphins are a very successful species—their physiology has evolved in a way that allows the animal to be perfectly adapted to its environment—but they were unable to even start on the road to civilization.

Dolphin "language," although sophisticated compared to almost all other animals on the planet, has not developed in a way that can lead to social development beyond a rudimentary level. Nor, with the bodies they possess, could they have constructed the aquatic equivalent of buildings. Also, they cannot record any knowledge they acquire in the way humans do using writing, nor easily cultivate their territory or manage other animals, which means they are constantly at the whim of fluctuations in food supply. Finally, dolphins could not have developed weapons and so could not have engaged in what is an extremely important developmental factor for any civilization—war. So, extending this example, it might be fair to say that any intelligent aquatic animal has only a slim chance of developing a civilization, as we understand the concept.

So, any planet that brings forth life must have enough land to allow animals to develop and for the correct ratio of plants and animals to arise in order to create a balanced ecosystem. Furthermore, any world that is entirely covered in water will almost certainly not harbor life any more advanced than the equivalent to Earth-style fish and relatively simple aquatic animals and plants, although in theory it might be possible for a special system to emerge and flourish on a world almost entirely covered in water, which contains animals and water-based plants rather than land-based plants.

So, turning to the other strand of the arguments to determine the nature of alien life, what about environmental conditions?

I mentioned elsewhere that bacteria have been found in some extreme environments on Earth—in radioactive waste, thousands of feet beneath the seabed, in hot springs, and in the frozen wastes of Antarctica. Bacteria are of course extremely hardy creatures; more sophisticated animals could not have evolved within environments as harsh as those in which bacteria appear to thrive.

This places limits upon possible environments on alien worlds harboring any form of *highly developed* life-form. First, the temperature of the environment must lie somewhere in the region of 0° to 40°c. Enzymes cannot operate at temperatures much higher than 40°c and begin to denature.* Coupled with this, the environment must not be flooded with intense radiation, as this damages biochemicals and inhibits many of the chemical reactions required for mechanisms that control the functioning and continued growth of cells.

The environment on a particular planet cannot be too harsh or else any form of complex multicellular beings could not have evolved. On the other hand, the environment has to offer a chal-

---

*Water freezes at 0°c (at a pressure of 1 atmosphere), and therefore biochemical processes, which all take place in liquid water, would cease. A caveat to this is that on a planet with a different atmospheric pressure, water will freeze at a different temperature, but, as we will see, environments with very different atmospheric pressures to that of Earth present their own problems for the evolution of life-as-we-know-it.

lenge to living things so that natural selection can operate and evolution can occur.

If we consider first the atmosphere, what are the limitations? Would it be possible to have a successful ecosystem on another planet that supports DNA-based life but does not have an atmosphere similar to that of the Earth?

The answer is almost certainly no. The reasons for this are complex. On Earth we live in a balanced ecosystem in which plants need carbon dioxide to facilitate photosynthesis, which produces oxygen. All animals use oxygen, which is transported by the blood and carried to the cells of the body, where it is involved in almost all the biochemical mechanisms that maintain our bodies. No alien creature could evolve or remain alive on a planet without interacting with other organisms. All organisms must be part of an ecosystem, and any ecosystem must include gaseous cycles similar to the oxygen–carbon dioxide cycle on earth. Such alien systems would have to integrate organisms similar to our plants and animals. An alien world may have an ecosystem based upon animals and some other kingdom—perhaps some form of animate mineral or living rock—but the same situation applies.

This does allow scope for an atmosphere with different *proportions* of the gases we have in our atmosphere, but of course precludes too high a concentration of any gases DNA-based life would find toxic. An illustration of this would be the fact that only some very rare bacteria can survive in an atmosphere in which there is too little oxygen or too much of a "toxic" gas such as methane.

What of other environmental considerations? What of atmospheric pressure and gravitational fields? How would these affect the diversity of life on an alien world?

On a planet where the atmospheric pressure is higher, it is possible that intelligent creatures could have evolved that look very different from humans. The layout of the respiratory systems of such creatures would probably be very different because the pressure of the gases they breathe would not be the same as it is here, which means that the processes allowing gases to diffuse into their version of a circulatory system would operate at different potentials. But this does not exclude the possibility of such creatures reaching a high level of development.

More difficult to resolve is the effect of high or low gravity. The strength of the gravitational field on an alien world will greatly affect the body types and the behavior of creatures living there. All land-based mammals on Earth fall within a relatively narrow range of size—there are no mammals two hundred feet long or insect-sized. But if the force of gravity was, say, fifty times greater than it is here, the creatures that developed there would be very much smaller than here, and far less mobile. In extreme conditions we could imagine intelligent creatures that were almost flat.

Conversely, planets with low gravity would bring forth creatures that were much larger, but lighter, and they would probably move by natural methods of flight as much as they would around the surface.

However, a great difference in physical size to what we witness on Earth creates its own problems. Think back to the aircraft analogy for a moment. At either end of the scale there is a definite limit to the size of aircraft that could be of practical use to humans. In the same way, very large creatures present design difficulties for nature.

One of the major problems for very large animals—and there are many such difficulties—is the fact that they would need huge hearts to supply the volume of blood needed to keep their bodies functioning, and big hearts need big lungs. As we know from the design of animals on earth, this is not a strict limitation, but for an animal to evolve into the dominant species and to establish a civilization, they also need a large brain that is not simply devoted to running a massive body. And the brain of a large creature would need a large head and still more blood to supply the cells with oxygen, which needs a massive heart and lungs, and so we enter a vicious circle.

Some claim that the "success" of the dinosaur refutes this argument. It does not. By this argument, "success" derives from the fact that the dinosaurs were around for many millions of years. Yet success is not based purely upon longevity; it must take into account the role an animal plays in the ecosystem. Humans are the only animals to have created a civilization, the only species to control their environment in any large-scale way.

So, even if we only consider DNA-based life, it is clear there could be a range of different shapes and sizes for extraterrestrials, but these have to fit into certain limits, and it may be that on planets with approximately the same gravity as Earth, the species that formed a civilization would be at least vaguely similar to humans. But what of the details, what exobiologists call "parochial characteristics," or "cosmetic" differences? What about the number of limbs, sensory apparatus, or coloring? Is it likely that we will one day encounter a two-headed, five-legged green thing that we are meant to befriend?

Whether or not an extra pair of limbs or a third eye would be favored within an ecosystem on another world is open to debate. There may be advantages in having these things. But as we saw earlier, within any environment, nature will always go for the most efficient option. If the advantages of a third eye or an extra pair of ears outweighs the demands produced by the extra weight, the extra blood requirements, the development time (both in terms of evolution and within the womb), then it could happen. If not, then evolution is unlikely to allow such creatures to dominate, and they would easily be made obsolete by better "models."

But how far can we go with this argument?

There are those who do take the anthropomorphic argument to its limit and suggest that the most likely design for a successful life-form that has evolved to the point of developing a civilization will be similar to ours—that they will look like us (or us like them). But why?

Consider the number of limbs a creature has. Do we need more than two legs? It is possible that a creature developing on a world with high gravity would need three legs or more to allow it to move more efficiently under the strain, but then it could be argued that a biped could have evolved on the same world that has two much better legs, allowing it to move around more readily than its multiped rival.

Does any creature need two heads? We have two of almost everything else, why only one head?

Quite simply, two brains would require too much blood for the same size heart, and we are back in the same cul-de-sac as with the size of lungs versus the size of the animal. But convergence would not allow for such a consequence anyway, it would enforce the most efficient solution—a biped with one head is better than a large, ungainly biped with two.

So, a three-legged, two-headed beast may never get through the evolutionary net. Indeed, it may be argued that this latter body type is unlikely because among the great diversity of life on Earth, there has never been a two-headed anything. However, strictly speaking, this is not an empirically sound argument, because we are trying to deduce outcomes for an alien world where (within the limits required for DNA-based life) conditions could be quite different.

So, to conclude, what can we determine from these arguments? First, most scientists believe that life in the universe is plentiful, and many would put money on the idea that our civilization is just one of many. Less convincing is the argument that there could be intelligent life based upon anything other than DNA-led biochemistry. If there are creatures that have flourished and become highly evolved by this route, we may never encounter them and would almost certainly never be able to communicate with them.

Turning to the narrower arena of life based upon DNA, we still have a truly mind-boggling range of possibilities—just look at the marvelous blend of life on Earth in all its varied glory. If we are to consider DNA-based life, it will form and evolve on a range of planet types, but there are certain restrains, certain environmental limits. These, along with caveats to the way evolution could progress on alien worlds, narrows the field somewhat and can lead us to believe that the most likely shape for an intelligent DNA-based alien approximates to the humanoid form. But details could produce very different creatures. They could perhaps use different chemicals to carry oxygen to their cells, and their sun could emit a slightly different range of radiation. Combined, this would mean that such aliens would be a different color to any human we have seen on earth, because it is the color of the hemoglobin in our blood and the amount of the melanin in our skin that determines our color.

It is even conceivable that we may one day encounter little green men.

🄴 Mendel's Monsters

> Right thinking people should check the procreation of base
> and servile types . . . all that is ugly and bestial in the
> souls and bodies of men.
>
> H. G. WELLS, *Anticipations*

Human mutants and man-made monsters are one of the mainstays of science fiction and horror, and have remained so from ancient myths to the movie *Gattaca*; they are the stuff of nightmares. Ever since Mary Shelley created the fictional Frankenstein's monster, coinciding as it did with the birth of modern science, people have wondered if such things could be possible—could humans make other humans? Can humans manipulate the very stuff of life to "manufacture" other living things? Can we alter our bodies at a fundamental level or imbue life into inanimate matter?

In the days when these questions were first raised, biology was in its infancy and the science of genetics was totally unknown. In the same way the alchemists dreamed of transmuting base metals into gold but did not have a nuclear reactor with which to accomplish it, early thinkers in the realm of biological mutation had none of the technology and little of the knowledge needed by modern day genetic engineers. But today we have the technology, and our understanding of the fundamental genetic processes of life is increasing each day. The nuclear physicist is no longer an alchemist working in his private laboratory; today they have the key to the basic forces in Nature and the potential for great good and great evil. In the same way, genetic engineers have their fingers on the biological button offering a world of wonder and benefit, or the way to the worst technonightmares.

As we will see, the question of morality is an issue of major concern in the world of genetics, and a contemporary dilemma for a growing number of ordinary people. We may now have the technology to manipulate our genes, to produce both angels and monsters; but should we?

The tremendous leaps we are now witnessing in the science of genetics represents one of the greatest revolutions in human understanding since ancient times, its pace almost supernaturally fast. According to one observer: "Genetics will touch our lives in the next century as powerfully as the development of silicon chip technology did during the 1980s. Just as virtually every home now contains machines with the chip, so nearly all of us will soon have some reason to welcome or curse the advance of genetics."[1] So important is the development of ideas in the field of genetics and their application to everyday life that "genetic engineering" is sometimes called the fourth medical revolution.

The first of these revolutionary changes was the realization, some two hundred years ago, that hygiene was essential for good health. The second was the invention of anesthesia, which allowed for far safer surgery during the middle of the nineteenth century. The third was the discovery of ways to combat bacteria and viruses—the use of antibiotics and the invention of the vaccine. And the fourth is the recently developed process whereby doctors can pinpoint genes and replace "bad" with "good."

In much the same way that the science of nuclear physics has come to be viewed as a mixed blessing to humanity, the field of genetics is a classic, double-edged sword. The nuclear physicist has given us an alternative power source to counter the gradual depletion of our natural resources, and may one day provide the means for interplanetary travel. But the knowledge we have gained from nuclear research has produced two atomic explosions, one catastrophic accident, and the potential to destroy human civilization many times over.

In the same way, genetics provides huge opportunities as well as a catalogue of anxieties for the future. On the one hand, within a few decades scientists hope to have intimate knowledge of the estimated five thousand genetic diseases suffered by humans, and be able to replace faulty genes, manufacture transplant organs, and favorably manipulate whole swaths of the Earth's gene pool to our collective benefit. On the other hand, there is the danger that cloning could be terribly misused and lead to the creation of man-made mutants, human clones, social eugenics, and the horrors these things could provide. Genetic engineering is truly a Pandora's box. But as we shall see, it cannot possibly be left closed. In fact, it is already half open.

At the root of genetics is the simple fact that all life-forms pass on characteristics to the next generation through a genetic blueprint stamped into every cell in every living creature. The blueprint is different for each individual. This is what distinguishes a squid from a racehorse or Tom Cruise from Bill Clinton. The blueprint is called the "genetic code" and it is made up of a sequence of tiny units called *genes* (100,000 of them in each human cell), which are themselves made from bits of a very large biochemical DNA, which stands for *deoxyribonucleic acid.*

The study of genetics goes back a long way. Our modern understanding of the basic concepts behind it can be dated to an Austrian monk, Gregor Mendel, who during the 1850s realized that characteristics were inherited from generation to generation by means of what he called *discrete factors.* These are what we today call genes.

Mendel realized that each individual inherited two complete sets of genes, called *alleles,* one from each parent, and that these alleles change during reproduction but are passed on unaltered from parent to offspring. What makes us different is that, because each parent also has two alleles, each of their children has a fifty percent chance of getting one or another allele from each parent, or a twenty-five percent chance of receiving any particular combination of genes. This "shuffling" of genes produces variations in characteristics ranging from our coloring or build to our susceptibility to certain diseases, or even such traits as sexuality or certain psychological characteristics.

Although Mendel conceived the idea of inheritance of genetic material, he had no idea how the process worked on a chemical level. It was not until 1953, almost seventy years after his death, that the mystery was solved when James Watson and Francis Crick, work-

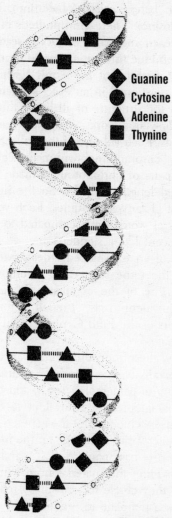

Guanine
Cytosine
Adenine
Thynine

Figure 21.1

ing in Cambridge, discovered that genes consisted of a very complex molecule, DNA—made of two strands, the famous double helix.

DNA is found in every cell of every living thing. And although it is a vast molecule, amazingly, it is made from just four small chemical units, or "bases," called A (adenine), T (thymine), G (guanine), and C (cytosine). There are hundreds of millions of A's, T's, C's, and G's in each molecule of DNA, scattered throughout its structure. They combine three at a time to form specific three-letter codes (what we can think of as "words") which allow individual aminoacids to be positioned in an exact sequence to form proteins. Proteins make up the structure of all living things.

To visualize how this works, imagine a cell as being a set of huge encyclopedias. Each volume of the collection of encyclopedias is equivalent to a "chromosome" within the cell. Each human cell has twenty-three pairs of chromosomes, made up of an incredibly long, tightly coiled length of DNA. So the human "encyclopedia set" would consist of forty-six volumes. Each volume in this library would be billions of words long; compared to which, *The Lord of the Rings* would seem like a slender tome.

Each volume of an encyclopedia deals with a whole range of subjects, and in the same way, each chromosome controls every physical characteristic of the organism. So, in this analogy, hair color might be equivalent to say, "French Colonial History"; height, "Chinese Emperors of the Fifth Century"; nail shape, "The Life of William Shakespeare."

In our analogy, these individual entries in the encyclopedia are equivalent to individual genes. And, of course, real encyclopedia entries are made up of paragraphs, words, and letters. So, extending our model, the paragraphs are equivalent to the large sections of DNA that make up specific parts of the genes; words are the counterpart of the three-letter words that code for the individual aminoacids; and the letters are the base pairs, A, T, C, and G.

As well as providing the material from which genes are made, DNA also acts as the vehicle for growth by replicating itself within the cell. It acts as a template to produce new copies of itself. It is as if the encyclopedias in our collection could be endlessly photocopied and distributed; any number of exact copies of each can be made by generating copy after copy based upon the original. But as with photocopies, tiny mistakes occur during the copying. These "mistakes" are called *mutations*, and they can either be beneficial or detrimental to our offspring. But, crucially, these mutations occur all the time. We all possess mutated genes—in fact at a basic level, if our parents' genes were not mutated during reproduction, we would all be biologically identical.

By the late 1960s, a decade and a half after Crick and Watson's ground-breaking work on the structure of DNA, scientists had a pretty clear idea of how the words in the encyclopedia collection could be formed. They realized that out of the four bases—A, T, C, and G—only three were needed at any one time to correctly position an aminoacid to form a protein. This meant that with a supply of four letters, there would be sixty-four ways in which three-letter combinations could be created. But the rate of development in genetics has accelerated exponentially since then.

During the 1980s a new "Big Science" project, costing almost as much as the Apollo missions and involving many more scientists around the world, was established to attempt to map every gene in the body. This is called the Genome Project, and it is now reaching its concluding phase.

Along the route taking us from its foundations in the early 1980s, the Genome Project has had its dissenters and opponents, but it has also unearthed a vast array of information about genetics, and in particular the genetic makeup of humans. Now, hardly a day passes without the newspapers of the world heralding a new breakthrough in the study of one disease or another, thanks to the discoveries of the global genetics community. In fact, we are now in the peculiar position of being able to diagnose far more diseases than we can possibly cure, which leads to problems of its own—do people really want to know they have a disease if nothing can be done about it?

This is but one of the mounting and increasingly obvious moral issues that has come out of the incredible and almost too-rapid advances being made in the science of genetics. But if we are to enjoy the undoubtedly huge benefits that could come from genetic manipulation, then we also have to learn to deal with the potential dangers and problems it throws up.

The good and the bad of genetics seem to get equal attention from the media. Just as the optimists shout from the rooftops their hopes and claims for the future and their successes of the moment, we hear of the frightening scare stories that come with it. What are the two sides of this coin? What are the fantastic hopes and the terrifying nightmares?

Let's first look at the great opportunities the study of our genetic makeup can bring.

At the top of the list must be the breathtaking advances already being made in the field of *gene therapy*. As the name implies, gene therapy is a medical treatment reliant on the potential to alter the genetic makeup of our bodies. As I discussed earlier in this chapter, the human body consists of over 100,000 genes positioned in various places in the twenty-three pairs of chromosomes present in each cell.

Many diseases known to humanity have nothing to do with our genetic characteristics. The most obvious examples are the many thousands of contagious diseases. Usually their only genetic component is our varying natural ability to resist infection, and the efficiency of our defenses to fight it. But there are some five thousand diseases already known that are directly affected by the presence of certain genes in our genome.

There are two types of genetic disease—those precipitated by what is called a "dominant gene" or collection of dominant genes, and those caused by a "recessive gene" or a collection of these working together.

A dominant gene is one that acts individually. It could come from either parent, but it will generate a characteristic on its own without the complimentary gene from the other parent (remember, we obtain a set of genes from each parent). These genes can be quite innocent. For example, the gene for brown eyes will always "dominate" over that of any other color. So if an individual receives a gene for brown eyes from one parent and a gene for blue eyes from the other, they will certainly have brown eyes.

If an individual acquires a dominant gene for a disease, they will have a far greater chance of developing that disease later in life than if they did not have this gene or if the gene was of the recessive variety—the gene can simply "switch on" at some point and precipitate a series of biochemical changes that will lead to a disease. An example of this type of serious illness is Huntington's disease (HD). This is a horrible degenerative illness that affects about one in five thousand people. It is always fatal, and triggered by a solitary dominant gene. And in 1983 this gene was the first to be located that precipitates a hereditary disorder. We still have no cure for HD, but at least individuals can now be screened to determine whether they have the gene and are therefore likely to pass it on to their offspring.

Recessive genes work in pairs. In other words, an individual may inherit a gene that could precipitate a characteristic, but this characteristic, "good" or "bad," will only appear if the complimentary gene is also "switched on." In many cases, people inherit a recessive gene that is mutated in some way so that it does not function properly. This will mean that it does not do its usual preventative job—say producing a special chemical that stops some other biochemical process occurring, which would trigger illness. Alternatively, people can inherit a gene that does work properly but actually sets a disease in motion by allowing other complex biochemical pathways to open up. In the case where such genes are dominant, then the individual has a high risk of developing the disease; if the gene is recessive, then its complimentary gene has to be "knocked out" by some unknown factor—such as carcinogenic compounds in the environment, exposure to radiation, hormonal disturbances, smoking, the list is a long one.

Until recently, these diseases have only been treatable by surgery or intensive drug programs, if at all. Illnesses such as HD remain completely incurable, and as mentioned before, the greatest benefit to be derived from our genetic knowledge of them is to help prevent them from being passed on. But it is hoped that in the not-too-distant future doctors will be able to treat an increasing collection of diseases based upon an understanding of where genes are and how they interrelate. With gene therapy, it is now just possible to implant healthy genes into the body of a patient with an inherited disease in the hope that the body will start to copy the healthy version and stem the deficiency, or to counter the damaging effects of the gene that has malfunctioned. So far, only a very limited number of diseases have responded to this technique, but methods are improving all the time and geneticists are spreading their nets further.

The first step in this process has been the effort to discover which genes do what. With 100,000 genes to choose from, and the fact that almost all diseases or any other human biointeraction with the environment is a vastly complex matter involving many genes, the task is incredibly difficult. But gradually, as we learn more about the genome and the Genome Project reveals many of the functions and the interplay of genes, we are learning more about how the entire system operates.

A recent success in gene therapy has been the treatment of a disease called ADA (adenosine deaminase deficiency), which affects the functioning of the immune system. Another has been the use of gene therapy to combat a debilitating illness called SCID (severe combined immunodeficiency). Today, geneticists are active in attempting to use this technique to fight a range of potentially fatal diseases including an ever-growing array of cancers, cystic fibrosis, hemophilia, and AIDS.

One of the great stumbling blocks so far has been the difficulty of delivering these "good" genes into the appropriate part of the patient's body and encouraging the body to take the new genes and replicate them naturally and correctly. A great deal of work is being done on this aspect of the treatment and it is very gradually improving. Some workers are even experimenting with the idea of producing a new forty-seventh chromosome which would carry any genes that were required. The chromosome would be made outside the body, tailored to specification, then implanted where it would be replicated just as geneticists have done with single genes.

This idea is still a long way from practical use and remains very much a theoretical concept. It offers huge advantages over the simpler process of single gene implantation but is also immensely more difficult to achieve. As we will see, such advanced ideas offer huge potential for medicine, but also bring to mind such extreme possibilities that alarmists can find reason for worrying about perversions of sophisticated science. If we can tailor whole chromosomes for medical jobs, could we not produce genetic monsters?

On only a slightly more prosaic level, there is the question of personal involvement in genetic engineering, because this science, developing as it is at such an extraordinary pace and taking us all into unknown territory, offers vast potential for the individual, as well as for terrible misuse.

If we know the detailed structure of the genome, then it will soon be possible to map the entire genetic makeup of a fetus in the womb. This will put us in a position where parents can decide if this is the child they want. This ability offers all sorts of opportunities, both beneficial and alarming. We all want a healthier world, but most people would consider it a misuse of knowledge were they to decide, say, that a fetus is not "perfect" enough.

Today, doctors can genetically screen for many diseases before a couple decided to start a family. There are some seriously debilitating diseases that run in families, and by detecting an individual's susceptibility to such diseases, potential parents can weigh up the dangers and make their personal decision. This is considered a very positive outcome of genetic research, but exactly the same information, exactly the same discoveries and precisely the same techniques, also offer scope to tailor fetuses to personal requirements.

Most people would not want to interfere in the development of their unborn child or terminate a pregnancy if a relatively minor problem was spotted, but others would. And how far does this thinking take us? There are many parents who only want a son, if technology could tell them early on that the fetus was female, they might make the decision to abort. Equally, they may discover that the child will be born deaf, or have a high risk of developing a certain cancer by the age of thirty—what decision do they then make?

Current public opinion seems to be firmly against improving the attributes of a fetus. In a 1993 *Daily Telegraph*/Gallup poll in Britain, 78 percent of respondents were against the notion of parents "designing" their children. But in the U.S., one survey found that 43 percent of people approved of the idea of gene therapy to boost a child's intelligence.[2]

Another aspect of this is the danger that genetic information falls into the wrong hands. Should insurance companies have unlimited access to an individual's genome? Should someone applying for life insurance declare that they have a gene that gives them an enhanced susceptibility to a serious illness?

In the future these questions will be everyday problems we will all have to face one way or another. We will all be aware of our genetic profile; indeed, within a decade we will probably all carry a "genetic passport" that profiles our personal genome. We will all be drawn into the debate about privacy at the most personal level—the very internal structure of our bodies. We can only hope that as society develops these new technologies we will also produce adequate ways of dealing with them on a social and ethical level.

But society also has to find ways through a subtly different moral maze. Could our enormously enhanced knowledge of genetics lead us one day to eugenics? The considerations of the individual and the decisions they may have to make about their own unborn children presents a form of *individual eugenics*, but what of *social eugenics*?

As a civilization, we have toyed with this dangerous idea many times: the quote at the beginning of this chapter comes from a book by H. G. Wells advocating social eugenics, and the obvious example of the application of eugenics was the Nazi's program of *selective breeding* during the 1930s and early forties—an attempt to produce a "super race." This simplistic program had nothing to do with high-tech genetic engineering but was purely based upon the false assumption that by choosing "perfect" parents, "desirable" features could be engineered to pass on to the next generation. We know now that genetics does not work like that, and that there is no way to control the inheritance of certain particular features over others without manipulation at a genetic level. But with the latest technology at our fingertips, and the progress that is sure to be made during the next decade, we can now do just that—eradicate what some would consider to be "negative" attributes and promote "positive" ones by altering the genome of a fetus, or by cloning.

At first glance there seems to be little wrong with the fundamental principle of eugenics. As I said earlier, we all want a healthier world, and few would deny that a planet populated by more intelligent and more socially responsible people would be an improvement. But there are many problems with the concept.

First, who makes the decisions? Who has the right to say which fetus should live and which should die? Second, characteristics or attributes that would be perceived as "undesirable" by eugenicists sometimes produce remarkable results. Take two obvious examples—Stephen Hawking and Beethoven. Hawking developed the degenerative neurological disease ALS at the age of twenty-three, yet he has gone on to become one of the most influential physicists in the world and one of the all-time best-selling authors. Beethoven became profoundly deaf as an adult, yet he composed some of the most sublime music ever heard. If specially appointed individuals with the power to dictate who lives and who dies were to base their ideas purely upon empirical evidence garnered from a genetic profile, then both Hawking and Beethoven would have been terminated.

Finally, there is the problem of control. If eugenics were ever to become policy, the culling of genetically imperfect fetuses could be allowed to go to ridiculous extremes. Would the drive for perfection prevent, say, the birth of someone who was susceptible to obesity later in life? Or someone who may be predisposed to body odor?*

Perhaps even more exiting than gene therapy is the potential offered by the latest genetic research to hit the headlines—cloning.

Cloning has been a science fiction favorite for many years but has only recently begun to emerge as a respectable and potentially epoch-changing science. The possibility of cloning has been considered almost as long as the human race has had knowledge of how genetics works at a basic level, and soon became a subject that penetrated into the public consciousness. Indeed, cloning has been the subject of several novels. Probably the most famous is *The Boys from Brazil*, in which the real-life character Josef Mengele, the Nazi physician and experimenter, clones Hitler's genes in an attempt to produce a master race. A more chilling but less well known example of cloning in science fiction appeared in the novel *Solution Three*, written by Naomi Mitchison, the sister of the great British biologist and philosopher of science J.B.S. Haldane. This is set in a postnuclear apocalypse world in which the human race is all but de-

---

*Interestingly, geneticists now have the ability to detect this characteristic. Ian Phillips from the University of London has recently discovered a mutation in a gene that produces a defective enzyme that is unable to break down a chemical called trimethylamine, which produces the characteristic "rotting fish" smell associated with body odor.

stroyed and cloning is used to repopulate the planet—with disastrous consequences.

In more recent times there was an attempt to hoodwink the public and the scientific community into believing that human cloning had been accomplished. In 1978 a freelance journalist, David Rorvik, wrote what he claimed to be a nonfiction account of how he had helped an eccentric millionaire clone himself. It was, of course, a rather crude hoax, and was exposed, but only after Rorvik had earned an estimated $400,000 in royalties and advances from the book.

What is most significant about this last example of cloning in literature was that it sparked off a violent reaction from the scientific community—several geneticists went on record as saying that the science of cloning was "not even on the horizon, let alone workable."[3] Partly because of this, for many years cloning remained in the scientific wilderness. Since Rorvik, there have been some on the fringes of the scientific community who have further marginalized themselves by making serious claims that cloning is possible and declaring that they have succeeded. For one reason or another, these have all been ignored and even openly dismissed by the scientific community.

Then, in July 1996, the world was shaken by the news that a little known British scientist, Ian Wilmut, working in what had for long been considered a scientific backwater—the study of embryology—became the first person to successfully clone a mammal from adult cells. The sheep, Dolly, had been "created" in the lab.

This breakthrough was immediately splashed across the newspapers of the world and soon became the subject of books and television documentaries. But what exactly is cloning and how could such work as Wilmut's have the potential to change the course of civilization and push society further toward some of the greatest ethical dilemmas we have ever had to face? How does this knowledge lead us to the potential to create monsters imagined by Shelley, and central to the plot of a thousand and one science fiction stories?

Cloning is the process in which the genetic material of an egg is completely removed and replaced with the entire genome (a complete set of genes) from a donor, the individual to be cloned.

Simple types of cloning have been possible for some time; indeed, before producing Dolly, Wilmut and his team had successfully cloned two sheep, named Megan and Morag. But these had been produced by splitting a single embryo, creating two identical copies of the same sheep. These techniques differ from the cloning experiment that produced Dolly in two distinct ways. First, until Dolly, only relatively simple organisms have been cloned by implanting genetic material into a host egg, rather than splitting an already fertilized egg. The list of cloned creatures included bacteria, plants, and, in some rare cases, animals such as frogs. The second and even more important factor distinguishing Wilmut's achievement from all other previous experiments is that he used genetic material taken from an *adult* animal, a mature sheep, and implanted this into a recipient egg.

Now, obviously, this has huge implications, not just for science but for all our lives in the future. For the very first time, a scientist had brought to reality the "science fiction" idea of cloning—taking genetic material from an adult animal, removing the material already present in a recipient egg, and replacing it with this "foreign" material. What this means, if we extrapolate the science only a tiny degree, is that an adult human could have genetic material removed. This could be placed in a host egg and a new human grown in the lab. This human would not be merely a twin or close relative of the original, it would be an identical, exact copy, possessing exactly the same genome.

The process enabling this to happen may be broken down into seven clear steps:

1. Remove some donor cells. Place these in a glass dish and label it *Clone*.
2. This is what may be considered the most crucial step—the eggs need to be put into a state of "hibernation." To do this, they are starved of nutrients, which means they stop multiplying and enter a state in which they can "reprogram." What this really means is that they are "tricked" into thinking they have returned to an embryonic state and become ready to multiply once the new

DNA has been added to them and the old DNA removed. No one really knows how the starving of cells sends them into this state.

3. Give a special cocktail of drugs to a large group of women to push them into a state of "superovulation" in which they produce large numbers of eggs in a short space of time.

4. Harvest the eggs and remove their nuclei. The nucleus is the nerve center or "brain" of the cell and contains the twenty-three pairs of chromosomes, which constitutes the genome of the human. This stage involves dextrous practical application because we must not lose or damage the rest of the cell's material within the outer cell membrane—the cytoplasm.

5. Place a nucleus from the removed cells of the donor to be cloned into the recipient cell in place of the removed nucleus. Fire a pulse of electricity into the cell. This facilitates the acceptance of the new material in the recipient cell. Again, no one knows why a pulse of electricity will do this, but it does.

6. Place this fused egg into the womb of another woman, who acts as the surrogate mother for the clone.

7. Wait nine months. The expected success rate for mammals has been placed by Wilmut's team at about one in 300. This figure is based upon his own experiments using sheep. No one knows what the success rate would be in human cloning.

As soon as news of Wilmut's cloning of Dolly was announced, fears were voiced about the possible application of this work to the cloning of humans. Fortunately, the United Nations have agreed to ban any form of research that involves the use of what are called human *germ* cells—ova and sperm taken from human beings. This limits geneticists to manipulation of cells known collectively as *somatic* cells, cells other than those involved in reproduction. But according to some, this ban is flawed in a number of significant ways.

First, those of a cynical disposition suggest that such a ban, precipitated by a moral judgment on the part of the West, might only hold for a time, and that as soon as a nation realizes the commercial potential of experiments involving human cloning, the ban will be eased. They point to other attempts at prohibition based upon ethical criteria and the way these have eroded (for example, sanctions against Iraq, the banning of land mines, and others). But more important, some observers have pointed out that it might be ethical and wise for the West to ban human cloning experiments using germ cells, but what is to stop similar "rogue" states—those, say, who are currently experimenting with germ and nerve gas warfare systems or selling arms to unscrupulous dictators—from gaining access to this technology and producing a real-life *Boys from Brazil* scenario?

The international community is gradually becoming aware of the dangers involved with cloning as well as its enormous potential. When Ian Wilmut was called to account by the House of Commons Science and Technology Committee, he declared: "We would find this sort of work with human embryos offensive. We can see no clinical reason why you would wish to make a copy of a person."[4] But as the scientist, Professor French Anderson, director of the Gene Therapy Laboratories in the U.S., has recently pointed out: "There is a real danger that our society could slip into a new era of eugenics. It is one thing to give a normal existence to a sick individual; it is another to attempt to 'improve' on normal—whatever 'normal' means. And the situation will be even more dangerous when we begin to alter germ cells. Then misguided or malevolent attempts to alter the genetic composition of humans could cause problems for generations."[5]

In a recently published article entitled "Cloning? Get Used to It," the geneticist and science writer Colin Tudge goes further, saying: "We should face reality. There are no biological laws, apart from the underlying laws of physics, and technology might achieve anything that does not break these bedrock laws. We should also recognize that the potent new biotechnologies are already outside humanity's present control. Individual countries may have their laws, mores and customs, and influence technology up to a point through the flow of research funds. But the U.S. in particular is driven by market forces, and while most Americans are known to deplore the idea of human cloning, that same majority supports a Constitution that defends the rights of minorities effectively to do as they please.

If one percent of Americans want cloning, and it's known they do, then most of the other 99 percent will not gainsay them."[6]

Meanwhile, in Britain the debate over the uses of this new work is heating up. The Human Fertilization and Embryology Authority bans the use of experiments on human fetuses over fourteen days old, which means that any clone that is produced has to be destroyed after this time. Ruth Deech, chairman of the HFEA, has said: "We would never grant a license for any treatment that would result in the production of an actual cloned baby."

But those opposed to the whole idea of cloning are deeply suspicious of the scientific community. Peter Garrett, research director of the antiabortion charity, Life, claims: "We know what a nightmare world is just around the corner if we once accept any manipulation of human life in this way. They want to lull us into allowing the cloners to get to work by proposing a fourteen-day limit on the clones' lives initially. Then we get all the usual utilitarian talk about potential benefits of research using laboratory clones. Having got us used to the idea, they will quickly relax the age limit and away we will go, full sail ahead, in a year or two's time."[7]

Perhaps because the first cloned mammal was produced there, the country most interested in keeping up research into this new science is Britain. Recently, the British government rejected an agreement signed by nineteen other European countries, including Italy, France, and Spain, to place strict limits on research. And although the U.S. Senate similarly rejected an anticloning research bill, the present administration in the U.S., and in particular the President, is deeply opposed to clone experimentation.

No one really knows how different human cloning would be to cloning a sheep. Of course, humans and sheep are both mammals, so there should be few fundamental differences in the methods, and if sheep cells are activated by electrical surges and hibernated by starving them, then this will also happen with human cells. Cloning is a highly technical procedure and requires skilled scientists and refined technological backup, but it is not in the same league as the utilization of nuclear technologies or even development of effective germ or chemical weapons, along with the delivery systems and defense mechanisms they need.

To start a clandestine human cloning facility, those involved would need a team of experts (available at a price), a sophisticated biochemical laboratory (again, readily available with sufficient funding), a supply of human tissue and recipients, donors and surrogate mothers, and a great deal of luck on their side. The facts are clear: for a sufficiently motivated and sophisticated state with modest funding, cloning is a technological possibility. But what would they do with such a facility?

The potential is almost limitless. The most obvious use would be to clone the country's leader. It is easy to imagine an egomaniac like Saddam Hussein or past leaders of the Idi Amin or Galtieri school being interested in such a thing. Another obvious use is the production of genetically altered individuals for specific purposes. By combining cloning with techniques drawn from gene therapy, it should be possible to generate clones possessing certain characteristics altered along the way. Using this method, it would in theory be possible to produce "superhumans"—beings "manufactured" to fit requirements.

However, before we get too carried away with these ideas, there are some serious limitations to the entire process of genetic engineering. It is certainly possible to clone mammals; at the same time, the techniques to implant desired genes and to manipulate the existing genome is becoming more and more sophisticated. These genetic skills can be used to alter characteristics covering a wide variety of human attributes—from eye color to height; they may also play a part in adjusting how an individual responds to nonphysical stimuli, because it is now believed that our genetic makeup is in part responsible for some psychological aspects of human beings.

Now, the important caveat here are the words "in part," because we are not just simply a collection of genes, and this becomes particularly significant when we look at the psychological aspects of a human being.

A few years ago there was a great fuss created by the announcement that a "gene for homosexuality" had been found. At other points during the past few years geneticists have located what newspapers have dubbed the "alcoholic gene," the "violence gene," and others. But what geneticists mean when they announce such findings is that they have found a gene which predisposes us to certain attributes—both good and bad. Of equal importance to the genetic structure of our bodies is the environment in which we mature. This is especially important when we consider such characteristics as our sexuality, our tastes, and our ability to control moods.

It may be that within a very short time the potential to produce tailor-made clones will be available, but the mere manipulation of genetic material is not the whole story. Even if we consider the benign use of cloning, officially sanctioned and within a controlled environment, the outcome of such experiments would not be entirely predictable. We may be able to make an exact physical copy of someone, but would they really grow into the same person? The answer is: of course not.

If we take billionaire A who wishes to have themselves cloned, he or she could donate genetic material, the experiment is conducted, and the clone born. Even if they were brought up under incredibly contrived circumstances, that child would experience totally different things from the original donor; after all, the original donor was not a clone.

A further, more subtle point to consider is that the development of an embryo in the womb may not be solely due to the nature of the genes in its cells. No one is sure how the cytoplasm (the material outside the nucleus of the cell) affects the development of the fertilized egg; or indeed how the biochemistry of the mother and her experiences during pregnancy impacts upon the progress of the fetus.

However, let's take the cloning scenario several stages further, into what would now be considered the supernatural, a realm we might call *superscience*. Consider some of the more extreme possibilities genetic research offers.

Some neurophysiologists are currently toying with the idea of creating a "secondary brain" for newborn babies. The idea involves implanting at birth a microchip into a baby's head. This chip can record every experience the brain has (as it happens), and stores the same information as the natural brain. In other words, the chip is a "backup brain." When the human dies, the chip is removed and plugged into a new body, enabling its owner to live again with all the information derived from their "first life."

If we couple this idea with the technology of cloning, it is conceivable that a human could live many lives, and be the same person physically and mentally—the chip would simply go into a new body built from the old one and they would be "complete" again.

Of course, religious people may have a thing or two to say about this. They would argue that the person was not "complete"; they may have the same memories and the same body as the original, but what of their soul?

Many scientists dismiss the entire notion of the soul as pure fiction. They argue that what we mistakenly believe to be a soul is just an aspect of our personality—and this in turn is a mere product of the gestalt that results from the complexity of the human brain. In other words, we have such complex brains that when we are alive we have self-awareness and a personality and this is "us" or our "soul." Kill the brain, and the soul or personality (whatever you want to call it) ceases to exist. But if the complexity of the brain can be preserved, along with the actual information derived from experience, and stored, it could be argued that the "personality" can also be restored and preserved within a new body.

Naturally, cloning, without the added invention of storing our minds as well as our bodies, presents many problems for religious people. For instance, if humans can make another human without the need for natural reproduction—can this being really be considered human?

Such ideas lead us into a moral and ethical minefield. It raises a plethora of questions, all of which are extremely difficult to answer. For example: Where does the soul derive? Is the soul created at the moment of conception, as some religions profess? If this is the case, what is to be made of creatures born but not conceived? Does the activation of the cell by means of an electrical pulse, one of the key stages mentioned above, constitute "conception"? And if it does, is a soul produced at that moment, just as some believe happens during natural conception?

These matters will occupy a growing number of people in the future. There is little doubt that somehow a human clone will be produced within the next decade; the only mystery is through which process. Will such a being come from a "hostile" regime, from a clandestine experiment in the West? Or will western governments slowly relax the rules and allow this to happen—perhaps in an effort to preempt a pariah state getting there first?

As clones begin to appear, more and more people will start to wonder about the meaning of the soul, what is special about natural conception, what the interaction of mind, body, and soul could be. Today this is really almost entirely the intellectual territory of philosophers. But tomorrow it will, by necessity, become an issue argued over by everyone, as much a part of our lives as a discussion about the ecology of the planet or the morality of abortion is now.

22 Earth Magic, Leylines, and Circles in the Corn

> Our civilization . . . has not yet fully recovered from the
> shock of its birth—the transition from the tribal or "closed
> society," with its submission to magical forces, to the "open
> society," which sets free the critical powers of man.
>
> KARL POPPER, *The Open Society and Its Enemies*

Since the time before civilization began, humans have believed in what has come to be known as *earth magic*, the notion that they can in some way communicate with the natural forces at work in the world.

It is now very fashionable to believe that we humans are part of a greater *web*, a network of animate and indeed inanimate matter. This holistic viewpoint embraces the idea of a gestalt within Nature, a greater force emerging from a benevolent fusing of individual parts.

To the rational observer the foundation of this broad concept must be the fact that, in some mysterious way, some part of a human being can communicate or link up with the natural forces at work in the world in which we live. This is of course a central tenet of many of the paranormal phenomena discussed in this book. Many believers in telepathy think the ability to read other's thoughts works by some form of web, a hidden link between the minds of all humans. Those who believe we are being visited by alien beings point to the idea of a *cosmic web* that in some way links beings across the universe. The alchemists, and followers of the ancient Hermetic tradition, thought (and some still believe) that the secrets of the universe may be manifest in the everyday, that there is again a pattern or net connecting the forces of nature with the beings living on this planet.

For the scientist, many of these ideas sound nebulous and vague, without any tangible root. This is not helped by the fact that like so much of the paranormal, single, strong ideas are turned into a lukewarm mishmash of muddled thinking because of hidden agendas and a willingness to accept hocus-pocus. However, some enthusiasts of the paranormal accuse scientists of harboring what they believe to be similar ideologies. Take as an example Bell's experiment, in which two particles from a common source are fired at a device that then sends them in opposite directions at the speed of light.

As we saw in Chapter 6: Visions From a Future Time, if one of these particles is then "altered," it changes the character of the other. For example, if its *spin* is changed, the second particle, traveling in the opposite direction at the speed of light, is also altered at precisely the same instant. This appears to transgress the laws of physics, in particular Einstein's theory of relativity, which puts a speed limit—the speed of light—on anything in the universe.

The current interpretation of this experiment is that if the two particles have been linked at any time during the history of the universe, they will always be able to communicate with each other. The physicists who discuss and analyze this interpretation do so using mathematical tools, and they are usually clear-thinking individuals who have little time for the vagaries of the occult and metaphysical realm. But researchers into supernatural phenomena point to such concepts as mirroring their own nonmathematical theories and explanations.

They suggest that there are mysterious unknown forces at work in Nature. However, as I have pointed out in other parts of this book, it would appear unlikely that any major force could exist that we have no idea about, unless it was a very weak force that did not interact with other known forces. The reason for this is simple: if it is a strong force, we would have noticed it already. But if, by some strange chance, we had not noticed it and it was suddenly discovered, such a finding would destroy science; an inconceivable thought because science *works*. You would not be reading this book if not for the fact that science (technology) had produced it.

The remaining possibility, according to misinformed occultists, is that some strange force we have not yet discovered does operate, but only in another part of the universe. I have argued against this already in Chapter 20: Our Brethren Among the Stars? The law of universality rules out this possibility. Besides, if this strange force only operated somewhere else in the universe, it would be of little use in explaining many of the paranormal experiences that are supposed to occur here on Earth.

Earth magic takes many forms. As I have said, it may be interpreted as the key to almost all paranormal phenomena, certainly telepathy, clairvoyance, remote viewing, astral traveling, perhaps even ball lightning, ghosts, and poltergeists. But it also finds a home in explanations for a rostrum of other occult and metaphysical ideas, including the reason for standing stones, leylines, divining, pyramid-building, and crop circles.

Leylines are one of the most frequently written about and discussed manifestations of earth magic. The lore surrounding the phenomena is ancient, but like many ideas of the occultists, it had been forgotten or lost for perhaps thousands of years until it reemerged early in the twentieth century and gained a new following among New Age enthusiasts.

The name *ley* was first coined by an amateur researcher into ancient paths and road systems, an Englishman named Alfred Watkins, who wrote a book on the subject called *The Old Straight Track*. Watkins's concept of the ley was that all road systems and pathways zigzagging the countryside followed ancient patterns. In other words, all roads, old or modern, followed the same route and were built along the same pathways. He gave no reason for this curious alignment and certainly saw no mystical or occult connection between the ancient paths humans created and those designed and constructed by present-day engineers.

From Watkins's idea came the metaphysical concept of leylines being *lines of force*, a principle that was then applied to explain a whole range of occult ideas. For example, investigators of paranormal phenomena use the idea of lines of force or *channels of energy* to explain dowsing. This is a technique with which certain people appear able to locate underground water using dowsing rods that were traditionally made from forked hazel twigs but are now produced from almost any material, most commonly wire cut into an

L or a Y shape. The idea is that the mind of the dowser is in some way hooked up with the flow of natural energy along leylines.

Enthusiasts point to the fact that animals are sensitive to patterns within the natural forces at work on Earth. They cite the theory that migrating birds use lines of magnetic flux to help them navigate, and refer to reported cases in which animals appear to be sensitive to approaching storms or impending earthquakes and tremors. They suggest that humans still retain a primitive instinct for the lines of force around the earth and use this to dowse. But as we saw in Chapter 18: We Are Made of Stars, the hypothetical force that astrologers claim facilitates the interpretation of future events, charting of personalities, and elucidating the compatibility of humans has never been detected, cannot be squared with the known forces of nature, and really has no tangible link with the ways in which birds navigate or salmon are able to return to their birthplace to breed—phenomena that may be explained by empirical science.

The evidence for and against dowsing is plentiful but contradictory. There are numerous documented stories of individuals being able to locate water using dowsing rods, but just as many have produced embarrassing results. During the 1970s a well-known French dowser claimed that in a single afternoon he had succeeded in finding the precise point to sink a well for a house he had bought in England. Yet, as we saw with research into telepathy and telekinesis, scientists have never been able to pin down this talent. In 1913 a group of scientists who set out to determine the mechanism through which dowsing could work were disappointed to find that a group of well-known dowsers failed to notice they were standing above an underground reservoir that produced fifty thousand gallons of water per hour. Many years later, in 1970, a study financed by the Ministry of Defense in Britain produced less than exciting results. The report claimed that "the map dowsers could not even match the accuracy of volunteers taking guesses at the locations."[1]

So what is the scientist to make of claims linking dowsing and its apparent link with so-called leylines? If there are some people who are able to locate water (and in some cases deposits of other substances such as minerals and ores), then this talent may derive from a heightened sensitivity toward some known forces and energies. Perhaps there are some rare individuals who can detect parts of the electromagnetic spectrum that for most of us lie beyond the range of our everyday senses. It is even conceivable that if there are lines of force forming a grid around the world, then the energy that travels between the linked points is a form of electromagnetic radiation, and that this is in some way detected by an unknown part of the human brain remaining active in some people, but dormant since prehistoric times in the rest of us.

However, some enthusiasts of the paranormal have taken things much further than connecting leylines with what may be natural hidden talents such as dowsing. They suggest that leylines were used by ancient peoples in the construction of roads and in the determination of sites for ritualistic monuments, that they chose these paths and places because they believed they could somehow tap the energy traveling along the leylines for religious, spiritual, or emotional empowerment.

This is far removed from Alfred Watkins's original, comparatively prosaic ideas, but the researchers of earth magic claim that there are striking connections between the standing stones at Avebury (interpreted by some as the nerve center or nexus of the world's leyline system), Stonehenge, standing stones in France, and in other parts of continental Europe. Others take this still further and attempt to connect these systems with the lines found at the Plain of Nazca in Peru, known to have been constructed by the Incas between 600 and 1200 B.C. Still more unsupportable connections have been made between the pyramids at Giza, the ancient temples at Machu Picchu, the stones on Easter Island, and even the position of the legendary lost continents of Atlantis and Mu.

The latest application for the idea of leylines is to connect them with UFOs. Some claim that the grid system acts as a guide for alien visitors, that UFOs access the power that flows along the lines to propel their craft within the Earth's atmosphere. Others believe the ancients built the standing stones and their roads along these natural routes to emulate or honor our alien neighbors.

In the early 1960s a former RAF pilot named Tony Wedd became the first to popularize this hypothetical link between leylines and UFOs. He believed the pilots of alien craft used the leylines to help them navigate and claimed to find a vortex of energy where many lines met near his house in the west of England. He then went on to declare that he had been contacted by an alien called Attalita, who passed on instructions to build alien machines. None of these have yet appeared in the malls of Earth.

Others picked up the baton inadvertently passed on by Watkins and metamorphosed by people like Tony Wedd. The 1970s saw a deluge of books on leys and their importance to both a secret ancient history and UFO lore. The best known of these were *The Flying Saucer Vision* and *The View Over Atlantis*, both by UFOlogist John Michell. Both these books had sexy titles but contained almost nothing of any value to support the claims of the author—the idea that aliens use leylines and that there is a link between the lost civilization of Atlantis and the lines of some intangible natural energy crisscrossing the globe.

Of course there are strange connections between the positions of stones in parts of England, and all sorts of odd shapes and configurations can be mapped out to link the pyramids with Stonehenge or the supposed location of Atlantis, but such connections can often be made by studying what are otherwise totally unconnected points of reference. This does not mean there is any distinct mystical value to be placed upon them.

Nevertheless, claims and counterclaims continue to confuse the issue of leylines and their links with the paranormal. To support the skeptics is the fact that even poor, much maligned Alfred Watkins's rather innocent and modest claims have been shown to be fallible. He wrote a second book, called *Archaic Tracks Round Cambridge*, in which he described sixty-two lines around the city created by lining up dozens of ancient and modern sites and monuments. But in 1979 a group of geologists from the Institute of Geomantic Research in Britain found that only nine of the sixty-two really did line up.[2]

On the other hand, what many believe to be good evidence to support the importance of leylines has come from a variety of studies made during the years since Watkins's death in 1935. Before extending his ideas to the esoteric edge and linking leys with almost anything under the umbrella of the paranormal, the author John

Michell found that there are many stone rows dotted around the world, but concentrated especially in the west and the north of England and parts of northern France. Although initially less dramatic than stone circles, some of these stone rows appear to line up for many miles across the country. The most important of these is the Devil's Arrows in Yorkshire. This is a line of three stones that fall into an almost perfect line for almost eighteen kilometers (twelve miles). The line linking them also passes through four Neolithic earthworks, including Nunwick, one of the most prominent in northern England.

The enthusiasts claim that this is proof that ancient peoples were able to utilize an innate ability or sensitivity for a mysterious form of energy flowing along leylines. Indeed, computer simulations of the Devil's Arrows created by mathematicians Michael Behrend and Robert Forrest showed that the chances of the alignment coming about by chance were negligible.

But there is another way of explaining this linking of ancient landmarks which does not rely upon the occult. It could be that ancient peoples did indeed want to place these stones and earthworks in a line for some lost religious purpose, but that need not have arisen from any form of supersensitivity toward unknown energies. Could they have aligned their stones in conjunction with a star, or the sun at a point in its arc that was of some particular significance to them, a significance probably lost forever? It is very common for researchers of the paranormal to jump quickly to conclusions drawn from the occult, and thus overlook the cleverness of ancient peoples. Such an attitude displays a condescension not unlike that of extreme skeptics who scoff at every aspect of the supernatural. In fact, it's quite possible that ancient people were able to line up monuments and objects of significance across a distance of eighteen kilometers without the use of mystical forces, and the reasons for the effort may not be rooted in anything more mystical than their particular religious ideology.

While the enthusiasts of the paranormal have become increasingly obsessed with the apparent mystical meaning of leylines, public imagination has been captivated by another strange phenomena, which, according to some interpretations, also comes under the umbrella of earth magic—the appearance in fields of crop circles.

Media excitement with crop circles began in August 1980 when a Wiltshire farmer named John Scull claimed that a circle of flattened corn sixty feet in diameter had appeared overnight in one of his fields. The local newspaper, the *Wiltshire Times*, ran a report, and within days the national media had leaped on the story and photographs of the mysterious circle in Farmer Scull's field were plastered over the front pages.

During subsequent summers, more and more of the strange circles appeared in fields. Most of these were concentrated in the western counties of England, especially Wiltshire and Hampshire. And as more and more appeared, media interest grew and theories to attempt to explain the phenomenon abounded. But for some time, the basic facts concerning what was happening in the fields of western England remained sketchy and often contradictory.

In those early days of crop circle reports, no one had actually seen one being formed, and the earliest any investigators could reach the site was the morning after the event. There were plenty of claims from local residents, stories of strange lights seen in the vicinity of the circles at night, odd sounds emanating from the fields. Investigators arriving on the scene sometimes spoke of mysterious electronic effects inside the circles, of their apparatus being disturbed by high levels of static electricity. Others declared that dowsing equipment had gone haywire inside crop circles and that photographic equipment had malfunctioned. The stories grew more and more elaborate as time passed. Some would also point out that the tales became more colorful as the interest of journalists escalated.

In the midst of this mounting fascination, several well-timed books appeared. The most successful of these was *Circular Evidence*, which became an international best-seller. Published in 1989, it was written by two UFO enthusiasts, Pat Delgado and Colin Andrews. Delgado was a space engineer who had worked on the British missile project in Australia and then for a time at NASA; Andrews was a highly qualified electrical engineer. Both wrote frequently for such journals as *Flying Saucer Review* and became interested in the paranormal aspects of crop circles as well as the apparent physical anomalies they presented. Their book was widely condemned by scientists but found a large and enthusiastic audience among believers in the occult and the UFO investigation fraternity, although, to be fair, quite a few followers of UFO lore doubted many of the author's claims.

Even by the mid-1980s, when Delgado and Andrews were researching *Circular Evidence*, the facts about crop circles were still scant, but allowed for a great range of often colorful theories. It was known that the corn was not broken at the base, but simply bent. It was also observed that the crops were forced into a swirl rather than a circular pattern, as though a vortex at the center of the circle was flattening the stems like a spinning top or the conventional image of a tornado. Some claimed the effect was caused by rutting hedgehogs, others that a strange weather effect was behind them.

Dubbing themselves *cereologists*, Delgado and Andrews pointed to the fact that during the time from the first appearance of the circles in 1980 to the writing of their book, the markings had become increasingly elaborate. They declared that this was evidence the crop circles were being produced by an alien intelligence as part of an effort to communicate with humanity. This idea was then expanded upon by other authors, gradually drawing in other aspects of occult lore.

But still others disagreed, and suddenly there seemed to be as many theories as there were subdivisions of UFO legend, each group managing to adopt crop circles for themselves. There were those who believed that the circles were created by what they called "pandimensional beings"—what we are to believe are beings from other "dimensions." And this intriguing theory is still alive and well today, long after most sane individuals have grown convinced of rational (but nevertheless fascinating) explanations for the phenomenon of crop circles. As recently as the summer of 1997, a photographer who specializes in aerial images of crop circles, Lucy Pringle, claimed that "crop circles are the work of beings in a 'parallel world' " and warned that entering a crop circle might represent a health risk to pregnant women because of "microwave emissions."[3]

Others subscribe to the view that a race of intelligent beings live inside the Earth—this is linked to the concept of the *hollow earth*, and believers claim that flying saucers are not from other worlds but from this secret civilization that lives in an alternative world beneath our feet, a world that may be reached through portals at the Earth's poles. Puzzling though it may seem, supporters of this idea believe the inhabitants of the hollow earth are behind the appearance of crop circles and that they are using the technique to communicate with us.

Not surprisingly, the number of accounts of crop circles grew almost exponentially as the media picked up on the story and ran with it throughout the 1980s. Books and television programs on the subject began to appear. All at once, everyone was a crop circle spotter, a dedicated cereologist. Helicopter trips were arranged for tourists who were willing to pay a hundred pounds per hour to hover over West Country fields and skirt the hillsides where fresh circles appeared on a daily basis. Farmers began to open up their fields for a fee-paying public that was charged extra for each photograph they took or video film they shot. Film and TV crews found that access to farmland depended upon how much the networks were willing to pay, and landowners vied with one another to offer the best and most sensational new pattern in the corn.

Then, just as the furor reached a crescendo, in August 1991, two people reported seeing a crop circle in the making, and press excitement reached a new peak. Gary and Vivienne Tomlinson were walking along a track beside a field close to the village of Hambledon in Hampshire when they heard and saw what they later claimed to be a corn circle in the process of being formed.

"There was a tremendous noise," they reported. "We looked up to see if it was caused by a helicopter but there was nothing. We felt a strong wind pushing us from the side and above. It was forcing down on our heads—yet incredibly, my husband's hair was standing on end. Then the whirring air seemed to branch into two and zigzag off into the distance. We could still see it like a light mist or fog shimmering as it moved. As it disappeared we were left standing in a corn circle with the corn flattened all around us. Everything became very still again and we were left with a tingling feeling."[4]

But then, only a month later, the hopes of the enthusiasts heightened by this eyewitness account of corn circle formation seemed dashed when two men came forward announcing that they had been responsible for creating the circles all along.

Doug Bower and Dave Chorley, two retired artists, demonstrated how they had faked dozens of crop circles during the previous ten years, even leaving their DD trademark close to some of their creations.

When this news broke, there was an immediate and impassioned response from enthusiasts and skeptics alike. The enthusiasts were initially devastated, but then fell back on the idea that the artists were themselves fakes, that the CIA, MI5, and FBI were all conspiring to dispel the idea that the circles were part of a huge plan for aliens to make contact. The skeptics made as much capital from the news as they could. British newspapers began to run "fake-a crop-circle" competitions, whereas only a year or two earlier they had offered £10,000 rewards to anyone who could present a clear and convincing explanation for the phenomenon.

Although Bower and Chorley received their fifteen minutes of fame from this stunt, much to the delight of the enthusiasts, it soon became clear there were some major flaws in their claims. First, it was obvious that two men working with simple tools could not have produced the hundreds of crop circles that had appeared regularly each year, some as far from the West Country of England as Australia and the U.S. But then other teams of fakers went public and circles were soon discovered that carried their hallmarks—coded messages cut into the corn close by. Sometimes the hoaxers even added their names: The Bill Bailey Gang, Merlin & Co., and others.

But there were more serious problems with the claims of the fakers. It was slowly dawning that there were "natural" crop circles that no one had faked. Some were so perfect it was almost impossible for the doubters to explain how they could have been the handiwork of practical jokers. Other circles were found in remote places stumbled upon by chance—hardly the place a self-publicizing joker would choose. Coupled with this was the testimony of the Tomlinsons, who had not been involved in the debate until the night they stumbled upon a crop circle being formed, and had no hidden agenda or even any particular interest in crop circles.

However, the most convincing argument that the fakes were only part of the story came from the fact that crop circles had been reported hundreds of years before any of the players in this modern game were born.

The first documented case occurred in Assen, Holland, in 1590, but the most famous incident took place in a field in Hertfordshire in 1678. The farmer who owned the land reported finding the circle in one of his corn fields, and wrote a pamphlet about it (these were the days before newspapers). He suggested that the markings were produced by what he called a "mowing devil." The pamphlet was illustrated with a cover depicting a devilish figure cutting a swath through the field with a scythe. Intrigued by this case and other reports, a seventeenth century scientist, Robert Plot, investigated the phenomenon and came to the conclusion that the circles were produced by air blasts fired at the ground.

It is a depressing fact that in the seventeenth century, an age of great superstition and widespread belief that the occult controlled everyday life, Dr. Plot could study the phenomenon with empirical detachment; yet in our own time, so many people have leaped at the supernatural to find answers.

Thankfully, amidst the hysteria generated by the enthusiasts of all things paranormal, and the cynicism of the skeptics, the press, the farmers cashing in, and the proud fakers, there were still a few people in the early 1990s who were willing to look at the phenomenon objectively. One of them, Dr. Terence Meaden, head of the Tornado and Storm Research Organization and editor of the *Journal of Meteorology*, may well have come up with an explanation for how genuine crop circles (as opposed to the faked circles of Bower and Chorley) could have been created.

Meaden proposes that the circles are formed as a result of what he calls a *plasma vortex*. If a hill obstructs a gust of wind, the wind can eventually meet stationary air on the lee side of the hill. This creates a vortex, or spiraling column of air, which then sucks in more air and atmospheric electricity. This hovers close to the ground until it encounters a corn field, flattening the stems into the now familiar spiral pattern. The static electricity generated in the vortex— produced from high concentrations of ionized air—could be the reason for the high-pitched whirring noise (a sound like a helicopter) reported by witnesses who have been near a corn circle as it formed. It may even account for the reports of odd electrostatic effects within the circle sometime after it's formed. Some residual charge may disturb machinery and photographic equipment.

In an effort to quantify the forces at work in crop circle formation, others have taken Meaden's work further. Professor John Snow, working at Purdue University, and Professors Yoshi-Hiko Ohtsuki and Tokio Kikuchi from the University of Tokyo, have all visited prominent sites in England and studied the vortex effect in the lab. They produced computer simulations and created virtual crop circles with exactly the characteristics observed in fields in Wiltshire and Hampshire.

The Japanese team took the research to another stage and produced a real vortex on a small scale, and used it to create circles in metal sheets. They then persuaded the Tokyo underground railway to allow them to conduct a larger scale experiment in a subway tunnel, along which they passed ionized air. In the confined conditions of a subway tunnel, the effect of air and high concentrations of electrical energy from the tracks did indeed produce a set of small circles in the dust close to the live lines.

So, what should be concluded from the evidence gathered about crop circles in recent years, and how does it relate to what occultists call earth magic? Indeed, what is earth magic?

Crop circles are clearly a complex phenomenon, and no one explanation covers all of those found. In recent years, more and more complex patterns have been appearing in fields around the world. The last years of the millennium have produced a bumper crop of amazing designs, including the Star of David, snowflakes, Florentine needlework patterns, Maltese crosses, even torus knots and symbols from the Kabbalah. Unless you are willing to believe that these are produced by bored aliens or it is some part of a larger secret scheme to initiate humans into the great galactic brotherhood, these are obvious very clever and quite beautiful fakes. But simple crop circles still appear spontaneously in farmers' fields. Naturally, these are increasingly ignored as the more glamorous models take center stage, but they are there nevertheless, and are almost certainly produced by natural means.

Research into the meteorological anomalies that could account for these natural crop circles is continuing, but it faces pressure from the cynics, who would like to discredit anything to do with crop circles. This is a shame, and actually hands a pyrrhic victory to the crank element that, together with the press, has generated the hysteria and overblown hype surrounding crop circles.

How do crop circles fit into the larger picture of earth magic? Well, as we saw at the start of this chapter, what the enthusiasts of the paranormal call earth magic may well be a collection of known forces acting in strange ways and interacting with human beings in a perfectly natural rather than supernatural way.

We are constantly buffeted by forces and energies, some of which scientists understand clearly, while others remain only partly clarified. Crop circles are a perfect example of how odd physical effects can be produced by these forces. If we discount the fakers and showmen (just as we should when analyzing all aspects of the paranormal), crop circles represent a phenomenon that, on the surface, defies explanation. But as is always the case, this is merely due to a dearth of information. Dig deeper and an explanation unfolds. In the case of crop circles, they are the result of a natural but rare process that creates an effect only seen occasionally until recent times.

People living in the eighteenth century, say 250 years ago, placed supernatural meaning on why their loved ones mysteriously fell ill and died. Only when humanity learned of the existence of bacteria and viruses were we able to substitute the concept of evil spirits with the reality of microorganisms. Suddenly, the demons and devils that killed us were known to be nothing more than other physical creatures with whom we share the planet.

There is an earth magic, a natural magic that can cast its spell and perform its wonders. It's called biology. At other times it takes on the guise of physics, chemistry, geology. It is the natural wonder of existence, and it is miraculous enough.

23 Remote Viewing and the Psi Detectives

Fighting is the most primitive way of making war on your enemies. The supreme excellence is to subdue the armies without having to fight them.

SUN TZU, FOURTH CENTURY B.C.

I was once involved in a remote viewing experiment. In 1997, I was the science consultant for a series for the Discovery Channel called *The Science of the Impossible*, and one of the programs was devoted to "Mind and Matter." Short of an "actor," they decided I would do for a sequence in which an experienced remote viewer would try to determine where in the world I was and what I was doing.

The arrangements were necessarily convoluted. The idea was that I would be taken to a secret location. I would only be told where I was going on the morning of the shoot. Meanwhile, in Virginia, remote viewer Joe McMoneagle, a Vietnam veteran who had discovered his talent after being wounded, was to be filmed live trying to get into my mind and to describe what I was seeing across the other side of the world, in England. The only thing he was told was that (much to my disappointment) I would not be leaving the country.

Why I could not have been whisked to the Bahamas, I don't know. Instead I was treated to Stanstead airport in sunny Essex. This is an impressive glass and granite homage to the 1980s boom, a beautifully designed building which is also one of the largest open structures in Europe. We arrived on time and walked around the airport, had a coffee and waited for a private guided tour that was to take us onto the runways and around a few aircraft. Meanwhile, the second crew started filming Joe trying to "see" where I was.

But then we hit a problem. The tour guide, the public relations executive for the airport, was late, and so far I had not seen a single plane—not even as we had parked and entered the building. As far as I was concerned, I could have been anywhere. Then, via a mobile phone, we heard that Joe had begun to draw, and as he produced the pictures, they were faxed to a mobile fax machine the producer had in one of the film crew vehicles. By the time our guide arrived and escorted us out to the runway and I saw my first plane, the test was over and Joe had said that he was finished with the location.

So how did Joe McMoneagle's illustrations match up with reality?

Well, things were confused by the mix-up at Stanstead airport, but even so, I can't honestly say the experiment was terribly convincing. Joe made a few striking connections. The most surprising was that just at the moment I turned to the producer to draw attention to the amazing granite floor in the building, Joe had described the place in which I was standing as "made of lots of stone and glass." He then went on to draw archways and canopies of stone; the roof of the airport is held up with metal supports, which could be described as "arched." He said we were near a main road and a railway line, that there was a church or place of worship nearby. But most important, he did not mention aircraft, flying, or anything to do with an airport.

The interpretation of this experience was as confused as the events. The producer claimed that it was really a success because Joe could not have known I was at an airport, since I hadn't seen a plane until after he finished viewing. Joe had pointed out that there were two major highways nearby, and an underground railway that we had passed as we entered the airport. Most significantly, he discovered there was a chapel on the lower ground floor of the airport.

However, for me there were problems with these interpretations. First of all, I'd been totally unaware of the underground railway we had passed, although I admit I may have been aware of it subliminally. I was quite unaware of the chapel, yet Joe was supposed to have picked that up from me. If this was true, why had he failed to pick up the aircraft I hadn't seen? Finally, the fact that he correctly realized there was a major road close by is hardly an amazing feat of supernatural skill—it is difficult to get away from roads in the 1990s in Britain.

My feeling was that Joe thought I'd been in a church or a cathedral. His drawings consisted of stone arches, glass, and ornate decoration, and he was pushing the idea that I was in a city and there was something to do with worship at the center of the experience. My interpretation was that he thought I was in St. Paul's Cathedral or Westminster Abbey.

Remote viewing (or R.V.) is a boom industry, and for some it is the most lucrative application of supposed powers linked to telepathy and the use of what some believe to be powerful mental energies. It is also an application taken seriously by senior executives of multinational companies who employ remote viewers, by the military establishment of many nations around the world, and by government agencies and police forces, who frequently use people known as *psi detectives*, who have helped solve serious crimes.

There have been some impressive examples of remote viewing and psi detection during the past three decades. A great deal of secret work was conducted on both sides of the Iron Curtain during the cold war, but much of this is still kept under wraps. More readily available are the details of how certain individuals appear to have helped police around the world in their work.

Two of the most successful exponents of psi detection are from Holland, Peter Hurkos and Gerald Croiset. When he was thirteen, Peter Hurkos fell from a tree and suffered a concussion. Soon after, he started to realize that he could locate lost objects by just thinking about them, and had strange "impressions" of distant locations and activities which, according to witnesses, could be matched with the behavior of individuals he'd never known and with whom he had no other contact. From the late 1950s, he helped the police forces of several countries solve serious crimes, determined the location of murder victims, and gave descriptions of perpetrators, including a period working with American police hunting the Boston Strangler. Hurkos was especially effective at passing on data he gathered from the technique of *psychometry*, in which a psychic appears to "sense" information about an individual by handling some of their possessions, usually clothes or items that had a particular significance for the missing person.

Gerald Croiset's successes were, if anything, even more impressive. He has never charged for his services and is not happy traveling, usually helping police with their investigations over the telephone. In this way, he helped the New York Police Department find the body of a four-year-old girl and the identity of her killer. In 1967 he received a call from a journalist named Frank Ryan, who wanted him to help find a teenage girl, Patricia Mary McAdam, who was missing from her home in Scotland. Croiset told Ryan that the girl was dead, where she would be found, and even added details such as the fact that the body lay close to a car wreck and that a wheelbarrow was leaning against the car. Strangely, the area was located, the wrecked car and even the wheelbarrow were found, but no body. To this day Patricia Mary McAdam's body has never been discovered.

Another famous case yielding what appears to be impressive results for the psi detective involved the hunt of the Yorkshire Ripper. During the late 1970s there was a series of gruesome murders concentrated primarily around cities in the north of England, especially Sheffield and Bradford. The victims were always young women, many of them prostitutes, and the murderer dismembered the bodies, earning his nickname because of the way he butchered the bodies in a similar fashion to the Victorian murderer, Jack the Ripper.

The hunt for the Ripper lasted years and became the biggest murder case in modern British history. During the latter stages of the investigation, an amalgamation of several regional police forces turned to psychics for help. The famous medium, Doris Stokes, gave information to the police which turned out to be of little use; but then another female psychic, Nella Jones, volunteered her services. She told police the murderer lived in a large house in Bradford, in northern England, that the house was number 6, but she could not pinpoint the name of the street. She claimed the man was called Peter, that he was a long distance lorry driver, and that the name of the company he worked for was embossed on the door of the cab. She claimed the company name began with the letter C, but she could not make out the full name.

Early in 1981 the Yorkshire Ripper was caught and jailed for life. His name was Peter Sutcliffe, he had been a long distance lorry driver who lived in a large house, number 6, Garden Lane, Bradford, and he had worked for a company called Clark Holdings.

Similarly impressive results are sometimes obtained through remote viewing, and used by the military as well as government agencies such as MI5, the CIA, and the KGB.

The Russians appear to have been the first off the mark with research into remote viewing and the uses of mental powers for military purposes. Some claim that research began in the 1920s and continued until the Berlin wall came down and communism crumbled. However, there is plenty of the usual hyperbole surrounding accounts of anything to do with military use of telepathic energies. A journalist with the magazine *Encounters*, recently stated: "First developed in the Soviet Union and then adopted by the West, psi makes some use of the ninety percent of the human brain that is normally left unused."[1]

But even if we ignore such bland statements unsupported by any form of evidence, it is certainly true the Russians and the other major powers have been very interested in the possible practical application of psi since at least the 1940s, and despite the collapse of communism, there are almost certainly research establishments still investigating its use for military purposes. And why not? It makes perfect sense to study such things. Compared to the development of conventional weapons, the money spent on research into the paranormal would be almost insignificant, perhaps a few million dollars. Yet, if a mysterious form of telepathic power could be isolated and controlled, it would be one of the most effective weapons any nation could wish for.

Some claim that such powers have been used; but even the most enthusiastic supporter of such ideas as psi admit that, at best, the application of these powers is inefficient and produces only patchy results.

The most popular use for mental powers by the military is as an aid for spying, and many people formerly involved with psi espionage have come forward in recent years. Some are still bound by security to keep sensitive material to themselves, but others, such as those who worked for the KGB and other Communist Bloc agencies, have been more forthcoming. As well as this source, information about the fascination with parapsychology in former Warsaw Pact countries has been known about since the 1960s, from investigators and research groups in the West, some of whom gathered firsthand experience by clandestine means during the cold war. According to some authorities, these sources alerted the Pentagon to the potential of research into psi powers.

In 1968 two investigators of paranormal activity, American Sheila Ostrander and Canadian Lynn Shroeder, visited the USSR and detailed their findings in a book called *Psychic Discoveries Behind the Iron Curtain.* Soon after its publication, they claimed they were each visited by agents from the FBI and the Royal Canadian Mounted Police, although this could have been a convenient story to enhance sales of the book.

By this time, the U.S. government had detailed knowledge of the Russian interest in psi research, and in particular, claims that the Russians had people who could influence the behavior of others and to remote-view secret installations. Even so, it took some time for this to be made public, and it was not until the 1980 issue of the *U.S. Military Review* that Lieutenant Colonel John Alexander was sanctioned to declare his belief that Russian psi research had developed to the point where they could interfere with U.S. military operations. This was followed by an official recommendation that the Pentagon set up a counterinvestigation. In fact, such an organization had already been in operation for many years.

The research that the Russians had apparently developed into a practical tool was based on a technique that has become known as "Sleep-Wake Hypnosis." This is a trick used by professional hypnotists and showmen that, some claim, enables them to induce hypnotic effects at a distance. The power to do this on stage is one thing, but it's been claimed that some practitioners have the ability to turn on and turn off hypnotic states at distances of thousands of miles.

These claims have yet to be studied thoroughly by investigators outside the military, so little information is available in the public domain. It could be that secret institutions have made some headway in the study of Sleep-Wake Hypnosis, but there is no documentary evidence available to support this. All we have to go by at present are the performances of professional hypnotists who insist they are not using any form of trickery, yet are able to produce what seem to be impressive results. One famous hypnotist who appears regularly on British television, Paul McKenna, has been filmed raising and lowering the heart rate of a volunteer seated in another room. However, it is important to note that to date this particular claim has not been verified rigorously by scientists under laboratory conditions.

Yet, to the believers in the practical uses of psi, the implication from the Russian research is that their military have had the ability to influence the actions of unaware individuals across the world for almost two decades. But even this extraordinary claim is not enough for some. It has been suggested recently that the Russians have mastered a technique through which they are able to create, store, and transmit at will what is dubbed "negative energy," and to pervade an area as far away as the West Coast of the United States with this energy.

One of the most extraordinary accounts of what some believers think is the use of this mysterious energy comes from a "mystic" named Michael Bromley, a self-styled "Celtic shaman." Bromley claims to have the ability to sense negative energy, and has been employed on numerous occasions by security agencies around the world to help them plan events and to know in advance when and where there is likely to be trouble. He claims to have been successful in pinpointing key areas in Los Angeles where he thought problems might arise during the 1984 Olympic Games held in the city, and was able to tell police in advance about a security guard he believed would attempt to rape one of the athletes performing in the games. He impressed the local police (and, it was revealed later, the FBI and the CIA) when he told them before the start of the games that the district of Westwood was particularly vulnerable and awash with negative energy.

On the opening day of the games a young man drove a car deliberately at a line of twenty people in Westwood. At his trial, the offender told the jury that he had been hit by what he called "waves of energy."

For Michael Bromley and others, the 1984 Olympic Games was

particularly significant because it was a time of increased pre-Glasnost political tension, and the Russians had refused to take part, just as the U.S. had not participated four years earlier at the Moscow Olympics. Bromley was convinced that something sinister was going on during the 1984 games. "I lived outside Los Angeles," he reported. "Quite independently people were calling me from the city saying they felt waves of energy coming into the area. The Soviets didn't come to the games. All the phone lines from that country to America were engaged a lot of the time. I realized that phone lines were carriers of energy. Now if it was possible to send *psychic* energy down those lines . . . I believe the Soviets were projecting negative energy into Los Angeles. I know it sounds farfetched, but we do it every day. It's the same as wishing someone well, sending their love, or wishing them harm. The Russians have been carrying out scientific experiments into parapsychology for decades."[2]

For the scientist, this statement illustrates many of the misguided attempts at explanation offered by enthusiasts, and what is often muddled thinking concerning any attempts to bring the occult into the arena of scientific investigation. It is worth looking at in some detail.

First, the whole account is an odd blend of emotive statement and wild speculation. I would like to give Michael Bromley the benefit of the doubt and assume he is not combining these elements to deliberately manipulate the reader, that he is sincere about what he believes. However, this technique of mixing stirring, thought-provoking one-liners with suggestive remarks is used frequently by politicians and speech-makers who are trying to convince an audience. So, for example, we have a description of how Bromley was being called by people from Los Angeles reporting "waves of energy." He follows this sentence immediately with: "The Soviets didn't come to the games." Then we have a long discourse about how energy could be transmitted via telephone lines, followed by: "The Russians have been carrying out scientific experiments into the paranormal for decades." This has the effect of reinforcing a link between the two statements that is not really there.

And what are we to make of the theory postulated by Michael

Bromley, that the negative energy sent by the Russians could be transmitted through telephone lines?

He presents no evidence for this hypothesis. He may think he has, because he says things like: "I know it sounds farfetched, but we all do it every day. It's the same as wishing someone well, sending their love, or wishing them harm." This is not evidence, this is bland statement actually contradicted by fact. Yes, we may all send our love to people every day, but does it actually do any good? We do not live in a world where things happen because we wish they would. There are undoubtedly forces known as positive and negative *thinking*, but these have nothing to do with projecting our desires or wishes upon the world or other people. Positive and negative thinking are emotions or drives within us that dictate how confident, optimistic, or indeed pessimistic we feel, and this in turn affects our performance. If we were really able to influence events by wishing alone, would it not be an everyday occurrence to see the sick rise up from their hospital beds because loved ones wished it so?

And what about the telephone lines? They do indeed carry energy, electrical impulses with tiny electrical potentials. But how can we accept a leap from a documented, practical scientific principle such as the use of electrical impulses to communicate around the world, to some vague notion Bromley proposes, such as the use of telephone wires to carry an unknown force?

For Michael Bromley, the phone lines carry energy, therefore the Russians must be sending negative energy down those lines. How does he know this? Because the lines were engaged! Does it not seem more likely that the phone lines were jammed with people wanting to talk across the world, disappointed people who had been hoping to meet their friends from Russia?

Which then leads us to another problem with this story—where is the motivation for such an act? Why would the Russian military waste so much effort on upsetting the people of Los Angeles during the Olympic Games? It might be argued that during the cold war, agencies in the Soviet Union and in the West did some crazy things for little gain, but is this really the most likely explanation for why people felt what they called "waves of energy"? Is it not more reasonable to suggest that these waves of energy were imagined, that someone contrived the idea that because those dastardly Commies were not attending the Olympic games in America, they must be transmitting negative energy; after all: "The Russians have been carrying out scientific experiments into the paranormal for decades." Once the idea was circulated, others started to believe it, and in the summer heat and with the excitement of the games fueling the hysteria, even criminals trying to find excuses for their murderous intentions realized they could tap into the zeitgeist.

But there is actually more to this story than meets the eye. People have been fascinated by the idea that the Soviets have been very interested in the occult for a long time, and this undoubtedly played a part in starting the rumors in Los Angeles and sending people to their telephones to call up Michael Bromley. As I mentioned earlier, at least one highly successful book has documented some of the experiments conducted by Russian scientists and parapsychologists working for the military, and it is worth noting that 1984 was a time of heightened political friction between the two superpowers, a time reminiscent of the cold war hysteria of the 1950s and sixties.

There is a rich history of attempts to apply psi powers, not just as a tool for the spy, but also in efforts to exert some sort of mental control at a distance, of which Paul McKenna's media-friendly efforts to utilize Sleep-Wake Hypnosis is only one example.

In 1977 a psychic investigator named Andrija Puharich told a surprised audience at one of his lectures in London that the Russians had for years been developing a device that could control minds at a distance. This, he claimed, was based upon the work of the great physicist, Nikola Tesla, the man who had made practical the use of alternating current.

According to Puharich, around the turn of the century Tesla had developed a method of sending an extremely low frequency wave (an ELF wave) which has a frequency of four to fifteen pulses per second (4-15 Hz) through the core of the Earth. The idea was to use huge transmitters positioned diametrically on the surface of the planet that would set up a *stationary wave* (a set of waves all vibrating at the same frequency to produce a single wave front) able to penetrate the Earth's core and emerge at a pre-determined spot somewhere in the world.

The plan for Tesla stationary wave machine

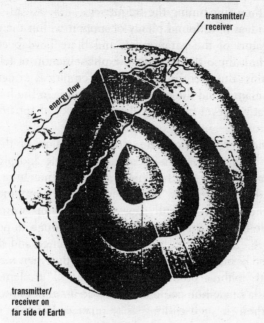

transmitter/
receiver

energy flow

transmitter/
receiver on
far side of Earth

Figure 23.1

The use of a frequency between 4 and 15 Hz was no mere coincidence. This is the frequency range of alpha rhythms produced by the human brain. These rhythms are usually linked with relaxation, but have also been identified as the frequencies most usually enhanced in the brains of fakirs during feats of extreme physical endurance (see Chapters 3 and 7).

In an attempt to explain how the stationary wave generated by Tesla's transmitters could affect the thoughts and behavior of a target, supporters suppose that a *resonance wave* is established in the brain of the recipient of the signal. This wave is created when two waves vibrate in harmony—an everyday example is the way an opera singer can shatter a glass because their voice vibrates at the same frequency as the molecules in the crystalline structure of the glass. The implication is that a controller using the Tesla device could alter the brain waves in a target subject by establishing a resonance wave and then fine-tuning the frequency.

These ideas have found plenty of support within the community of investigators of the paranormal, and there have been many attempts to link these studies with the phenomenon of telepathy and psychokinesis. But there is still no clear empirical evidence to support these claims, and the concept that rhythms in the human brain, operating at whatever frequency, lay at the root of apparent psi powers is unproven.

In some ways this scenario is an illustration of the problems anyone faces in trying to find genuine legitimate and provable links between science and the supernatural. Tesla's machine is theoretically sound. Extremely low frequency waves can be produced, and, again in theory, if sufficiently powerful generators were used, it is conceivable that a stationary wave powerful enough to penetrate the Earth's core could be constructed. Where science and the paranormal do not harmonize is in the linking of this man-made wave to the natural rhythms produced in the brain. The elements are all there—Tesla's machine could be built, the brain does produce alpha rhythms, there is such a thing as resonance; but they do not link up except in the minds of the enthusiasts.

The western powers have been just as keen as the Russians to attempt to exploit any possible practical use for psi powers. In a 1992 Symposium on UFO Research, U.S. Major General Albert N. Strubblebine III chaired a seminar on remote viewing in which for the first time a high-ranking military official revealed the degree to which the U.S. government had used RV for military purposes.

Stubblebine himself headed a research group called Psi Tech, which was set up by two respected physicists, Hal Puthoff and Russell Targ at Stanford Research Institute in California, and jointly funded by the CIA and the U.S. Navy. At the talk, he claimed RV training took about one year and that selected individuals with appropriate discipline and commitment could do amazing things using the power of their minds. "Time is no object," he claimed. "I can go past, I can go present, I can go future. It is independent of location so I can go anywhere on earth . . . I can access information at any location I choose." He then went on to describe how in

1991 his group had helped a large American corporation assess the effect of the Gulf War on the price of oil. His group came up with an answer by, he claimed, "looking inside Saddam Hussein's head."[3]

To the dispassionate observer, there are many problems with this account. We can assume that such a group was established, but the successes they claim to have achieved are highly dubious. The overriding reason for this is that if such a group had been able to truly "go past, present, or future and any location on earth," there would be no need for any conventional spies—the U.S. government would know everything there was to know about any foreign power. In fact, such a group could easily have become billionaires and now rule the world!

We must assume that Major General Stubblebine III is, like many of us, prone to hyperbole and exaggeration. Caught up in the enthusiasm of his delivery, he must have failed to add caveats to his claims, not least of which is the fact that even the most successful remote viewers are often vague, get things wrong as often as they get things right, and are unable to give precise details except on extremely rare occasions.

The second point concerns Stubblebine's story of his involvement in Gulf War espionage. This is unconvincing on two counts. First, when a corporation funds a project, there is pressure on those employed—in this case, the remote viewers—to come up with something, anything, that might justify a check. This understandably casts everything they do in a suspicious light. But the second and far more important issue comes from the simple fact that anyone with any awareness of world affairs and global finances would be able to produce a coherent report detailing the effects of the Gulf War on global economics. You certainly don't need to have psi powers to do that, and one of the last people to be of help would have been Saddam Hussein.

Yet, beyond all this, in order to validate or refute this research we must look at the work of the two men who started the project, Hal Puthoff and Russell Targ. According to their own accounts, they conducted over a hundred RV experiments during the late 1970s and claim to have achieved remarkable results.[4] In at least one set of experiments, they described a remote viewer who was able to find precisely a location not simply when a "sender" was there, but before the site was even chosen.

To the scientist, this alone would imply that something suspicious was going on with their procedure—that there must be a security leak somewhere. But putting this aside, other researchers have tried unsuccessfully to duplicate Puthoff and Targ's results. The most thorough and well-publicized of these was a set of experiments conducted by two other physicists, David Marks and Robert Kammann. They found that their subjects produced results no better than would be expected from chance or simple guesswork.

In a book the pair wrote in 1980 entitled *Psychology of the Psychic,*[5] Marks and Kammann described their experiments, how they then contacted Puthoff and Targ to request access to their data, and how, to their amazement, their request was refused.

To deny other scientists access to one's work is almost unheard of, and immediately sets alarm bells ringing in the ears of other researchers trying to verify results. Consequently, Puthoff and Targ's refusal to release their data gave Marks and Kammann renewed enthusiasm to track down what could be happening in RV experiments.

After extensive investigation they came to the conclusion that in all the tests Puthoff and Targ performed, the remote viewer was either totally inaccurate or, when they had succeeded, they had been provided with subliminal cues or unconscious hints by those involved in the experiment. By matching the transcripts of the conversation between the viewer, the "sender," and the experimenters, Marks and Kammann showed how these clues are picked up.

In *Psychology of the Psychic,* they present many examples of how this can be done. Suppose a viewer is told that he will be required to identify three locations: 1, 2, and 3. The viewer is told that one location is a building, another is an open area, a third is a road, but they have no idea of the order in which they will come during the tests. However, in the transcript of the dialogue, there are a series of cues the experienced viewer can use. For example, the experimenter says before one of the tests: "Third time lucky." During another of the three, he says "Okay, take it slowly, we've got a long day ahead of us." With these hints, the viewer knows which site is 1, 2, or 3.

After persisting for five years and facing renewed requests from others in the field, Puthoff and Targ did eventually publish their findings, in 1985. Upon detailed analysis of these documents, independent investigators found that there were many cues given in their transcripts, particularly during the most successful remote viewing tests.

Other researchers have conducted their own investigations similar to the work of Marks and Kammann and have obtained concurring results. In a paper published in *Nature* in 1986, coauthored by Dr. Marks and an independent colleague, Dr. C. Scott, the researchers came to the scathing conclusion that "remote viewing has not been demonstrated in the experiments conducted by Puthoff and Targ, only the repeated failure of investigators to remove sensory cues."[6]

But is this really the whole picture? Naturally, like all areas of the paranormal, there is a high proportion of misinterpretation, misunderstanding, wishful-thinking, and out and out fraud involved with remote viewing and psi detection: But is it possible that a few individuals do not rely upon cues, do not fake their abilities, and are able to pick up incomplete and often hazy images from distant locations? Is it possible that some special individuals are able to picture scenes and people in the past, present, and future? What are we to make of some of the remarkable success stories of people such as Nella Jones? Do such people have a genuine talent, and if so, how does it operate?

Remote viewers claim they can "see" distant locations, they can "get inside the heads" of the person they see through. But as we saw in Chapters 3 and 4, the energy needed to do such a thing would destroy the cells of the brain. A safer option may be the use of some form of "channeling." This is an idea that has gained popularity during the past twenty-five years and is based upon the principle that some talented people can "tap into a mental network," some mysterious "level of human consciousness" in which all humans are linked.

Parapsychologists are fond of trying to involve quantum mechanics in their explanations of how this could work, but these links are tenuous in the extreme. As I have said elsewhere, well-meaning but untrained investigators are often too keen to drag out Bell's experiment and other seemingly odd aspects of quantum theory to explain what could be happening with telepaths, clairvoyants, and indeed, remote viewers.

It is conceivable that remote viewers may be able to utilize some strange talent involving receiving information passed through a wormhole, such as the process described in Chapter 6: Visions From a Future Time. In other words, they could be receiving data through a wormhole linking two places on Earth. To do this, they would require the extraordinary ability to not only access random information as clairvoyants seem to do (people who receive uninvited images of future events), but to actually manipulate the wormhole so they could see and sense anywhere in the world they wish to probe.

For even the most open-minded scientist, this last hypothesis appears so farfetched as to be impossible. For this idea to work successfully, it would mean that Nella Jones and other successful psi detectives, along with the rare individuals who have produced successful remote viewing results unaided by cues and strong hints, were able to manipulate the very matter of the universe and such fundamental entities as wormholes, without consciously realizing it.

This is no explanation, simply a wild idea, an attempt to match up the meager facts unearthed by parapsychologists with the latest ideas on the fringes of science. But until we have more information about how the mind can process information, how subliminal information may seep into our subconscious and be dredged up by the human brain, and indeed until we know a lot more about how the universe functions at the most fundamental level explored by quantum mechanists, some very rare cases of remote viewing and psi detection remain totally unexplained.

24 The Gods Themselves?

There were vast heaps of stone . . . There . . . under my eyes, ruined, destroyed, lay a town—its roofs open to the sky . . . Further on, some remains of a giant aqueduct . . . there traces of a quay . . . Further on again, long lines of sunken walls and broad, deserted streets . . . Where was I? Where was I? . . . Captain Nemo . . . picking up a piece of chalk . . . advanced to a rock . . . and traced the word . . . "Atlantis."

JULES VERNE, *Twenty Thousand Leagues Under the Sea*, 1869

Most of us love the idea of Atlantis, and upward of three thousand books and articles have been written on the subject during the past two hundred years. There are many reasons for this interest. Psychologists might argue that the image of Atlantis holds a mirror to our world, that it has a similar emotional energy to the biblical Garden of Eden. For others, Atlantis represents an idealized version of our future as a society, a culture closer to our human roots, but technological and global.

In some respects Atlantis is all things to all people. For some it is nothing more than a myth, to others it is a lost continent that may one day be found. For a smaller group it was a place visited by aliens, perhaps even established by extraterrestrials who then passed on their knowledge to the Egyptians. In this last scenario Atlanteans are the forebears of our technological existence, the progenitors of ancient knowledge, keepers of what the alchemists and Hermeticists called the *prisca sapientia*.

But what lies at the root of these ideas? Was there ever an Atlantis, and if so, what sort of place was it? And who were the Atlanteans?

The story of Atlantis has come to us from the Greek philosopher Plato, who lived during the fourth century B.C. Renowned for his masterpiece, the ten-volume *The Republic*, devoted to political structure and the nature of government, he also wrote a pair of dialogues (books in which two or more characters argue over a subject) which dealt with philosophy, history, and science. These were entitled *Timaeus* and *Critias*. In these texts Plato's teacher, Socrates, holds an imaginary conversation with three of his friends. One of these is Plato's maternal great-grandfather, Critias, who had heard a story about a place called Atlantis from his grandfather, Critias the Elder. He had been told the story by his father, who had learned of the legend from the great Athenian thinker and law-giver, Solon, who died 130 years before Plato's birth. Solon claimed to have come by the tale when he visited Egypt, where an ancient priest and guardian of the Temple of Sais had been privy to ancient records documenting all remaining knowledge of the lost continent.

According to these ancient texts, Atlantis was a vast land mass populated by peace-loving demigods who presided over a global culture that existed some nine thousand years before Solon's time—about 11,500 years ago.

According to Plato: "There was an island opposite the strait you call . . . the Pillars of Hercules, an island larger than Libya and Asia combined . . . On this island of Atlantis had arisen a powerful and remarkable dynasty of kings, who . . . controlled, within the strait, Libya up to Egypt and Europe as far as Tyrrhenia [Italy]. This dynasty . . . attempted to enslave at a single stroke . . . all the territory within the strait."

Originally a noble people descended from the gods, the Atlanteans eventually became greedy and sought to dominate realms beyond their boundaries, invading neighboring states and oppressing less-developed cultures. Finally Zeus became angry and destroyed them at a single stroke, the grand palaces and the golden wall swept aside and sunk beneath the waves as the land of Poseidon's children was deluged by the ocean.

"At a later time," Plato tells us, ". . . there were earthquakes and floods of extraordinary violence, and in a single dreadful day and night . . . the island of Atlantis . . . was swallowed up by the sea and vanished; this is why the sea in that area is to this day impassable to navigation, which is hindered by mud just below the surface, the remains of the sunken island."[1]

Plato's *Timaeus* and *Critias* also details the political structure and society of Atlantis. The Atlanteans were ruled by ten kings, descendants of five pairs of twins, the offspring of the union of the mortal woman Cleito and the god Poseidon. The kings met at intervals of five years to make far-reaching and long-term decisions using a system of votes. Their meetings took place in the great capital city, which was surrounded by a golden wall. The city had hot springs, temples, exercise areas, and, oddly, a racecourse.

For many historians, the account outlined in Plato's dialogues is an example of a technique he used often, most famously in *The Republic*, portraying his philosophical ideas through morality tales and stories that highlighted ethical issues. With such details as a racecourse, the description of the Atlantean capital sounds suspiciously Greek, which would imply that the original tale as handed down to Plato has been greatly elaborated and embellished by the author to make it fit the requirements of his own culture and to convey his own ideology. The structure of the story is also an ancient one—the notion of a people acquiring too much power and becoming corrupt before being taught a lesson, or in this case, destroyed by the all-seeing, all-knowing gods. But this of course does not mean the story is completely contrived. It might be simply that Plato used an ancient legend for his dialogues and altered the details.

Plato's pupil, Aristotle, took the view that in writing *Timaeus* and *Critias*, his master had expanded upon a kernel of truth in order to create a myth to convey his philosophical teachings, calling Plato's work, "political fable." But despite the fact that Aristotle became the colossus of teaching throughout the world for at least 1,500 years after his death, and his ideas were the cornerstone of philosophy and science until the Enlightenment, there remained a large contingent of people who did not see *Timaeus* and *Critias* as a mere "fable."

Even by the time of the philosopher, Gaius Plinius Secundus, known as Pliny the Elder, who wrote the encyclopedic *Natural History* in the year 77 A.D., four centuries after Aristotle, the idea of Atlantis within the world of philosophy and history was ambiguous, and opinion about it divided. Some scholars held the view that Atlantis had been a real place, lying opposite the Pillars of Hercules, while others saw it as simply a myth.

Plato is the only original surviving source of the Atlantis story, and all other accounts are based entirely upon it. And, because there is so little to go by, it is almost impossible to judge the political structure or the form of society that may have been adopted by the Atlanteans. Consequently, those fascinated with the subject have concentrated on finding the location of the lost continent in the hope that one day an expedition will unearth the great walled city and reveal the secrets of the demigods who some believe once ruled much of the world.

Throughout ancient times, and indeed until the late fifteenth century, the Atlantic was a largely uncharted ocean. Roman and Dark Age historians and geographers described a vast array of unexplored islands and isolated lands throughout the Atlantic, almost all of which turned out to be fictitious. These included the islands of the Seven Cities, the Fortunate Isles, St. Brendan's Isle, and a mysterious place named Hy Breasil, which remained on mariner's charts the world over until the late nineteenth century.

Plato had described Atlantis as lying beyond the Pillars of Hercules, by which he meant the Strait of Gibraltar, the gateway from the Mediterranean to the Atlantic. It was only natural, then, that this information should lead cartographers and explorers to place the mysterious lost continent of Atlantis somewhere in the Atlantic Ocean. But where exactly?

During the first few decades after America was discovered by Columbus, European philosophers thought that this new landmass might be the remains of the lost continent. In the 1550s the Spanish historian, Francesco López de Gómara, believed that some of the features of what was then known of America and the West Indies fit the description Plato had offered in *Timaeus* and *Critias*, and the English statesman and philosopher, Francis Bacon placed Atlantis in the New World in his masterpiece *Nova Atlantis*, published in 1618.

But gradually, as America was explored and mapped, it became clear that it had nothing to do with Atlantis. However, the idea that Atlantis was to be found somewhere in the Atlantic Ocean persisted. One of the most determined adherents to this view was the American author and historian Ignatius Donnelly, who was convinced that the Azores are the only remains of Atlantis above sea level. In his book

Atlantis: The Antediluvian World, published in 1882, Donnelly propounded the theory that Atlantis sank beneath the Atlantic waves and that a feature called the Mid-Atlantic Ridge, discovered in the 1870s (of which the Azores are the volcanic peaks), was a major geographical component of the continent.

Donnelly went to his grave believing he had solved the mystery of the location of Atlantis, but in the 1960s the study of plate tectonics showed that his theory could not be correct. Plate tectonics describes how the present configuration of the Earth's crust was produced by shifting plates (large segments of the Earth's surface), which have created features such as mountain ranges and ocean ridges including the Mid-Atlantic Ridge. Rather than this range and the Azores being the remains of a sunken continent, as Donnelly believed, the ridge was generated by the movement of the Earth's tectonic plates in relatively recent times.

Today, Donnelly's idea is dismissed by scholars and scientists. The Greek historian Professor A. Galanopoulos, has said of the idea: "There never was an Atlantic land bridge since the arrival of man in the world; there is no sunken landmass in the Antarctic: the Atlantic Ocean must have existed in its present form for at least a million years. In fact it is a geophysical impossibility for an Atlantis of Plato's dimensions to have existed in the Atlantic."[2]

Yet, to this day many believe the theory that Atlantis was a large landmass in what is now the Atlantic Ocean, and some contend that there is particularly compelling evidence to support the argument that it was located close to the West Indies.

In 1968 a diver in the Bahamas known to the locals as Bonefish Sam met an American zoologist and keen amateur archaeologist, Dr. J. Manson Valentine, who was visiting the islands. Bonefish Sam showed him an underwater anomaly that he thought might be of archaeological interest.

The anomaly is about a kilometer off Paradise Point in the Bahamas and consists of what could have once been a wall (interpreted by some as a road) made from large stones lying under about six meters of water. The stones, now known as the Bimini Road, are each estimated to weigh between one and ten tons, and several dozen of them are aligned to run for about half a kilometer in a straight line before ending in a sharp bend.

Dr. Valentine described the structure as "an extensive pavement of rectangular and polygonal flat stones of varying size and thickness, obviously shaped and accurately aligned to form a convincing arte-factural pattern . . . Some were absolutely rectangular and some approaching perfect squares."[3]

Occultists became very excited by this discovery. This was mainly due to the clairvoyant, Edgar Cayce, who during the 1920s and thirties said he'd entered a trance in which messages from "higher authorities" informed him of mystical connections. Before he died in 1945, he was quoted as saying: "A portion of the temples [of Atlantis] may yet be discovered under the slime of ages of sea water near Bimini . . .Expect it in '68 and '69, not so far away!"

Initially this does seem a rather startling prediction, but it should be borne in mind that the underwater feature was called the Bimini Road because of its proximity to North Bimini Island in the Bahamas and was a well-known tourist spot, but had been overlooked by scientists until Bonefish Sam's revelation in 1968. Cayce may also have known of the feature and used it in his prophesies. It is also possible that many people on the islands knew about Cayce (he was a very famous mystic in his day), and had deliberately chosen to inform the scientific world through Dr. Manson Valentine in order to fit the prediction.

But to the increasing satisfaction of the Atlantis enthusiasts, there was soon more ammunition for their arguments. In 1975, Dr. David Zink, author of The Stones of Atlantis, discovered what appears to be a block of stone that looks very much like concrete and is cer-tainly man-made because it contains a tongue-and-groove joint.

However, skeptics argue that this and other artifacts, including some anomalous marble pillars (marble is not found in the area naturally) are relics of shipwrecks. These arguments were strength-ened when a building found alongside the Bimini Road and origi-nally thought to be an Atlantean temple was shown to be nothing more exotic than a sponge store that had been built during the 1930s. This has led skeptics to encourage the theory that the Bimini Road was actually produced by a natural phenomenon, via an accepted geological process called "Pleistocene beach-rock erosion and cracking." Others tread a middle road, suggesting that the underwater feature is natural, but could nevertheless have been used by ancient peoples.

Today, the mystery of the Bimini Road remains unsolved, and it is the battleground for supporters and opponents of the many theories surrounding the reality and the location of Atlantis. In 1997 a group of British researchers from the Building Research Establishment (BRE) analyzed samples from the man-made block close to the Bimini Road and concluded that it is made from a form of concrete manufactured by an old-fashioned technique and certainly older than the modern-day Portland cement process, devised in 1820. Just how long before this is uncertain, so the block could have been produced in Europe anytime between the sixteenth and early nineteenth century, or it could be far more ancient.

Using an electron microscope, one of the BRE team, Dr. David Rayment, head of the organization's Electron Microanalytical Unit, has found a strip of gold in the concrete block which shows clear signs of having been worked by a skilled craftsman. But though this is a fascinating development, it does not mean that the block was part of a building in Atlantis. As Dr. Kelvin Pettifer of the BRE's Petrographic Unit points out: "Much as I'd like to believe it, there is nothing in any of the samples that is enough to convince me. It could have been that the marble pillars and other man-made materials were intended for a cotton plantation mansion but ended up on the seabed following a shipwreck."[4]

Clearly, more research on these materials is needed before a definite conclusion can be reached. Carbon dating would be of little use in this case because the concrete block is man-made, but one possibility would be to try to find pollen grains or other organic material inside the core of the block which may have been deposited there when it was being produced. These could be matched with samples found in different parts of the world in an effort to determine where the block was made. And of course, the pollen grains or other natural materials could be carbon-dated.

Meanwhile, a Russian team, lead by Professor Viatcheslav Koudriavtsev, director of the Moscow Institute of Metahistory, which studies how natural catastrophes affect human development, are investigating the idea that Atlantis could have been positioned off Land's End, in Cornwall, England. They are sending a group of divers to explore a little known set of sunken ruins that lie a hundred miles off the western tip of Cornwall, clustered around Little Sole Bank, an undersea hill that rises to fifty meters below the surface.

Koudriavtsev has based his theory on a combination of Plato's descriptions in *Timaeus* and *Critias* and ancient Cornish myths that tell of a rich land on which stood the City of the Lions, containing no fewer than 140 temples. In these tales, the area now under the waves was called Lyonesse, a land featured in many ancient fables and legends, including *The Faeire Queen* by the Elizabethan poet, Edmund Spenser and Alfred Lord Tennyson's version of the Arthurian legend, *Idylls of the King*, published between 1859 and 1885 in the form of twelve poems.

The Russian professor has said that the idea Atlantis existed off the coast of Cornwall is backed up by plentiful research, claiming that his conclusions are "based upon fresh translations of the Greek texts, which have fascinated me since I was a student at Moscow University many years ago."[5]

Indeed, geologists do believe that a series of natural disasters set in motion by the melting of the last icebergs left over from the most recent Ice Age could have flooded a fertile plain that may have existed at the location marked by the Russian investigators. Dr. Geoff Kellaway, a local geologist, has called the Russian theory "not unreasonable. Billions of gallons of ice-age waters flooded fertile lands that could have supported civilizations. Mammoth teeth have been washed ashore. But the Celtic Shelf is a massive area—the Russians will be looking for a needle in a very deep haystack."[6]

Sadly, even if Professor Kourdriavtsev's team do locate evidence that Little Sole Bank represents a part of Atlantis, it will solve one mystery but offer another, because just miles away from this point, the Celtic Shelf falls away to a depth of four thousand meters, and much of the lost continent will be truly lost, having slid to the bottom of the Atlantic Ocean.

But these are just a few of the disparate ideas that surround the possible location of Atlantis based upon one interpretation of Plato's descriptions and combined with other seemingly connected ideas. But there are many who refuse to accept the notion that Atlantis was anywhere near the Atlantic. According to a growing body of experts who have in recent decades held sway over the official line in linking the myth with the reality of Atlantis, the lost continent was actually thousands of miles away from the Atlantic Ocean.

Close inspection of Plato's account shows many confusing contradictions and anomalies. First, it seems that he has exaggerated all the dimensions by a factor of ten. This has been noted by Professor Galanopoulos, who points out that in his description of the great capital of Atlantis, the Royal City, Plato ascribes a length of 10,000 *stades* or 1,135 miles to the city wall.[7] Even Plato questioned the validity of this figure in his transcription, and it does indeed seem excessively large even for a culture ruled by demigods. The Great Wall of China (which is, incidentally, the only man-made object visible from earth orbit) is 1,500 miles long, but a wall over eleven hundred miles in length would circumvent Greater London twenty times. If we reduce the length of the wall by a factor of ten we have a more reasonable number.

Plato also says that Atlantis existed 9,000 years before his time, which places it in an era when the rest of the world was still in the Palaeolithic period or Old Stone Age, at least 6,000 years before the origin of the Egyptian civilization. But, it is argued, the mythical flavor of the lost continent is eradicated if we again divide Plato's figure by ten, placing the high point of Atlantean culture 900 years before Solon. This then fits neatly with the description offered by Plato of battles between the Atlanteans and the embryonic state of Athens during the time the kings of Atlantis were attempting to expand their empire.

This mistake could have occurred, it is argued, because the Egyptian copyist mistook the ancient Egyptian symbol for 100 (a coiled snake) for the lotus flower, the symbol for 1,000. This would be analogous to us confusing the British billion (a million million) with the American billion (one thousand million).

A further confusion arises over a simple phrase in the original tale. Plato was told that the lost continent was "larger than Libya and Asia combined," but the Greek words for "greater than" and "between" are almost identical, which suggests that Plato should have described Atlantis as "between" Libya and Asia.

If for the moment we take these ideas as facts, it places an entirely different complexion on the tale of Atlantis, at once making it more prosaic and eligible for links with another culture that is known to have existed at the same time, but far from the Atlantic Ocean—the Minoan civilization of Crete.

Beginning in 1900 with Sir Arthur Evans, a succession of archaeologists have stuided the region encompassing the islands of the Aegean lying south of mainland Greece and have gradually pieced together a picture of what may have happened to a great civilization that once lived there around 1500 B.C.

The most southerly of the Greek islands is Santorini. Today it is actually a collection of three islands. The largest is the beautiful isle of Thera, which is a major tourist attraction with its black sand and crystal clear waters. About 3,500 years ago, Santorini was a single, almost circular island which was blown apart by a massive volcanic eruption thought to be four times more powerful than the eruption of Krakatoa in 1883.

The eruption of Krakatoa has been estimated as equivalent to one million Hiroshimas, and although this may be an exaggeration, the explosion that occurred on Santorini around 1520 B.C. must have been truly devastating. It is believed to have created hundred-foot waves that swept in all directions from the island, entirely engulfing another advanced community living a mere ninety-six kilometers north—the Minoan civilization that was then thriving on the island of Crete.

Evans, who discovered the Palace of Knossos on Crete, and with it unearthed the lost history of Minoan culture, did not link the destruction of this civilization with the volcanic eruption on Santorini; but others did soon find a link. As early as 1909 some scholars were suggesting that the ruins of the Minoans were in fact the lost royal city of the Atlanteans. By the 1930s the Greek archaeologist, Spyridon Marinatos, took this and made the link between the eruption on Santorini and the destruction of Knossos after he found more Minoan remains in the north of Crete, along with pumice, a frothy form of volcanic glass left over from a volcanic eruption.

Later, during the 1960s, ruins of an advanced culture were found on Thera, including the remnants of a massive circular channel on the edge of what remains of the original island, which matches Plato's description of channels circumventing the great metropolis of Atlantis.

Other possible links come from a comparison of the Minoan culture and the legends of Atlantis. According to Plato, the Atlanteans worshiped the bull, and it was discovered from the ruins at Knossos that the bull cult also lay at the heart of the Minoan religion.

If we take Plato's geography as misguided, along with his time frame being wrong by a factor of ten, then the evidence to superimpose Crete with Atlantis is compelling. This then suggests that the tale of Atlantis passed from Solon to Plato was actually an Egyptian legend based upon an event that may have taken place some nine hundred to a thousand years earlier, in the Aegean.

However, even this rather neat explanation has its critics. Recent archaeological findings using accurate dating techniques suggest that the volcanic ash from Santorini is at least 150 years older than the date assigned to the destruction of the Cretan palaces, implying that the volcanic eruption on Santorini did not destroy the Minoan culture after all.

Evidently, a great deal more work has to be done on the possible links between Santorini and Crete before a plausible hypothesis linking this area with Atlantis can be formulated; but for many, what appeared to be a promising connection is too flawed to be accurate, and they are actively searching for new solutions.

One of the best researched alternative theories of recent years has been the work of the writer Graham Hancock, who, along with others has proposed a quite different location for the lost continent.

Hancock, in his book *Fingerprints of the Gods*, and archaeologists Rose and Rand Flem-Ath in their book, *When the Sky Fell In*, also published in 1995, both propose that the site of Atlantis was in fact Antarctica.

Sticking to Plato's original dates for the existence of Atlantis, they subscribe to the idea that catastrophic displacement of the Earth's crust caused the extinction of an advanced civilization that existed on the edge of an extended continent of Antarctica about 11,000 years ago. Their contention is that at this time Antarctica was very much larger than it is today, that its most northerly coast reached at least two thousand miles farther north. They further contest that the reason no one has yet located Atlantis is because it lies beneath the frozen wastes of modern-day Antarctica.

In *Fingerprints of the Gods*, Hancock quotes the Flem-Aths as saying: "Antarctica is our least understood continent. Most of us assume that the immense island has been ice-bound for millions of years. But new discoveries prove that parts of Antarctica were free of ice thousands of years ago, recent history by the geological clock. The theory of 'earth-crust displacement' explains the mysterious surge and ebb of Antarctica's vast ice sheet."[8]

The link with Plato, they say, comes from the fact that some of the records of Atlantis were taken by survivors to the area around the Mediterranean, a group who millennia later seeded the Egyptian civilization, providing them with the technological expertise needed to construct the pyramids, embalm their pharaohs, and model the sphinx.

But does Atlantis need to have existed at all? After all, the only account we have to go by is Plato's testament, and although there may have been genuine elaborate legends hidden in lost documents in Egypt, perhaps in the library of Alexandria before its destruction, it is also possible that Plato's story is nothing more than a total fabrication.

Noting the remarkable similarities between artifacts found in different cultures that developed around the Atlantic Ocean as far apart as Africa and Europe, there are some who hold the view that there had to be a real Atlantis that existed some 11,000 years ago to account for this. But the work of such pioneers as Thor Heyerdahl and others have shown that there is no need for recourse to occult explanations for such things; that, indeed, people traveled more widely at this time than was believed previously.

Yet, the story of Atlantis does lie at the heart of a great occult tradition. The Theosophists (or Theosophical Society)—a group that flourished toward the end of the nineteenth century—were particularly enamored with the legend of Atlantis.

The Theosophy Society was established by Madame Helena Blavatsky in 1875. She wrote several books that have become classics of the alternative historical tradition, including *Isis Unveiled* and *The Secret Doctrine*. The Theosophists believed in what they called the *Akashic records*, what some describe as an "astral library"—a source of mystical knowledge tapped into by skilled mediums who then divulge secret knowledge to the rest of us.* From the Akashic records, Blavatsky and other Theosophists, most notably Rudolph Steiner, constructed an image of a train of seven civilizations or *root races* dating from the distant past, of which we were supposed to be the fifth.† The Atlanteans, who according to this idea are our immediate ancestors, the fourth race of humans, possessed an advanced technology, used flying machines, and had developed sophisticated medical techniques.

It is interesting to note that Theosophists writing in the late nineteenth century were fascinated with the potential of technology, and in some respects their descriptions of ancient lost civilizations bore a marked resemblance to Victorian western culture—the Atlanteans used airships and X-ray machines. This is even more interesting when we consider the fact that imprinting one's own culture upon alternative ancient scenarios is exactly what Plato did in his tracts describing the lost continent. It is also a common phenomenon among witnesses who claim to have see alien spacecraft—they describe them in a way that is fitting for their time (see Chapter 13).

One of the most prolific writers on the subject of Atlantis and the legends surrounding it was an expert of the occult, Lewis Spence, who produced over a dozen books on the subject, all based upon the fantasies of the Theosophists and Plato's original seven

---

*Edgar Cayce claimed to be able to access these records and believed they were the source for his predictions.

†According to Theosophic doctrine, the first root race were invisible, made of "fire-mist," and lived at the North Pole. The second lived in northern Asia and were almost invisible, but managed to see each other well enough to develop sexual intercourse. The third root race was the Lemurians, who lived in a place called Mu several hundred thousand years ago. The fourth were the people of Atlantis, and we are the fifth. The sixth root race will supposedly return to Lemuria, and after the seventh race has run its course, humans will leave Earth altogether and emigrate to Mercury.

thousand words on the subject in his two dialogues, *Timaeus* and *Critias*. Many of these books have become classics of the alternative tradition and remain in print courtesy of specialist publishers around the world today.

And in more recent times, the myth of Atlantis has found a new impetus among believers in the idea that our planet has been visited and perhaps even colonized in the distant past. By amalgamating some of the ideas of the Theosophists and the convoluted hypotheses of such writers as Eric Von Daniken, a large body of people claim to believe that the human race was seeded by aliens, that Atlantis was really the home of an advanced culture destroyed perhaps by a nuclear accident or wiped out by an AIDS-type disease. Ironically, some are trying to use the Atlantis story as a model for all that they perceive to be wrong with our culture. This is again exactly what Plato was doing two and a half millennia ago in Greece.

Believers in the idea that there have been advanced human civilizations that have existed and thrived here on Earth in ancient times point to the many and diverse legends incorporating advanced but lost civilizations in our own deep past, but these provide very flimsy evidence for such a bold claim.

The "evidence" may be broken down into three groups—ancient texts, ancient images, and ancient monuments.

The first of these come from a variety of sources and different cultures, including ancient India, China, Egypt, and South America. These texts often describe events that could be interpreted (again using contemporary culture as a template) as describing visitations by aliens, abductions, even colonization, and have been used by many enthusiasts in a growing collection of books that attempt to make links between extraterrestrials and ancient peoples on earth. An example is the accounts of Old Testament prophets such as Ezekiel, interpreted by writers such as Eric Von Daniken as coded descriptions of alien visitations, cosmic travelers who passed on secret knowledge, men who some identify with the original colonizers of Earth and possibly with the establishment of Atlantis.

A favorite of the occultists is what has been claimed to be Ezekiel's encounters with an ancient astronaut taken from the biblical passage that begins: "Now it came to pass in the thirtieth year, in the fourth month, in the fifth day of the month, as I was among the captives by the river of Chebar, that the heavens were opened . . . And I looked, and behold, a whirlwind came out of the north, a great cloud, and a fire unfolding itself, and a brightness was about it, and out of the midst thereof was the color of amber, out of the midst of the fire. Also out of the midst thereof came the likeness of four living creatures. And this was their appearance; they had the likeness of a man. And every one had four faces, and every one had four wings. And their feet were straight feet; and the soles of their feet were like the sole of a calf's foot: and they sparkled like the color of burnished brass."[9]

At first glance this might appear to describe something like an advanced flying machine, perhaps one built by aliens, or by the people the Theosophists imagined lived in Atlantis, along with occupants clad in spacesuits. But it should be recalled that the Old Testament was written by simple people who had little experience of the world, who lived in constant fear of the forces of nature and the wrath of their God. To them, something as natural as a whirlwind or a volcanic eruption could be personified, anthropomorphosized with images of strange beings. It is even conceivable that these could be descriptions of real men from a slightly more advanced culture dazzling simple peasants with chariots, bright ornamentation, and well-crafted weapons.

Linked with these texts are records preserved by ancient cultures as verbal accounts. A striking example comes from the West African Dogon tribe, which, according to some, knew of the existence of a star called Sirius B, though it can only be seen with the aid of a powerful telescope and was first photographed in 1970. In his book *The Sirius Mystery*, the writer Robert Temple claims the tribespeople knew this star was a part of what modern astronomers call a binary star system (a star that orbits another), in this case the much brighter Sirius. Astonishingly, the Dogon tribe even knew the duration of the star's orbit—around fifty years. The Dogon, he claims, learned about Sirius B from the ancient Egyptians some three thousand years ago, and others extrapolate further and believe that such astronomical knowledge possessed by the ancient Egyptians was passed on to them by a much older race—once again, the same highly advanced people of Atlantis. However, astronomers are convinced that the knowledge of the Dogon is nothing more than coincidence, and point to the fact that a high percentage of star systems are binary and that the figure of fifty years was a lucky guess.

The second type of "evidence" proposed by enthusiasts of these occult ideas is pictorial representation, ancient images that have survived from long-dead civilizations, particularly the Egyptians, who, enthusiasts believe, were the custodians of the artifacts surviving the destruction of Atlantis.

Some of these have been widely publicized as proof that we were either visited by advanced extraterrestrials or there was a race of technologically advanced humans who lived on Earth many thousands of years ago. Perhaps the most sensational is a drawing discovered in the ancient Mayan Temple at Palenque in Mexico. It shows a human figure seated in what looks astonishingly like a modern space capsule. The figure is squeezed into a small space jammed with levers and what could be interpreted as control panels, and coming from the rear of the contraption appears to be a plume of smoke and fire not unlike the vapors expelled from a NASA rocket.

This is not the only picture form the ancient world that depicts what could be interpreted as space technology. According to some supporters of the ancient technology theory, primitive man seems to be obsessed with spacesuited figures. One drawing seen in cave dwellings found in Val Camonica, Northern Italy, depicts what may be interpreted as cosmonauts or NASA astronauts. They are dressed in large suits and what look like helmets and visors. Another interpretation may be that the drawings were actually showing nothing more exotic than the hunting gear worn by primitive people during a period now recognized as a mini Ice Age. Similar drawings have been found at ancient American Indian sites in North America, in Urbekistan, and in Tassili in the Sahara.

But for those who want to believe that an ancient people ruled the Earth using advanced technology tens of thousands of years ago, the most important link they have to the past is the towering edifice of the Great Pyramid at Giza, the circle of stones at Stonehenge, and other sites around the globe.

One of the original Seven Wonders of the Ancient World (and

the only one remaining today), the Great Pyramid is a truly amazing feat of engineering. Known to have been constructed during the third millennium B.C., it contains upward of one million blocks of stone, each weighing about 2.5 tons. It measures 230 meters (756 feet) on each side (equivalent to four blocks of Fifth Avenue), and was originally 147 meters (482 feet) high.

Staggering engineering achievement the Great Pyramid may be, but orthodox archaeologists are able to describe in detail how it was built, using tens of thousands of slaves who dragged the stones from boats that had brought them from quarries in the Lower Nile. They have plotted the route of roads specially designed and constructed to transport the stones, and have shown how Egyptian engineers had the mathematical and engineering skills to construct a building that is not only huge, but demonstrates sophisticated number relationships between the length of its sides, its height, and the area of the base.

A monument that required comparable engineering genius is Stonehenge. For several decades it has been associated with theories trying to link its construction with alien visitors or ancient humans who possessed technological ability approximately equal to our own today.

Stonehenge is to be found thirteen kilometers north of Salisbury, England. It was started a few hundred years before the Great Pyramid at Giza, around 2800 B.C. But unlike the Great Pyramid, the Stonehenge site evolved over a period of almost 1,800 years. Conventional archaeologists have identified four different phases of construction, with Period I beginning around 2800 B.C. and period IV ending about 1100 B.C.

Theories concerning the use of the site and the way such an edifice could have been constructed by primitive tribespeople are varied and plentiful. Again, enthusiasts of the ancient astronaut theory suggest that Stonehenge is one of many sites situated on ley-lines—hypothetical lines of "force" or natural energy that intersect at key points (see Chapter 22).

Although many books, articles, and television programs have been produced debating the idea that Stonehenge is in some way linked with the ancient people of Atlantis, or is perhaps of cosmic significance to extraterrestrials, once more, conventional archaeology can offer a clear picture of how this incredible edifice was built. A growing collection of scholarly works have gone to great lengths to explain the methods employed by the ancient Britons and the techniques they employed using the materials readily available at the time.[10]

But whatever the arguments over who built the pyramids and Stonehenge and why, the simple fact remains that any reliance upon an occult explanation is at best an insult to human ingenuity. To many, not just empirically minded or skeptical scientists, the occultist's attempts to dismiss the achievements of our ancestors is at once demeaning and crude. But beyond this, the reasoning of people like Von Daniken and other supporters of the idea that the ancients of traditional history could not have done the things they did is one of the worst examples of flabby thinking.

In *Chariots of the Gods?* Von Daniken makes claims such as: "Is it really coincidence that the height of the Pyramid of Cheops multiplied by 1,000 million corresponds approximately to the distance between the Earth and the sun?"[11]

Well, first, the answer is surely yes; but let us give this particular author enough rope to do with as he will. The distance between the Earth and the sun is 93 million miles. If we multiply the height of the great pyramid by one thousand million we arrive at a figure of 98 million miles. This is an approximation indeed, off by no less than six percent. So what does it prove? Does it suggest that the ancient Atlanteans, or perhaps visitors from another planet, calculate to within a margin of error of six percent? Nothing else about the Great Pyramid is off by even one-thousandth of this figure.

On a television program called *The Case of the Ancient Astronauts*, made to debunk Von Daniken in 1978, the producers drew an analogy between the author's pronouncements and what a wrongheaded archaeologist of the future might thing about our culture. Suppose, they said, in the year 5330 A.D., an impressive-looking ancient monument was unearthed in a site known to have been where a civilization once existed, a city thought to be called "Washington." Archaeologists of the time calculate that the height in miles of a needlelike construction in the center of the ruined city when multiplied by forty gives the distance in light-years to the second nearest star, Proxima Centauri. Would the obvious conclusion be that the ancient Americans were too stupid to have made this comparison themselves and that the Washington Monument was designed and built by ancient alien visitors?

So where do all these conflicting ideas leave us on the subject of Atlantis? Did it really exist? If it did, what sort of place was it, what sort of society did the Atlanteans have? Was their technology comparable to ours, or was Atlantis simply an isolated island kingdom, home to a culture just a little more advanced than its neighbors?

The problem with any suggestion that the Atlanteans had an advanced technology is the matter of what happened to the traces. Hancock's fingerprints of the gods are the encoded remnants of a great culture he believes can be seen in certain ancient materials and cultural heritage, but would there not be much more remaining? What would our society leave behind? Would people eleven or twelve millennia in the future be able to say with certainty that an advanced civilization once lived on earth?

I think the answer to this is an unequivocal yes. We are a global society and our marks are everywhere. If our culture was to be utterly destroyed, some traces would remain if people of the future were to look hard enough. We have left our stamp in the depths of the oceans, on the highest peaks, and on the surface of the moon and other parts of the solar system. Atlantis could not have been a truly global civilization, and therefore it could not have developed to the level we have. In fact, Atlantean society could not have been any more advanced than that of, say, Europe before the seventeenth century, when global exploration became commonplace.

And what of the relics, the fingerprints of their time on Earth? For the believer in an alternative history there is no concrete proof, no ten-thousand-year-old human skull with a steel plate, no artificial hip joint from five thousand years before Christ, no laser gun carbon-dated to the time of the ancient Egyptians.

The closest we have come to such a discovery occurred in 1936. Archaeologists working in Iraq stumbled across what is now known as the "Baghdad battery." This is a tube a few inches long which consists of all the components of a working cell minus the battery acid itself, but known to be at least two thousand years old. Researchers have made an exact replica of the device, and using fruit juice to substitute for battery acid, they produced half a volt of electricity.

No one has been able to explain the origin of this curious object, and it is still believed to be genuine many decades after it was first discovered. It may be a relic from an ancient technological society, but it is strange that such a find is entirely isolated. Advances in technology are always interlinked. It is extremely unlikely that a car, for example, could be built unless the society in which it is made has the support system to produce it—techniques to produce metal sheets, gasoline or some other fuel, materials for the tires, not to mention the machine tools to build the individual components.

It might be that the Baghdad battery was constructed by an unknown genius, a Leonardo da Vinci of his time who stumbled across the technique and built it from scratch; perhaps we will never know.

Yet, despite the marked lack of evidence, there is nothing intrinsically wrong or contradictory about the idea that a reasonably advanced civilization could have sprung up and flourished for a while, perhaps tens of thousands of years ago. Indeed there are no real intellectual objections to the idea that we could have been visited by aliens in the dim and distant past and that an advanced culture was seeded by colonizers or perhaps a small group stranded here. But equally, it is a quite unnecessary hypothesis.

And what of the location of the lost continent? If it did exist, the smart money is still with the idea that the Minoan civilization was the origin of the Atlantis myth, the story handed on to the ancient Egyptians, who wrote the account before it became elaborated by successive generations leading to Plato; just as it has been elaborated further still in recent centuries, and will almost certainly continue to be in the centuries to come.

And this is how it should be. Good stories never go out of fashion.

# Afterword

It seems that around every twenty years there is a renewed interest in the paranormal. The 1930s was a big time for the supernatural, with such figures as Aleister Crowley then at his peak. The 1950s saw the first blossom of the UFO myth and the earliest tales of close encounters as popularized by George Adamski. In the early 1970s the supernatural surfaced again and amalgamated at times with science fiction, giving us a succession of popular novels, films, and TV series—*The Exorcist, Carrie, UFO,* to name but a few. Now it has all come around again in a far more sophisticated form with some new elements thrown in that could not have been imagined in previous incarnations—genetic mutations, the Internet, cybertechnology.

So, why does the subject of the paranormal intrigue and captivate us so? Why do we tune-in to the *X-Files,* rent videos of the Roswell autopsy, flock to see *Men in Black* and *Gattaca,* buy the novels and wear the T-shirts?

It is partly a generational thing. The fact that the paranormal reappears in the public imagination about once every two decades is no coincidence, and for many young people, the investigations of Mulder and Scully are totally new and original. But it runs deeper than that. In each of us there is a yearning for something larger than life, something beyond the mundane.

Perhaps, as life becomes more comfortable, we need to find something extra, something beyond ourselves. Most of us cannot find this extra element in our "real" lives, so we look for it elsewhere—we escape.

Like most people, I really wish many of the ideas of the occultists were actual. It would be such fun if ghosts existed (other than as hypothetical played-back images). It would be tremendously exciting if aliens really were here and we could communicate with them. How much more entertaining life would be if we could develop large-scale, usable telepathic powers. I'm not so keen on spontaneous human combustion, but precognition could be handy if used properly, and time travel would be a dream come true. Sadly, for most of these, the evidence points the other way. There may be some effects on a small scale with phenomena such as telepathy and PK. We certainly can control our own bodies to a degree, and some gifted individuals may be able to enhance this ability in others. There are almost certainly alien civilizations not far from Earth (in cosmic terms), and they may well have visited, but there is no huge conspiracy or cover-up hiding a silent invasion. In the final analysis, *Independence Day*, *Alien Nation*, or *War of the Worlds* are fantastic entertainment, but once you leave the cinema, close the book, or turn off the TV, they have no existence beyond memory.

Yet, none of this stops us from striving to discover, and I am not one of those trained as a scientist who totally dismisses the paranormal (as I hope you will have already gleaned). It is important we all keep open minds, but also fully functioning ones. There is something off-putting about the smugness of a small minority of New Age practitioners, a smugness borne of ignorance and dysfunctional mentality, that gets none of us anywhere.

By all means, keep questioning, prodding, and investigating. Keep watching the skies—if for nothing else but to find a window onto the eternal.

 References

Chapter 2: Is There Anybody Out There?

1. As quoted in Paul Davies, *Are We Alone?* (London: Penguin books, Orion Productions, 1995), 23.
2. Frank Drake and Dava Sobel, *Is Anyone Out There?* (London: Souvenir Press, 1993), 56.
3. Ibid.
4. Adrian Berry, *The Next 500 Years: Life in the Coming Millennium* (London: Headline, 1995), 239.
5. Davies, xi.

Chapter 3: The Mind's Eye

1. James Alcock, *Parapsychology: Science or Magic?* (London: Oxford, 1985), 86.
2. Carl Jung and Wolfgang Pauli, *The Interpretation of Nature and the Psyche* (New York: Pantheon, 1955).
3. Carl Jung, *Man and His Symbols* (London: Aldus Books, 1964).
4. Rupert Sheldrake, *A New Science of Life* (London: Blond and Briggs 1981).
5. J. B. Rhine, *Extrasensory Perception* (Boston: Bruce Humphries, 1934).
6. Chuck Honorton, quoted in "Roll Up for the Telepathy Test," *New Scientist*, 15 May 1993, 29–33.
7. Terry White, *The Sceptical Occultist* (London: Arrow, 1988), Chapter 2.

8. Susan Blackmore, quoted in Terence Hines, *Pseudoscience and the Paranormal* (New York: Prometheus Books, 1988), 82.

Chapter 4: Moving Heaven and Earth

1. John and Anne Spencer, *The Encyclopaedia of the World's Greatest Unsolved Mysteries* (London: Headline, 1995), 259.
2. L. E. Rhine, *Mind Over Matter* (London: Macmillan, 1970).
3. Spencer, 261.
4. As quoted in Richard Milton, *Forbidden Science: Exposing the Secrets of Suppressed Research* (London: Fourth Estate, 1994) 46.
5. Hans Eysenck and Carl Sargent, *Explaining the Unexplained* (London: Weidenfeld & Nicolson, 1982) 102–103.

Chapter 5: The Fire Within

1. John Fairley and Simon Welfare, *Arthur. C. Clarke's Chronicles of the Strange and Mysterious* (London: HarperCollins, 1987), 160.
2. Jenny Randles and Peter Hough, *Spontaneous Human Combustion* (New York: Bantam, 1993), 152.
3. Ibid, 238.
4. John E. Heymer, *The Entrancing Flame* (Little, Brown, 1996), 170-171.
5. Michael Harrison, *Fire from Heaven* (Scoob Books Publishing, 1990).

Chapter 6: Visions From a Future Time

1. Lyall Watson, *Supernature* (London: Sceptre, 1974), 272–73.
2. Quoted in Hans Eysenck and Carl Sargent, *Explaining the Unexplained* (London: Weidenfeld and Nicolson, 1982), 13.
3. C. G. Jung, *Synchronicity: An Acausal Connecting Principle* (London: Routledge & Kegan Paul, 1972).
4. Lawrence Krauss, *The Physics of Star Trek* (New York: HarperCollins, 1996), 65–83.
5. Quoted in Eysenck and Sargent, 139.
6. John Gribbin, *In Search of Schrödinger's Cat* (London: Corgi, 1984) 229.

Chapter 7: The Agony and the Ecstasy

1. *Seven Hundred Chinese Proverbs*, translated by Henry. H. Hart (Palo Alto: Stanford University Press, 1937), 62.
2. Daniel 3: 27.
3. Leslie LeCron, *Self-Hypnosis* (New York: New American Library, 1964), 19.
4. Quoted in *Mind Over Matter*, edited by George Constable et al. (New York: Time-Life Books, 1988), 118.

Chapter 8: A Chance of a Ghost

1. G. Gorer, *Explaining English Character* (London: Cresset, 1955).
2. Gurney, Myers, and Podmore, *Phantasms of the Living* (London: Kegan, Paul, Trench, Trubner and Co., 1918).
3. Frederick Myers, *Human Personality and the Survival of Bodily Death* (London: Longmans, Green and Co., 1927).
4. Ibid.
5. Colin Wilson, *The Occult* (London: Hodder and Stoughton, 1971).
6. Jenny Randles and Peter Hough, *Encyclopaedia of the Unexplained* (London: Michael O'Mara Ltd, 1995) 141.
7. Andrew Davidson, "The Spectre Inspector," *The Sunday Telegraph*, 5 May 1996, 3.
8. H. Sidgwick, E. Sidgwick, and A. Johnson, "A Report on the Census of Hallucinations," *Proceedings of the Society of Psychical Research*, 10, 25–422, 1894.
9. Terence Hines, *Pseudoscience and the Paranormal* (New York: Prometheus Books, 1988), 61.
10. Ronald K. Siegel, *Fire in the Brain: Clinical Tales of Hallucination* (New York: Plume, 1993), 11.
11. James Gleick, *Chaos* (Cardinal, 1988), 7–8.

Chapter 10: Time and Again

1. Professor Arthur Buller, *Punch*, 19 December 1923.
2. As quoted in John Gribbin, *Companion to the Cosmos*, (London: Weidenfeld and Nicolson), 1996.
3. Barry Chapman, *Reverse Time Travel* (London: Cassell, 1995), 107.

4. Frank Tipler, "Rotating Cylinders and the Possibility of Global Causality Violation," *Physical Review* 9D (1974), 203–206.
5. H. G. Wells, *The Time Machine: An Invention* (London: Heinemann, 1895).

Chapter 11: Into the Light

1. Ernest Hemingway, "Indian Camp," published in *In Our Time* (London: Cape, 1926).
2. Steve Jones, "View From the Lab," *Daily Telegraph*, 5 June 1996.
3. Dr. Raymond Moody, *Life After Life* (London: Mockingbird Books, 1975).
4. Kenneth Ring, *Life at Death: A Scientific Investigation of the Near-death Experience* (New York: Conard, McCann and Geoghegan, 1980).
5. Robert Monroe, *Far Journeys* (London: Souvenir, 1986).
6. *The Mind in Sleep: Psychology and Psychophysiology*, edited by A. Arkin, J. Antrobus and S. Ellman (Mahwah, New Jersey: Lawrence Erlbaum Associates, 1978).
7. K. Millar and N. Watkinson, "Recognition of Words Presented During General Anaesthesia," *Ergonomics*, 36 (1983), 585-94.
8. Susan Blackmore, *Journal of Mental Imagery*, Vol. 11 (1987), 53.
9. Carl Sagan, *Broca's Brain* (New York: Random House, 1979), 143.
10. Susan Blackmore, "Birth and the OBE: An Unhelpful Analogy," *Journal of the American Society for Psychical Research*, 77 (1983), 229–38.
11. J. D. Cowan, "Spontaneous Symmetry Breaking in Large-Scale Nervous Activity," *International Journal of Quantum Chemistry*, 22 (1982), 1059–82.
12. Susan Blackmore, "Visions from the Dying Brain," *New Scientist*, 5 May 1988, 43–45.

Chapter 12: The Healing Touch

1. Anonymous, *The Week-End Book*, (London: HarperCollins, 1925), 158.
2. William Nolen, *Healing: A Doctor in Search of a Miracle* (New York: Random House, 1974), 98–99.
3. Jerome Burne, "Do You Believe?," *FOCUS*, December 1993, 36–41.
4. Cassandra Jardine, "Have Faith—and Ease Those Ills," *The Daily Telegraph*, 31 May 1996, 20.
5. R. Gracely et al., "Placebo and Naloxone Can Alter Post-surgical Pain by Separate Mechanism," *Nature* 306, 264–65.

6. L. Watkins and D. Mayer, "Organization of Endogenous Opiate and Non-opiate Pain Control Systems," *Science*, 216, 1185–1192.
7. John Sweeney, "Gigglers For God," *Life Magazine (The Observer)*, 3 March 1996, 30–33.
8. Burne, 36–41.
9. Matthew Manning, *In the Minds of Millions* (London: Allen, 1977).
10. S. Ostrander and L. Schroeder, *Psychic Discoveries Behind the Iron Curtain* (Englewood Cliffs, New Jersey: Prentice-Hall, 1971).
11. A. A. Mason and S. Black., "Allergic Skin Responses Abolished under Treatment of Asthma and Hayfever by Hypnosis," *Lancet* (1958), i, 1129–35.

Chapter 13: Swept Off Their Feet

1. John E. Mack, *Abduction: Human Encounters with Aliens* (New York: Simon and Schuster, 1994).
2. Stephen Rae, "John Mack," *The New York Times*, 30 March 1994.
3. Ibid.
4. James Willwerth, "The Man from Outer Space," *Time*, 25 April 1994.
5. Ibid.
6. Patrick Huyghe, "In Her Own Words," *OMNI*, June 1995.
7. Tom Hodgkinson, "Why It Is Easier to Believe in Aliens than God," *The Guardian*, 11 July 1995.
8. Susan Blackmore, "Alien Abduction: The Inside Story," *New Scientist*, 19 November 1994, 29–31.
9. Ibid.
10. Eris Andys, "Aliens in Our Ocean," *Encounters*, May 1996, 61–63.
11. Jaques Vallee, interviewed in Keith Thompson, *Angels and Aliens: UFOs and the Mythic Imagination* (New York: Random House, 1991), 194.

Chapter 14: The Cult of the Cult

1. Eileen Barker, *New Religious Movements* (London: HMSO, 1992), 13.
2. Anthony Storr, *Feet of Clay* (London: HarperCollins, 1996), xiii.
3. Ibid, 12.
4. Tony Allen-Mills, "Caught in the Net," *Sunday Times*, 30 March 1997.

Chapter 15: Mojo Rising

1. From the journal of Georges de Rouquct, quoted in "Beyond the Land of the Zombies," *The Unexplained*, September 1997, 12.
2. Celia Hall, "Zombie Culture All in the Mind," *The Daily Telegraph*, 10 October 1997, 11.
3. Herbert Basedow, *The Australian Aboriginal* (London: Bodley Head, 1925), 36.
4. Wade Davis, *Passage of Darkness*, (London: HarperCollins, 1988), 94.
5. Peter Hough, "Death Wish," *The X-Factor*, Vol. 20, 548.
6. Davis, 196.

Chapter 16: Miracles and Wonder

1. Arthur C. Clarke, *The Lost Worlds of 2001* (London: HarperCollins, 1973), 116.
2. David Hume, "Of Miracles," *Essays*, Section X.
3. Ibid.
4. Jenny Randles and Peter Hough, *Encyclopedia of the Unexplained* (Michael O'Mara, 1995), 130.
5. John and Anne Spencer, *The Encyclopedia of the World's Greatest Unsolved Mysteries* (London: Headline, 1995), 152.

Chapter 17: Searching for the Secrets of Life

1. From *Corpus Hermeticum*, attributed to Hermes Trismegistus, quoted in: Jack Lindsay, *The Origins of Alchemy in Graeco-Roman Egypt* (New York: Frederick Muller, 1970).
2. Quoted in Lindsay, 126.
3. Ibid.
4. Carl Jung, *Memories, Dreams and Reflections* (London: HarperCollins/RKP, 1963), 147.
5. *Man and His Symbols*, conceived and edited by Carl G. Jung (London: Aldus Books Ltd., 1964), 210.
6. For more detail see Michael White, *Isaac Newton: The Last Sorcerer*, (London: 4th Estate, 1997).
7. Maynard Keynes, "Newton the Man," in *Royal Society, Newton Tercentenary Celebrations* (Cambridge: University Press, 1947) 27–34.

8. Paracelsus, *Alchemy, the Third Column of Medicine*, edited by A. E. Waite (London: 1897), 44.
9. Thomas Birch, *The History of the Royal Society of London* (London: 1756-57), Vol. 4, 347.

Chapter 18: We Are Made of Stars

1. Unnamed staff writer, "Watch This Space," *Mail On Sunday, Night and Day* magazine, 11 August 1996, 19-22.
2. Michel Gauquelin, *Dreams and Illusions of Astrology* (Buffalo, New York: Prometheus Books, 1979).
3. Amanda Cochrane, "Science, Art or Superstition?" *FOCUS*, November 1993, 16–21.
4. Linda Goodman, *Linda Goodman's Star-Signs* (New York: Bantam Books, 1968).
5. Ibid., 475.
6. Ibid., 203.
7. G. Abell and B. Singer, *Science and the Paranormal* (New York: Scribner's, 1981), 86.

Chapter 19: Fire From the Sky

1. As quoted in John and Mary Gribbin, *Fire on Earth: In Search of the Doomsday Asteroid* (New York: Simon and Schuster, 1996).
2. Robert Matthews, "The End of the World Is Nigh—Official," *The Sunday Telegraph Review*, 21 July 1996.
3. From Plato's dialogues, *Timaeus and Critias*, and quoted in Clube and Napier, *The Cosmic Winter* (Oxford: Blackwell, 1990), 70–71.
4. From Gildas, *The Ruins of Britain*, quoted in Clube and Napier, 108.
5. Windsor Chorlton, "Ice Age," *FOCUS*, December 1994.
6. Adrian Berry, *The Next 500 Years: Life in the Coming Millennium* (London: Headline, 1995).
7. Matthews.
8. John and Mary Gribbin, 234.

Chapter 20: Our Brethren Among The Stars?

1. Leslie E. Orgel, "The Origin of Life on Earth," *Scientific American*, October 1994, 53–60.

Chapter 21: Mendel's Monsters

1. Sean Ryan and Lois Rogers, "Who's Playing God With the Gene Genie?" *The Sunday Times*, 1 January 1995, 12.
2. Janet Fricker, "DNA's Genetic Time Bomb," *FOCUS*, May 1994, 26-30.
3. Gina Kolata, *Clone: The Road to Dolly and the Path Ahead*, (London: Allen Lane, 1997), 86.
4. *The Report of the House of Commons Science and Technology Committee*, August 1996.
5. French Anderson, "Gene Therapy," *Scientific American*, September 1995, 98B.
6. Colin Tudge, "Cloning? Get Used to It," *Independent on Sunday*, 25 January 1998, 25.
7. Roger Highfield, "Human Embryo Clones Could Help Save Lives," *Daily Telegraph*, 30 January 1998, 9.

Chapter 22: Earth Magic, Leylines, and Circles in the Corn

1. Russell Warren, "The Craft of Dowsing: Worthwhile or Weird?," *FOCUS*, January 1997, 51.
2. Nigel Peddick, *Leylines* (London: Weidenfeld and Nicolson, 1997), 27.
3. Christopher Oliver Wilson, "Bumper Crop of Circles," *Mail on Sunday*, 31 August 1997, 52–53.
4. Nigel Blundell, *Mysteries* (London: Bookmart Ltd./Amazon Publishing 1992), 68.

Chapter 23: Remote Viewing and the Psi Detectives

1. Marie Louise-Small, "Psi Spies," *Encounters*, May 1996, 54–57.
2. Jenny Randles and Peter Hough, *Encyclopaedia of the Unexplained* (London Michael O'Mara Books Ltd., 1995), 110.

3. "Report of the 1992 International Symposium on UFO Research," detailed in *Encounters*, May 1996, 55.
4. R. Targ and H. Puthoff, *Mind and Reach* (New York: Delacorte Press, 1977).
5. D. Marks and R. Kammann, *Psychology of the Psychic* (Buffalo, New York: Prometheus Books, 1980).
6. D. Marks and C. Scott, "Remote Viewing Exposed," *Nature*, 319, 444.

Chapter 24: The Gods Themselves?

1. Quoted in Jennifer Westwood, *Lost Atlantis* (London: Weidenfeld and Nicolson, 1997), 5.
2. Galanopoulos and Bacon, *Lost Atlantis* (London: Nelson, 1963), 75.
3. Quoted in: Dr. Karl Shuker, *The Unexplained* (London: Carlton Books, 1996), 155.
4. Robert Matthews, "Is this the Road That Will Take Us to Atlantis?," *Sunday Telegraph*, 6 April 1997.
5. John Harlow and Alistair Crighton, "Divers Find Atlantis Off Land's End," *The Sunday Times*, 28 December 1997, 12.
6. Ibid.
7. Galanopoulos and Bacon, *Atlantis: The Truth Behind the Legend* (London: Nelson 1965).
8. Graham Hancock, *Fingerprints of the Gods* (London: Heinemann Ltd., 1995), 492.
9. Quoted in Eric Von Daniken, *Chariots of the Gods?: Was God an Astronaut?* (London: Souvenir Press, 1969), 55.
10. For example: John North, *Stonehenge* (London: HarperCollins, 1996).
11. Von Daniken, 99.